DIET, NUTRITION AND CANCER

Proceedings of the 16th International Symposium of
The Princess Takamatsu Cancer Research Fund, Tokyo, 1985

DIET, NUTRITION AND CANCER

Edited by
YUZO HAYASHI, MINAKO NAGAO, TAKASHI SUGIMURA,
SHOZO TAKAYAMA, LORENZO TOMATIS,
LEE W. WATTENBERG, and GERALD N. WOGAN

JAPAN SCIENTIFIC SOCIETIES PRESS, Tokyo
VNU SCIENCE PRESS BV, Utrecht The Netherlands

Published jointly by:
Japan Scientific Societies Press
Hongo 6-2-10, Bunkyo-ku, Tokyo 113, Japan
ISBN 4-7622-9494-2

and

VNU Science Press BV
P. O. Box 2073, 3500 GB Utrecht,
The Netherlands
ISBN 90-6764-085-9

Distributed in all areas outside Japan and Asia between Pakistan and Korea by VNU Science Press BV

Printed in Japan

Organizing Committee of the 16th International Symposium

Yuzo HAYASHI
Division of Pathology, National Institute of Hygienic Sciences, Tokyo 158, Japan
Minako NAGAO
Carcinogenesis Division, National Cancer Center Research Institute, Tokyo 104, Japan
Takashi SUGIMURA
National Cancer Center Research Institute, Tokyo 104, Japan
Shozo TAKAYAMA
National Cancer Center Research Institute, Tokyo 104, Japan
Lorenzo TOMATIS
International Agency for Research on Cancer, Lyon Cedex 8, France
Lee W. WATTENBERG
Department of Laboratory Medicine and Pathology, University of Minnesota, Minneapolis, Minnesota 55455, U.S.A.
Gerald N. WOGAN
Department of Applied Biological Sciences, Massachusetts Institute of Technology, Cambridge, Massachusetts, 02139, U.S.A.

Preface

The ultimate goal in control of any disease is prevention, and so it is with cancer. As data accumulate, life style characteristics come into focus as major determinants of cancer risk in humans. Of these, diet is the most ubiquitous. Constituents of food have been identified that can cause cancer or enhance the likelihood of its occurrence. Food also contains substances that have been shown to protect against cancer. These data encourage the hope that a full understanding of the impact of dietary constituents on carcinogenesis will lead to important means of cancer control.

"Diet, Nutrition and Cancer", the topic of the Sixteenth International Symposium of the Princess Takamatsu Cancer Research Fund held in Tokyo in November of 1985 was particularly timely. It occurred at a period of increasing awareness of the major impact that diet can have on the occurrence of cancer. It was a moment at which an evaluation of existing data was critical to developing approaches to an exceedingly complicated and important area of cancer research.

The scientists participating in the Symposium represented an extraordinary variety of disciplines ranging from epidemiology to molecular biology. The papers covered a diversity of subjects and provided a valuable compendium of current research in this complex field. The presentations gave rise to a sense of cautious optimism that future productive research on the relationships of diet, nutrition, and cancer is feasible and could result in the development of useful strategies for cancer prevention.

Lee W. Wattenberg
Yuzo Hayashi
Minako Nagao
Takashi Sugimura
Shozo Takayama
Lorenzo Tomatis
Gerald N. Wogan

Preface

Contents

Modulation of Carcinogenesis by Dietary Components in Experimental Animals

Cancer Epidemiology of Mutagens/Carcinogens in Foods

Concluding Remarks and Future Perspectives

Contributors

Sheila BINGHAM	Medical Research Council and University of Cambridge, Dunn Clinical Nutrition Centre, Cambridge, AB2 1QL, United Kingdom
W. Robert BRUCE	Ludwig Institute for Cancer Research Toronto Branch, Toronto, Ontario M4Y 1M4, Canada
K. K. CARROLL	Department of Biochemistry, Health Sciences Centre, University of Western Ontario, London, N6A 5C1, Canada
Shoji FUKUSHIMA	First Department of Pathology, Nagoya City University Medical School, Mizuho-ku, Nagoya 467, Japan
Yuzo HAYASHI	Division of Pathology, National Institute of Hygienic Sciences, Setagaya-ku, Tokyo 158, Japan
Takeshi HIRAYAMA	Institute of Preventive Oncology, Shinjuku-ku, Tokyo 162, Japan
Iwao HIRONO	Department of Pathology, Fujita Gakuen Health University, Toyoake, Aichi 470-11, Japan
Dennis P. H. HSIEH	Department of Environmental Toxicology, University of California, Davis, California 95616, U.S.A.
Margaretha JÄGERSTAD	Applied Nutrition Chemical Center, University of Lund, S-220 07 Lund 7, Sweden
Laurence KOLONEL	University of Hawaii at Manoa, Cancer Research Center of Hawaii, Epidemiology Program, Honolulu, Hawaii 96813, U.S.A.
Masanori KURATSUNE	Nakamuragakuen College, Jonan-ku, Fukuoka 814, Japan
A. J. McMICHAEL	Division of Human Nutrition, CSIRO, ADELAIDE. 5,000, South Australia
Sanford A. MILLER	Center for Food Safety and Applied Nutrition, Food and

Drug Administration, The United State of America, Washington, D. C. 20204, U.S.A.

Minako NAGAO — Carcinogenesis Division, National Cancer Center Research Institute, Chuo-ku, Tokyo 104, Japan

Paul M. NEWBERNE — Department of Applied Biological Sciences, Massachusetts Institute of Technology, Cambridge, Massachusetts 02139, U.S.A.

Hiroko OHGAKI — Biochemistry Division, National Cancer Center, Research Institute Chuo-ku, Tokyo 104, Japan

Yoshinari OHNISHI — Department of Bacteriology, School of Medicine, Tokushima University, Tokushima 770, Japan

Shigeaki SATO — Biochemistry Division, National Cancer Center Research Institute, Chuo-ku, Tokyo 104, Japan

Michael B. SPORN — Division of Cancer Cause and Prevention, National Cancer Institute, Bethesda, Maryland 20205, U.S.A.

Grant N. STEMMERMANN — Japan-Hawaii Cancer Study, Kuakini Medical Center, Honolulu, Hawaii 96817, U.S.A.

Tsung-tang SUN — Department of Immunology, Cancer Institute, Chinese Academy of Medical Sciences, Beijing, People's Republic of China

Michihito TAKAHASHI — Division of Pathology, National Institute of Hygienic Sciences, Setagaya-ku, Tokyo 158, Japan

Steven R. TANNENBAUM — Department of Applied Biological Sciences, Massachusetts Institute of Technology, Cambridge, Massachusetts 02139, U.S.A.

Lorenzo TOMATIS — International Agency for Research on Cancer, 69372 Lyon Cedex 8, France

J.C.M. van der HOEVEN — NOTOX Toxicological Research and Consultancy v.o.f., 5231 DD 's-Hertogenbosch, The Netherlands

Lee W. WATTENBERG — Department of Laboratory Medicine and Pathology, University of Minnesota, Minneapolis, Minnesota 55455, U.S.A.

John H. WEISBURGER — Naylor Dana Institute for Disease Prevention, American Health Foundation, Valhalla, New York 10595, U.S.A.

Walter WILLETT — Channing Laboratory, Boston, Massachusetts 02115, U.S.A.

Gerald N. WOGAN — Laboratory of Toxicology, Department of Applied Biological Sciences, Massachusetts Institute of Technology, Cambridge, Massachusetts 02139, U.S.A.

Opening Address

H.I.H. Princess Kikuko Takamatsu

It is a great pleasure for me to welcome everyone who has gathered here today to participate in the 16th International Symposium of the Princess Takamatsu Cancer Research Fund. I would particularly like to welcome the scientists from abroad who have taken their valuable time to come to Japan to participate in this endeavor.

The title of this year's symposium is "Diet, Nutrition and Cancer". In recent years cancer has become a major disease in many countries. It is convinced that finding ways to cure cancer is most important task, however, finding ways to prevent cancer should be equally as important in conquering cancer. I hear that recent epidemiological and experimental studies show significant roles of dietary and nutritional factors in the development and modulation of cancer. I sincerely hope that the facts and ideas presented in this meeting by the scientists who are at the forefront in this field of cancer research will contribute to the progress of establishing a scientific basis for the prevention of cancer.

I wish to express my sincere thanks to Dr. Gerald N. Wogan of Massachusetts Institute of Technology, Dr. Lee W. Wattenberg of University of Minnesota, U.S.A. and Dr. Lorenzo Tomatis of International Agency for Research on Cancer, France for their valuable assistance in organizing this symposium as members of the organizing committee.

Dr. Gerald N. Wogan

Your Imperial Highness:

I have the honor and privilege to speak in response to your gracious opening address. On behalf of Drs. Wattenberg, Tomatis, and the other members of the organizing committee, and indeed all of the participants in the Symposium, I express our gratitude for the generous support which made possible this 16th International Symposium of the Princess Takamatsu Cancer Research Fund.

The topic of the Symposium, "Diet, Nutrition and Cancer" is a very timely one. As you have already mentioned, much evidence has accumulated through epidemiological and laboratory studies to indicate the important influences exerted by diet and nutritional factors in the development and modulation of various types of cancer. Some of this evidence has been judged sufficiently convincing that provisional dietary guidelines have been suggested to the general public with the objective of reducing risks for several important cancers, especially those of the breast and colon.

Despite the strength of associations on which these guidelines are based, they cannot in themselves establish specific etiologic bases for the observed effects. This deficiency has made them vulnerable to challenges, and will undoubtedly impede their general acceptance and, therefore, diminish their potential impact. Research endeavors directed at fuller elucidation of these factors and mechanisms through which they exert their effects comprise the topics to be discussed at this Symposium. These include evidence relating to the presence of mutagens/carcinogens in foods, including naturally-occurring substances as well as those formed during processing or ordinary cooking. Also to be discussed are dietary components which exert protective effects, thereby modulating the development of experimentally induced cancers; these protective agents also include naturally-occurring as well as synthetic chemicals. Epidemiologic studies evaluating the roles of diet, nutrition and exposure to food-borne mutagens/carcinogens as cancer risk factors are subjects of the remainder of the program.

In this, as in subjects of the previous Symposia of this series, the importance of contributions made to the field by Japanese scientists is clearly evident. From an historical viewpoint, some of the earliest and most convincing evidence linking nutrition to cancer came from studies on migrant Japanese populations in whom risks for stomach and colon cancers changed dramatically in the course of their migration to the U.S. via Hawaii, and with increasing westernization of their diet. The contributions of Segi, Kurihara, and other Japanese scientists to these and subsequent studies are well known. More recently, the discovery by Sugimura and his collaborators that mutagens can be formed in foods during cooking opened a new and exciting field of carcinogenesis research to which many Japanese investigators, including participants in this symposium, continue to make important contributions. It is therefore particularly fitting, Your Highness, that this meeting is being held here in Tokyo under circumstances which will allow informal as well as formal discussion of current research topics. For those of us from abroad, it represents an opportunity to renew old, and establish new, professional associations with our Japanese colleagues, and we are most appreciative of your interest and support in making it possible.

LECTURES

DIET, NUTRITION AND CANCER, Y. HAYASHI ET AL. (EDS.).
JAPAN SCI. SOC. PRESS, TOKYO/VNU SCI. PRESS, UTRECHT, PP. 3-10, 1986

KEYNOTE LECTURE

Diet and Nutrition as Risk Factors for Cancer

Gerald N. WOGAN

Department of Applied Biological Sciences, Massachusetts Institute of Technology, Cambridge, Massachusetts 02139, U.S.A.

Abstract: Involvement of diet and nutrition as risk factors for human cancers has been established by two general types of evidence. Epidemiological studies in human populations have identified associations between patterns of incidence for various forms of cancer and diet composition or food consumption patterns. Such associations are particularly evident in studies on migrant populations and in patterns of geographic localization of specific forms of cancer. Reduction in gastric cancer (and increase in colon cancer) incidence has been observed in immigrants from Japan to the U.S. *via* Hawaii with concurrent change in dietary habits. High fat consumption has been linked to increased incidence of breast and colon cancer, and evidence is accumulating that suggests increased intake of dietary fiber (or some specific component of it) may be associated with diminished risk of colorectal cancer.

Laboratory studies have demonstrated the existence of dietary constituents that might impact cancer risk in people consuming them. Substances in this class fall into two general categories: genotoxic carcinogens/mutagens; and protective factors which inhibit experimentally-induced chemical carcinogenesis in animals.

Numerous naturally-occurring carcinogens/mutagens have been identified. Examples include aflatoxins, cycasin, and bracken fern carcinogen(s), among others. Genotoxic substances associated with the use of intentional food additives are the N-nitroso compounds formed as nitrosation products from nitrite. Potential carcinogenic risk from food constituents also comes from mutagens found in foods as natural components, or formed in the course of cooking. Extensive evidence has been produced on a series of compounds formed by protein pyrolysis, which are not only powerful mutagens in bacteria, but also carcinogenic in animals.

Foods also contain a variety of nutrient and non-nutrient constituents which act as protective factors (anticarcinogens) when administered to experimental animals during the process of chemically-induced carcinogenesis. Many of these are present in cruciferous vegetables and also carotene-rich (green/yellow) vegetables. Of particular interest in the later category are the protective properties of retinoids against certain types of experimentally-induced tumors. Certain of these findings in experimental animals are also applicable to people. For example, evidence exists that consumption of green/yellow vegetables is associated with a reduction in risk of cancer at several sites in humans.

Possible impact of ingestion of carcinogens/mutagens on cancer risk is also being investigated. Progress in this area will be greatly enhanced by the continued development of analytical methods which enable quantitative exposure measurements on individuals within population groups. Thus, further efforts to extend quantitative evaluation of the importance of diet and nutrition as determinants of overall cancer risk may be accomplished with greater precision.

Considerable effort has been directed to studying the influence of environmental factors on the incidence of cancer. As evidence has accumulated, it has become increasingly clear that most cancers have external causes, and in principle should be preventable (*1*). In seeking evidence for the avoidability of cancer, epidemiologists have studied differences in cancer incidence between communities; changes in incidence on migration; and changes in incidence over time in the same populations. Diet has emerged as a major factor associated with patterns of incidence for certain forms of cancer, and much effort has been devoted on attempts to identify which constituents of foods might be associated with the observed effects. In general, the evidence suggests that some types of diets and some dietary components, such as high fat diets, salt-cured or smoked foods tend to increase the incidence of certain cancers, whereas others, such as green-yellow vegetables and citrus fruits tend to decrease it. Mechanisms responsible for these effects are only poorly understood and remain topics of active current research.

Laboratory investigations have attempted to elucidate mechanisms by which diet and nutrition might influence cancer risk. These studies have examined the ability of dietary constituents, nutritive and non-nutritive, to enhance or inhibit carcinogenesis or mutagenesis, and thereby have provided epidemiologists with testable hypotheses regarding specific components of the diet. Combined epidemiological and laboratory studies have produced a vast amount of information. This overview will briefly summarize some of the major findings in order to provide a background against which current research activities will be discussed by other participants in the context of the general topic of the Symposium. For convenience, the information has been organized along three general lines: factors which enhance cancer risks; those which reduce cancer risks; and the implications of these for the development of a general strategy for cancer prevention.

Factors Associated with Elevated Cancer Risks: Nutrient Intakes

In attempts to determine which constituents of foods might be associated with cancer, epidemiologists have studied population subgroups in various parts of the world, including migrant groups, to examine the relationships between specific dietary patterns or the consumption of certain foods and the risk of developing particular cancers. Such studies have repeatedly shown an association between dietary fat and the occurrence of cancer at several sites. High levels of meat consumption in various countries were found to be strongly associated with cancer of the large bowel (*2*). Subsequently, a large body of evidence has accumulated linking the occurrence of cancer of the breast as well as the colon to an elevated intake of total fat (*3*). Data on cancers of the prostate and endometrium are more limited, but they also suggest that increased risk is associated with high levels of total fat. In general it is not possible to identify specific components of fat as being clearly responsible for the observed effects, but total fat and saturated fats have been associated most frequently.

In some populations, in particular those of China, Japan, and Iceland, frequent consumption of foods preserved by salt-curing or smoking has been associated with elevated risks for several cancers, especially those of the esophagus and stomach. Here again, the agents responsible for the observed association have not been specifically identified, but the involvement of polycyclic aromatic hydrocarbons, N-nitroso compounds and other mutagenic and carcinogenic substances has been postulated (*4*).

Some of the strongest evidence that dietary factors can influence carcinogenesis has come from studies in laboratory animals (*5*). In several instances, it has been shown that dietary factors whose intakes show a positive correlation with incidence of specific cancer in humans also induce the same types of effects in animals. Evidence from animal studies thus complements data from epidemiologic studies in man, and also may provide information on mechanisms through which observed effects are produced. The literature on nutritional enhancement of experimentally induced cancers is very large and complex. However, several generalizations are warranted. In general, occurrence of cancer of various tissues, including mammary glands and the large intestine induced by a variety of chemical carcinogens is enhanced by excessive intake of fat, protein, and total calories. Enhancement is also brought about by deficiencies of vitamin A, zinc, and lipotropic factors (*6*).

Factors Associated with Elevated Cancer Risks: Non-nutrient Food Constituents

The human diet contains a great number of substances which induce mutations in bacterial or other bioassay systems and cancer when administered to experimental animals at appropriate dose levels. Many of these are normal constituents of plants that are consumed as food under various circumstances (*7*, *8*). Included in this group is the bracken fern, which at certain times contains toxic substances that induce tumors of the intestinal tract in rodents. An active compound, ptaquiloside, was recently isolated and chemically characterized. Possible cancer risks associated with consumption of bracken fern is the only one of this class of hazards to have been studied epidemiologically, as discussed elsewhere in this Symposium.

Other members of this broad class of naturally-occurring carcinogens include hydrazines found in the widely consumed false morel *Gyromitra esculenta*; methylazoymethanol, the aglycone of cycasin, a component of the starch of cycad nuts; pyrrolizidine alkaloids, present in many plants including several used as herbal teas and vegetables; and safrole, estragol, and methyleugenol, all methylenedioxybenzene derivatives present in many edible plants.

Mycotoxins represent another major class of non-nutrient contaminants of foods which constitute potential carcinogenic risks for people exposed to them. Several members of these very large groups of chemicals have been shown to be carcinogenic when fed to animals, including the aflatoxins, sterigmatocystin and luteoskyrin. The aflatoxins have been very extensively studied because of their potency as carcinogens for the liver and possibly other tissues of experimental animals, and also because of the high frequency with which they contaminate certain food crops (*9*). Extensive investigation of levels of intake of aflatoxins by populations experiencing different rates of liver cancer has produced an impressive association between aflatoxin ingestion and incidence of primary liver cancer. However, subsequent studies have also identified hepatitis B virus as an important risk factor for liver cancer. Recent evidence suggests that the agents may act synergistically, through as yet undefined mechanisms, to cause a very large increase in risk for individuals exposed to both.

N-Nitroso compounds comprise still another major class of food-borne carcinogens and mutagens. Nitrosamines and nitrosamides that have been tested for carcinogenicity are noteworthy both for their overall potency and for their versatility as animal carcinogens. Various compounds have experimentally induced tumors in virtually every organ and tissue of animals. Particular attention has been paid to the endogenous formation of N-nitroso compounds through the interaction of nitrite with nitrosatable amines in foods. The hypothesis linking nitrate and increased cancer risk rests on the postulate that nitrate is endogenously reduced to nitrite by bacteria, and that nitrosation produces carcinogenic and/or mutagenic products in the stomach. Evidence accumulated to date indicates that the levels of nitrosamines found in man's environment may be involved in the causation of human cancers. However, it has not been possible to demonstrate in the general population, a cause-effect relationship between exposure to low levels of nitrosamines and specific forms of cancer (*10*). Further research quantifying individual exposure through application of ultrasensitive analytical methodology may reveal such relationships, if they exist.

Another important source of food contaminants has been elucidated through the systematic screening of food extracts for the presence of mutagens. Using this approach, it has been established that mutagens are formed during the cooking of certain foods, and therefore represent a previously unrecognized source of potentially carcinogenic substances. Structural characterization of the isolated mutagens revealed them to be a series of heterocyclic amines of several classes—derivatives of quinoxaline, imidazole, indole, and carboline, respectively. A total of 11 mutagens of these types have thus far been tested for carcinogenicity, and it is significant that all have been shown to induce tumors of the gastrointestinal tract and other organs of rats and mice (*11*). These findings demonstrate the usefulness of the bacterial mutagenesis bioassays as screening tools, and their continued application may reveal the presence of additional mutagens in the human diet.

The combined impact of food-borne carcinogens/mutagens on cancer risk of individuals in the general population is unknown. Although many of these substances are probably consumed only under very limited circumstances, exposure to others, such as certain plant constituents, aflatoxins, nitrosamines, and heterocyclic amines is likely to be virtually inescapable. However, inter-individual variation in exposure levels may be very great owing to idiosyncratic dietary habits and other factors. Only development of ultrasensitive methodology applicable to individuals in biochemical epidemiological investigations will be capable of revealing individual risks. Such methodology, *e.g.*, for detecting carcinogen metabolites and macromolecular adducts is being developed, and will be discussed elsewhere in this Symposium.

Factors Associated with Reduced Cancer Risks: Nutrient Intakes

A growing accumulation of epidemiological evidence indicates that there is an inverse relationship between the risk of cancer and the consumption of foods that contain vitamin A, or more especially of its precursors, the carotenoids, contained in deep green and dark yellow vegetables (*4*). Most of the data does not reveal whether the protective action is attributable to vitamin A itself, to the carotenoids, or to some other constituent of these foods. In these studies, investigators have found an inverse relationship between estimates of carotene/vitamin A intake and cancers of various sites. The effect is particularly strong for cancer of the lung, but recent evidence indicates marked effects on cancers of the gastrointestinal tract and pancreas as well.

As noted earlier, studies in animals indicated that vitamin A deficiency generally increases susceptibility of animals to chemically-induced tumors, and that an increased intake of the vitamin appears to protect against carcinogenesis in most, but not all cases. Results from the studies on vitamin A have stimulated efforts to find analogs with greater inhibitory activity, less toxicity, and the capability of reaching target tissues at concentrations greater than those attainable with the natural vitamins. Many such compounds, the retinoids, have been synthesized, but are not normal constituents of the diet. Experiments to study inhibition of carcinogen-induced neoplasia of the breast, bladder, skin and lung have produced impressive results (*12*, *13*). The biochemical mechanisms involved in the inhibition of carcinogenesis by retinoids are not well understood, but their ability to modulate the proliferative effects of peptide growth factors and related molecules appears to be of particular importance in this respect (*14*).

Dietary fiber represents another factor for which there is evidence of protective effects. Considerable effort has been directed to studying relationships between intake of fiber-containing foods such as whole-grain cereals, certain fruits, and vegetables and the occurrence of various forms of cancer. Most epidemiological studies have examined the hypothesis that high fiber diets protect against colorectal cancer. Results of correlation studies and case-control studies have produced contradictory results. Disagreements may be attributable in part to the lack of standardized methodology for measurement of fiber levels, and it has been proposed that the observed effects may have been due to specific components of "fiber" such as non-starch polysaccharides. However, studies involving measurement of intakes of this component specifically also have produced contradictory findings in different populations, as discussed elsewhere in this Symposium. Thus, more work needs to be done to clarify the protective role of fiber against the risk of colorectal cancer.

Factors Associated with Reduced Cancer Risks: Protective Factors in Foods

Numerous non-nutritive components of food have been examined for their potential to protect against carcinogenesis. An impressive number of compounds, representing more than 20 different classes of chemicals have been found to have the ability to inhibit experimentally-induced cancers in animals. These substances fall into two broad categories—those that are effective against complete carcinogens, and those that act against tumor promoters. They may be further classified according to the specific stages of carcinogenesis at which they exert their effects (*15*). Thus, certain compounds act by inhibiting carcinogen formation from precursors as, for example, the inhibition of nitrosamine formation by ascorbic acid or α tocophenol. Another class of chemopreventive agents has been termed "blocking agents," which act through: inhibition of carcinogen activation (flavones); induction of detoxifying enzymes (BHT, phenols, coumarins); enhancement of glutathione-S-transferase activity (isothiocyanates); or by trapping reactive electrophilic derivatives of carcinogens (glutathione, methionine). All such agents block the carcinogenic process of the initiation stage. A further group of substances, termed "suppressors" act by blocking the processes involved in tumor progression past the initiation stage. This group would include retinoids, selenium salts, and protease inhibitors, among others. Thus, it is well established that chemoprevention of carcinogenesis in animals can be achieved by administration of a large number of chemicals. It remains to be determined whether these observations can be applied for effectively bringing about protection in humans.

Strategies for Cancer Prevention

Current strategies for prevention of environmentally related cancers are heavily oriented towards the identification of mutagens and carcinogens and the development of means to minimize exposure to them. In some cases, these efforts have met with considerable success, For example, in countries having the requisite technologies and regulatory machinery, processed foods are regularly monitored for aflatoxin and nitrosamine contamination, and in general, average intakes of them are very low compared to regions in which such controls are not practiced. Nevertheless, given the large number of known naturally-occurring carcinogens, together with those formed during ordinary cooking procedures, complete elimination of exposure to food-borne carcinogens appears to be an unattainable goal. Thus, every attempt must be made to reduce exposure to those which can be controlled in order to minimize the total burden of initiating agents and their collective contribution to cancer initiation.

Available evidence, including recent findings to be discussed at this Symposium, indicates that diet and nutrition can have major impacts on risks for important cancers. Although the underlying mechanisms responsible for these effects remain to be elucidated, modulating influences appear to result from effects on processes involved in the stages of tumor development later than the initiation phase. Thus, the modifying effects of diet and nutritional status are operationally separable from those of the carcinogenic insult resulting from exposure to initiating agents, and can therefore be exploited in developing strategies for cancer prevention which are complementary to procedures aimed at minimizing carcinogen exposure.

In the recent past, several organizations in the U.S., including the American Cancer Society, The National Cancer Institute and the NAS/NRC have formulated provisional dietary guidelines intended to reduce risks for several major forms of cancer (*14*). The main points of these guidelines, on which there was general concensus among the several agencies, can be summarized as follws:

1. Reduce fat consumption (from 40% of calories to 30% in the U.S.).
2. Include in the daily diet the following: whole grain cereal products; citrus fruits; and green/yellow and cruciferous vegetables.
3. Practice moderation in consumption of salt-cured (pickled) and smoke-cured foods, and of alcoholic beverages.
4. Maintain appropriate body weight.

In addition, it was recommended that efforts continue to be made to minimize contamination of foods by carcinogens from any source, and also to identify mutagens in foods and expedite their testing for carcinogenicity.

Groups formulating these guidelines acknowledge that the available data linking diet and cancer risk are inferential and incomplete, but consider their recommendations to be of sufficient possible benefit to justify their publication. They have also emphasized that these guidelines are provisional in nature, and offer no certainity of benefit, especially to individuals. Nonetheless, there is reason to believe that further elucidation of the molecular and cellular events involved in the process of tumor initiation and development will in turn provide a basis for further refinement of these guidelines, and formulation of even more effective preventive measures.

REFERENCES

1. Doll, R. and Peto, R. The Causes of Cancer: Quantitative estimates of avoidable risks of cancer in the United States today. JNCL, *66*: 1193–1308, 1981.
2. Armstrong, B. and Doll, R. Environmental factors and cancer incidence and mortality in different countries, with special reference to dietary practices. Int. J. Cancer, *15*: 617–631, 1975.
3. Carroll, K. K. Experimental evidence of dietary factors and hormone-dependent cancer. Cancer Res., *35*: 3374–3383, 1975.
4. Commitee on Diet, Nutrition and Cancer, NAS/NRC. Diet, Nutrition and Cancer. National Academy Press, Washington, D.C., 1982.
5. Hopkins, G. J. and Carroll, K. K. Role of diet in cancer prevention. J. Environ. Pathol. Toxicol. Oncol., *5*: 279–298, 1985.
6. Newborne, P. M., Rogers, A. E., and Nauss, K. M. *In;* C. E. Batterworth and M. Hutchinson (eds.), Nutritional Factors in the Induction and Maintenance of Malignancy, pp. 247–271, Academic Press, 1983.
7. Wogan, G. N. Naturally-occurring carcinogens. *In;* F. Homburger (ed.), Pathophysiology of Cancer, vol. 1, pp. 64–109, S. Karger AG, Basel, 1974.
8. Ames, B. N. Dietary carcinogens and anticarcinogens: oxygen radicals and degenerative diseases. Science, *221*: 1256–1264, 1983.
9. Busby, W. F., Jr. and Wogan, G. N. Aflatoxins. *In;* C. E. Searle (ed.), Chemical

Carcinogens, second ed., vol. 2, ACS Monograph 182, American Chemical Society, Washington, D.C., pp. 945–1136, 1984.
10. Bartsch, H. and Montesano, R. Relevance of nitrosamines to human cancer. Carcinogenesis, *5*: 1381–1393, 1984.
11. Sugimura, T. Carcinogenicity of mutagenic heterocyclic amines formed during the cooking process. Mutat. Res., *150*: 33-41, 1985.
12. Sporn, M. B., Dunlop, N. M., Newton, D. L., and Smith, J. M. Prevention of chemical carcinogenesis by vitamin A and its synthetic analogs (retinoids). Fed. Proc., *35*: 1332–1338, 1976.
13. Sporn, M. D. and Newton, D. L. Chemoprevention of cancer with retinoids. Fed. Proc., *38*: 2528–2534, 1979.
14. Sporn, M. B. and Roberts, A. B. Role of retinoids in differentiation and carcinogenesis. Cancer Res., *43*: 3034–3040, 1984.
15. Wattenberg, L. W. Perspectives in cancer research: chemoprevention of cancer. Cancer Res., *45*: 1–8, 1985.

DIET, NUTRITION AND CANCER, Y. HAYASHI ET AL. (EDS.),
JAPAN SCI. SOC. PRESS, TOKYO/VNU SCI. PRESS, UTRECHT, PP. 11–26, 1986

NAKAHARA MEMORIAL LECTURE

Application of the Mechanisms of Nutritional Carcinogenesis to the Prevention of Cancer

John H. WEISBURGER

American Health Foundation, Naylor Dana Institute for Disease Prevention, Valhalla, New York 10595-1599, U.S.A.

Abstract: Historically, the field of experimental chemical carcinogenesis began in Japan with Yamagiwa and has been a traditional subject of study since that time. In Prof. Nakahara's lifetime, he and his disciples have contributed much to an understanding of the basic mechanisms of carcinogenesis. Most types of human cancer are likely associated with chemical carcinogens. In part, we understand the mechanisms whereby carcinogens lead to neoplasia. The overall process involves agents with distinct properties, 1) those modifying the genome with specific consequences, and 2) those controlling the growth and development of latent tumor cells with such an abnormal genome. The genotoxic pathway and the subsequent promoting processes proceed by distinct mechanisms and thus have different consequences as regards health risk, especially with respect to dosage and time requirements for effective carcinogenesis.

Through multidisciplinary approaches, it has been established that cancer of the stomach and esophagus, prevalent in certain parts of the world, depend on the presence in salted, pickled, or smoked foods of specific chemicals, that are genotoxic and the structure of which is a function of local conditions. Salt can have a cocarcinogenic or promoting role. In much of the Western World, cancers of the colon, pancreas, breast, ovary, endometrium, and prostate are linked to nutritional traditions. The genotoxic carcinogens for several of these neoplasms may be formed during cooking, especially broiling or frying. There is evidence for extensive promoting processes, in turn, a function of the total dietary fat intake, through partially understood mechanisms. Additional modifying factors include cereal (bran) fiber, but perhaps not other types of fibers, that reduce the risk for colon cancer. Further modifying elements are discussed in this Symposium.

Fair understanding has been achieved of the underlying basic mechanisms, relative to the formation of carcinogens during food preparation and processing, and on the role of certain promoting or inhibiting elements such as fat, fiber, or components of fruits and vegetables. Certain of these elements are sufficiently well established for application to the prevention of specific cancers in various parts of the world.

Prof. Waro Nakahara in his lifetime contributed much to the varied and complex aspects bearing on the etiology of cancer. He and his colleagues discovered the powerful carcinogenicity of 4-nitroquinoline-1-oxide (4-NQO) and studied the mode of action of this agent (*1*). Prof. Nakahara also was one of the first to be aware that the summation effects and quantitative actions of chemical carcinogens given together or sequentially should be distinguished from initiation-promotion experiments. He developed elegant approaches to the use of submanifestational doses of either radiation or a chemical that, when followed by low-level exposure to another carcinogen, could induce cancer when either agent by itself did not. These germinal experiments provided the basis for understanding the effect of complex products such as tobacco smoke or broiled or fried meats or fish. Indeed, these products contain relatively small amounts of distinct carcinogens, co-carcinogens, or promoters. Tobacco smoke includes carcinogens of the polycyclic aromatic hydrocarbon class and those related to nitrosonornicotine and related products as well as some heterocyclic aromatic amines. The effect is potentiated by accelerators, such as fluoranthene, and by co-carcinogens and promoters belonging to the catechol and phenolic class. The amount of each component by itself would have only minimal effects according to the classification by Nakahara. Taken together, however, this phenomenon of syncarcinogenesis, as well as cocarcinogenesis and promotion, accounts for the overall carcinogenic effect in the addicted smoker. In the area of nutritional carcinogenesis likewise, recent discoveries by Dr. Sugimura (*2*), one of the foremost students of Dr. Nakahara, indicate that certain cooking procedures produce powerful mutagens that may well be the carcinogens for major types of human cancer: breast, colon, or pancreas. Here also, each pure component is present in small amounts, but the totality of the entire mixture provides a picture of the process of syncarcinogenesis, by which these substances may act as initiators. In the presence of a high fat intake, acting through mechanisms of co-carcinogenesis and promotion, the overall effect accounts for the prevalence of these types of cancer in countries with Western nutritional traditions.

The research of Prof. Nakahara dealt with a definition of the possible additive or synergistic action of small amounts of potent carcinogens. In the last 20 years, through important national efforts to delineate the risk of cancer from chemicals, a number of chemicals tested in lifetime studies in rodents showed definite evidence of carcinogenicity. However, even though these carcinogens were administered at sizable dose levels for two years or more, the incidence of specific cancers was quite small and sometimes affected only one sex of one species.

An important question to be resolved is whether such carcinogens are really contributing to the human cancer burden under the actual conditions of exposure. This area of risk assessment is now a subject of considerable research, and any possible comment would be more a matter of personal opinion than factual knowledge. We, no doubt, would benefit from having the views of a statesman such as Prof. Nakahara.

In occupational carcinogenesis, search for the environmental factors affecting an occupational group has pinpointed specific chemicals or processes as cancer risks. A totally different approach to the question of risk assessment not only considers the potential effect of specific chemicals but also examines the environmental factors as-

TABLE 1. Documentation in the Elucidation of the Etiology of Human Cancers

- I. Epidemiology
  - A. Geographic pathology
  - B. Special populations
  - C. Time trends
- II. Laboratory studies
  - A. Metabolic and biochemical epidemiology—population studies
  - B. Model studies in animals
  - C. Model studies in cell and organ cultures
  - D. Definition of mechanisms
- III. Development and validation of hypotheses
  - A. Established risk factors and their mode of action
  - B. Suspected risk factors and their possible role

sociated with the occurrence of specific types of human cancer. Utilizing this approach to systematically explore the environmental conditions found in those regions of the world with a high incidence of a given type of cancer compared with other conditions prevailing in geographical areas with a low incidence, scientists have discovered major leads to risk factors. In population studies, a sequence of determinations, made preferably by multidisciplinary teams of researchers systematically explore the key elements listed in Table 1.

The application of this approach has provided the basis for the concept that in any area of the world the prevailing cancer incidence and cancer death rates are for the most part due to factors associated with lifestyle. One is the use of tobacco products, especially cigarette smoking. Cigarette smoking is associated with cancer in the respiratory tract and contributes to cancer in the pancreas, kidney, and bladder through as yet unknown mechanisms. Increasing in use in the United States, but well established in Sweden and in a different context in India and parts of China, tobacco chewing and snuff dipping relate to cancer of the oral cavity and esophagus. Also, a habitual high intake of alcoholic beverages together with smoking enhances risk in the oral cavity and esophagus.

The other major lifestyle-related factor is nutrition. It is well documented that Western style nutrition with a fat intake of approximately 40% of calories is associated with higher risk of cancer of the breast, ovary, and endometrium, and probably also with cancer of the prostate and pancreas, and quite definitely with left-sided colon cancer. The carcinogens for these diseases are as yet unknown, but a likely working hypothesis implicates the mutagens and carcinogens formed during cooking. A totally distinct picture is relevant for gastric cancer and perhaps esophageal and liver cancer seen in the Orient. In this instance the customary salting and pickling and certain other traditional nutritional customs are associated with risk for these types of neoplasms.

The complex elements of nutritional carcinogenesis can be viewed by considering the underlying mechanisms of carcinogenesis.

Mechanisms of Carcinogenesis

The process of cancer causation and development involves a series of sequential points. There is good evidence that an early event in neoplasia is a somatic mutation, an alteration of the genetic material (*3, 4*). This key event can arise through a number of mechanisms that modify cellular DNA.

(1) through a direct attack by (a) radiation, (b) chemicals, or (c) viruses, which may result in the mispairing of bases during DNA replication, or by viral insertion of new DNA segments to yield abnormal DNA.

(2) through defective operation of DNA polymerase during synthesis.

(3) through errors introduced by DNA repair enzymes.

(4) through gene translocations, transposition, recombination, and amplification.

Abnormal DNA and genetic material obtained by any mechanism is only the first step in a sequence of molecular and cellular events yielding neoplasia. An abnormal cell population needs to achieve a selective growth advantage in the presence of surrounding normal cells. Cell duplication is a function of a number of endogenous and exogenous controlling factors operating by epigenetic mechanisms. Two such elements are (a) promoters and (b) inhibitors of growth, which either enhance or retard the process. In addtion, during the successive generations, early tumor cells can undergo phenotypic changes of expression, perhaps as a result of faulty steps in differentiation.

Promoters do not produce an invasive cancer in the absence of an antecedent cell change (*4–7*). Thus, in exploring the causes of any specific human cancer, a systematic search is needed for the agents leading to an abnormal genome, and any other agents involved in the growth and development of the resulting abnormal neoplastic cells and their further progression to malignancy (Table 2).

TABLE 2. Types of Agents Associated with the Etiology of Human Cancers

Problems or questions to be resolved for each type of cancer	
I.	Nature of genotoxic carcinogens or mixtures—can be chemical, viral, or radiation
II.	Nature of any nongenotoxic promoting or enhancing stimulus—can be chemical or viral
III.	Amount, duration of exposure, and potency for each kind of agent
IV.	Possibility of inhibition of the action of agents under I or II

Carcinogens in Nutritional Carcinogenesis—Stomach, Esophagus, and Liver Cancer

Salted, pickled, or smoked foods

In relation to the traditional intake of foods noted in the title, several specific chronic diseases will be discussed: (a) glandular gastric cancer, (b) cancer of the esophagus, and (c) hypertension and stroke. These three chronic diseases have an unequal distribution in various parts of the world and even within specific areas of countries (*8–16*). Time trends for these diseases exhibited a pronounced decrease in incidence in the last 50–60 years in some areas of the world. In other areas with a high incidence, a decline has been observed more recently, but in other parts of the world, no change has been noted. Within the limitations of data collection, such as incidence and mortality records, none of these diseases have exhibited an increase anywhere, except for esophageal cancer in parts of Africa.

The decline in incidence and mortality in the United States has been the subject of detailed research (*17*). The decrease in gastric cancer rates had a parallel decline in hypertension and stroke (*14*). This observation has been associated with pronounced changes in nutritional traditions. In many countries, prior to the advent, first, of commercial refrigeration and then later, of household refrigeration, foods were preserved for long-term storage by salting and pickling. Salting and pickling practices involved as much as 7–10% NaCl or brine solutions or mixtures of salt with saltpeter, with amounts of up to 5,000 ppm of saltpeter and a few percent of salt (*18*). These traditions have changed. When it was found that saltpeter or sodium nitrate was biochemically reduced to sodium nitrite, this preservative was substituted for sodium nitrate. Smaller levels of nitrite, initially of the order of 300 ppm nitrite and much more recently, levels as low as 50–80 ppm, were utilized.

Among others, Mirvish (*9*) has contributed sound data on the kinetics of formation of nitrosamines and nitrosamides as a function of such variables as concentration of nitrite, concentration of amine, substrate, pH, temperature, and other conditions. Of great relevance is the observation by Mirvish (*9*) that nitrosation of substrates could be blocked almost completely, depending on conditions, by vitamin C. This finding was extended to vitamin E (*19*). These observations have been utilized in practice to lower the formation and thence the presence of specific nitrosamines in foods such as bacon and other meat products consumed by the public. Also, Haenszel and collaborators (see ref. *10*) and Hirayama (*16*) have noted that yellow-green vegetables may have an inhibiting effect in gastric cancer, and by extension in mechanistic terms, even in head and neck cancers. Yellow-green vegetables are sources of vitamins C and E, but the possible protective effect of carotene-vitamin A in such foods should not be ignored, and certainly deserves more study (*20*).

There is agreement among all workers that gastric cancer in various parts of the world stems from the reaction of nitrite with certain substrates. The nitrite might stem from certain procedures of food preservation such as salting and pickling, a hypothesis we favor on the basis of a number of lines of evidence, or from the presence of high geochemical nitrate levels leading to foods containing high nitrate levels, or the formation mainly in the oral cavity or the stomach of nitrite, and thence leading to endogenous nitrosation mainly in the stomach. In addition, such nitrosation reactions likely can occur during the exposure of certain foods such as fish or meats to wood smoke.

In order to account for the high risk of gastric cancer in parts of Latin America, the groups of Correa (*21*) and Tannenbaum (*22*) have studied the nitrosation of fava beans and have identified a product stemming from this reaction that might be the carcinogen. Mirvish (*9*) has subjected fish to nitrosation and has developed an indication for the production of a nitroso derivative. The group of Weisburger (*17*) has reacted a number of different fish, particularly those customarily used in northern Japan, still a high risk region for gastric cancer, with nitrite and has noted the rapid formation of a direct-acting mutagen (without requiring S9 fractions) in the Ames assay. An extract of this mutagen when administered to Wistar strain rats in a small pilot experiment yielded cancer of the glandular stomach and associated preneoplastic lesions. In the meantime, the group of Sugimura (*23*) found an association between the development of mutagens from different kinds of soy sauce upon nitrosation and the presence of tyramine. They have demonstrated that the product of the reaction of nitrite with tyramine is a novel form of reactive, yet fairly stable, diazonium salt. Weisburger's group has preliminary evidence that the product stemming from the reaction of an extract of *Sanma* fish with nitrite also yields this kind of reactive product.

Briefly, *Sanma* fish purchased from a local Japanese food center was homogenized and extracted with ethanol. The ethanol extract was concentrated and the alcohol-free solution diluted with water, adjusted to pH 3 with HCl and cooled to 4°C. A concentrated solution of $NaNO_2$ was added and the mixture allowed to stand for 1 hr. The excess nitrite was destroyed by ammonium sulfamate. An aliquot of this solution was analyzed for mutagenic activity. The main portion was added to a solution of (1-naphthyl)ethylenediamine adjusted to pH 5. The mixture progressively assumed a reddish-purple color after standing a minimum of 3 hr, but usually overnight. The pH was adjusted to 7 and the desired coupled compound was extracted with ethyl acetate. The extract was concentrated, and the residue dissolved in methanol and streaked on thin layer plates of silicic acid (Whatman LK-6-F). Upon development in 1-butanol: 2 M ammonium hydroxide: ethanol (6:2:2 v:v), the slow R_f, a purple-red band with a motility of 0.46, was scraped off and eluted with methanol. The methanol extract was concentrated, and aliquots were injected for high pressure liquid chromatography (HPLC) using a Zorbax TMS analytical column and a gradient of water/methanol. The UV detector was set at 486 nm. Three peaks were obtained. Two of the peaks had low absorption at 486 nm. The main peak, however, had a UV-visible spectrum consonant with an azo dye structure. The fraction corresponding to this peak was concentrated, the residue dissolved in methanol, and this solution subjected to HPLC under the same conditions. Preliminary observations indicate that the main component in this fraction has a molecular weight of 446 daltons. The exact structure at this point is not known.

Carcinogens in Nutritional Carcinogenesis—Colon, Breast, and Pancreas Cancer

1. Fecapentaenes

The specific carcinogens causing large bowel cancer and its anatomic subsegments, ascending, transverse, descending colon, and rectal cancer, are not known.

Several possible types of carcinogens have been proposed in relation to colon cancer. Stools of some people were found to contain mutagenic activity, present to a greater extent in populations at high risk in the United States or Canada (*24*, *25*). Vegetarians in the United States or in Africa had no fecal mutagenic activity, and of the people tested in Kuopio, Finland, a low-risk population, a lower proportion had mutagenic activity (*26*). A family of mutagens, produced in the membranes of specific colonic bacteria, was identified (*27*, *28*). An example is a highly unsaturated, reactive chemical, (S)-3-(1,3,5,7,9-dodecapentienyloxy)-1,2-propanediol, quite unstable in oxygen. Its carcinogenic activity is currently under investigation.

2. *Heterocyclic amines*

Another lead stemmed from the discovery of Sugimura and Sato (*29*) that broiled fish or meat contained agents with powerful mutagenic activity. Several chemicals, isolated from fried meat or fried fish, belong to a new class of heterocyclic amines such as 2-amino-3-methylimidazo[4,5-f]quinoline (IQ), 2-amino-3,4-dimethylimidazo[4,5-f]quinoline (MeIQ), and 2-amino-3,8-dimethylimidazo[4,5-f]quinoxaline ($MeIQ_x$). These chemicals are highly mutagenic in bacterial systems, and are highly active in the DNA repair test in liver cells (*30*), properties that indicate carcinogenicity. Results from the extensive series of bioassays conducted in Japan in mice and in rats indicate that IQ and MeIQ have select target organs related to human nutritional carcinogenesis in the Western World, including the intestinal tract, pancreas, and bladder (*2*, *31*, *32*). In this Institute IQ was found to be a potent carcinogen for the mammary gland, of the same order of activity as the positive control 4-aminobiphenyl, a human carcinogen (*33*). Thus, the mode of cooking, especially of fish or meat, can be a source of widely distributed carcinogens with target specificities relevant to the types of cancer frequent in the Western World that show an increasing trend in Japan as Western nutritional habits are adopted.

Promoters in Nutritional Carcinogenesis

Type and amount of dietary fat

It is known that populations with a high risk for cancers of the colon, pancreas, breast, prostate, ovary, and endometrium, such as in the Anglo-Saxon countries, Western Europe and Scandinavia, are also people at high risk for coronary heart disease, with a generally high intake of total fat of about 40% of calories (*34–36*). An exceptional case is Finland, with a high fat intake, an elevated risk for heart disease, but a low risk for colon cancer. On the other hand, traditional Japan with a low fat intake of 10–15% of calories, had a low rate of coronary heart disease, and also of these dietary fat related cancers. These situations have been duplicated in carefully controlled laboratory models for certain of these cancers, and the fat effect is seen to translate to a promoting action. Whether or not overt invasive disease develops depends a great deal on epigenetic promoting factors. In the case of prostate cancer, *in situ* lesions have been found in diverse populations (*37*), but clinically invasive disease is observed only in populations consuming higher levels of dietary fat. Thus, the conclusion that promoting stimuli are important for the occurrence of invasive disease is strong.

Also, one of the best arguments for a role of fat in enhancing the risk for these cancers, and of heart disease, is the recent progressively increased incidence as the Japanese traditional low-fat regimen became partially westernized with 25% of calories currently stemming from fat (*16*). This observation also points to the need to recommend to all countries an adjustment in their dietary habits and traditions so that fat remains at about 20% of calories. In the Western World, this amounts to a 50% reduction of the fat calories currently in our diet.

The following discussion will emphasize the rationale for epigenetic actions for cancers of the colon, pancreas, breast, and prostate.

With the advent of good animal models for cancer of the breast, especially those of Huggins and Gullino, a number of groups have uniformly demonstrated that intake of diverse lipids, saturated and unsaturated, at 40% of calories, yields usually more tumors than an intake at 10% of fat calories (see refs. *34*, *38–40*). However, coconut oil and olive oil, discussed below, have a somewhat different effect. Later, Reddy and associates as well as Nigro's group demonstrated a similar relationship in a number of distinct, chemically-induced models for colon cancer (see ref. *41*), as did the groups of Longnecker and of Pour in models for pancreas cancer (*42*, *43*).

In the N-methylnitrosourea (NMU) breast cancer model and in the azoxymethane induced colon cancer model, the effect of olive oil at 40% of fat calories does not differ from 10% fat calories, in contrast to the use of oils such as corn or safflower or fats such as beef fat or lard (*44*, *45*). Interestingly, epidemiologic observations in the Mediterranean countries where olive oil is a large component of the overall fat intake suggest that the rates of heart disease and nutritionally-linked cancers are lower than in other Western countries (*46–49*). Also, C-8, C-10, or C-12 medium chain triglycerides (MCT) failed to increase the incidence of mammary gland and colon cancer in rats in dietary mixtures with 10% of calories as corn oil and 30% of calories as MCT, for a total lipid content of 40% of calories (*45*, *50*). Several studies with fish oils suggest that those products rich in C-20 and C-22 highly unsaturated fatty acids present opportunities for reduced risk in nutritional carcinogenesis (*40*, *51–54*).

In part, the underlying mechanisms have been worked out. The current status is summarized in a symposium volume (*55*). For colon cancer, the dietary fat level eventually translates to specific bile acid levels stemming, in turn, from fats controlling the endogenous production of cholesterol. The cholesterol is further converted, mainly in the liver, to bile acids. The latter are not carcinogens and are not converted to carcinogens through metabolism, but rather act as promoters for colon cancer through specific mechanisms yet to be established (see ref. *56*). The fatty acids stemming from a high fat intake may also exert a promoting action in the colon (*57–59*). The role of dietary fat in modulating breast, ovarian, and endometrial cancer is not yet fully defined by specific mechanisms, but endocrine balances appear to be involved (*54*, *60–64*). This, in turn, accounts for certain other epidemiologic observations delineating the risk for breast cancer such as the role of an early or a late pregnancy, ovariectomy, and the like. A more immediate effect of fat has also been discussed (*65–68*).

Different fats, saturated and unsaturated, had similar high promoting effects at the 40% of calories dietary level. An exception was the fully saturated coconut oil, a poor source of essential fatty acids (EFA). Studies, using mixtures of a saturated fat with low inherent levels of EFA together with relatively smaller, yet adequate amounts of EFA, indicated that, indeed, EFA play a role in the development of mammary gland cancer (see refs. *54*, *69*, *70*). The exact mechanisms have not yet been fully validated, but current views are that endogenous membrane and prostaglandin factors are involved.

In the colon cancer model, the effect of dietary fat can be antagonized by specific dietary fibers found in bran-containing cereals. These findings, in turn, were based on the epidemiologic observations that the colon cancer risk in Finland is much lower than in other Western countries (see below). Thus, the intake of cereal fibers is an excellent tool for reducing the risk of colon cancer, and the fat level is less critical for colon cancer, but most likely not for the other nutritionally-linked cancers in the Western World. Recently, calcium ion was suggested as a means to reduce colon cancer risk by lowering the solubility of bile acids and fatty acids by conversion to calcium salts (*71*, *72*). This area deserves more exploration since the effect was not found in a large prospective study of Japanese men (*73*).

In all the nutritionally-linked cancers, fat may function *via* promoting mechanisms. Those actions are highly dose-dependent, and reversible, at least up to a certain point of development of a neoplasm. Thus, lowering the pressure of promotion is expected to decrease the risk of disease relatively rapidly.

DISCUSSION

During his active career, Prof. Nakahara successfully provided tools for the study of chemical carcinogenesis. His internationally renowned research dealt with the environmental aspects of cancer, with modulating factors such as toxohormone, reminiscent of the contemporary interest in tumor necrosis factor, and with the mechanism of action of a number of newly discovered chemical carcinogens such as 4-NQO. Overall, in his career he set the stage in the Japanese medical and scientific establishment for the nationwide development of cancer research as a subject worthy of interest for young people. This influence and the significant worldwide epidemiologic investigations of Segi and his school have set the stage for major advances. Most cities and university medical schools in Japan can proudly point to investigators who have generated sound information not only with theoretical, fundamental relevance but bearing on etiologic factors for prevalent diseases such as gastric cancer or large bowel cancer. The facts generated form the basis for public health recommendations for lowering the risk of these important diseases. Much progress has been made in the last ten years, and it is sad that Prof. Nakahara, the pioneer of much of this progress, could only see the beginning.

The field of nutritional carcinogenesis has grown because the main participants recognized that it was important to structure research questions along multidisciplinary lines. Through field studies and geographical pathology, the precise environment constituting high risk for gastric, esophageal, or liver cancer in the Orient was found to be in sharp contrast with the environment in the Western World where the major nutritionally-linked cancers were those in the breast, ovary, endometrium, prostate, pancreas, and left-sided colon. Migrant studies pioneered by Haenszel, Wynder, and Hirayama demonstrated that genetics play a minor role since these investigations clearly indicate that the local nutritional customs dictated the type of cancer that would be seen, irrespective of the ethnic origin of a population. The exception, perhaps, is a risk for gastric cancer maintained in the migrants from Japan, a fact consonant with current concepts bearing on the etiology of this neoplasm. The traditional customs of pickling, salting, or smoking of foods, particularly meats or fish, almost invariably was associated with a high risk for gastric, head and neck, or liver cancer. Likewise, residence in areas high in geochemical nitrate such as parts of Latin or central America or eastern Europe seemed associated with these kinds of cancer, especially when other nutritional traditions indicated a diet often poor in protective elements such as milk (protein can be a nitrite trap) and especially of fresh fruits and vegetables (sources of vitamins C and E to detoxify nitrite or carotene or vitamin A to maintain differentiated function). Among others, Joossens and Geboers (*14*) pointed to the association between the occurrence of gastric cancer and that of hypertension-stroke. This association patently stems from the risk for hypertension under conditions of excessive salt intake, without compensation by protective potassium and calcium ions.

Studies, identifying specific carcinogens in several types of frequently eaten foods in high-risk regions, suggest strongly that effective public health recommendations can now be made. These suggestions include changes in nutritional traditions from childhood onward. Specifically recommended is an avoidance of salted, pickled, or smoked foods and a regular intake of fresh fruits and vegetables as sources of specific vitamins and other micronutrients. Along these lines in the United States, the Food and Nutrition Board of the National Academy of Sciences provides guidance in the form of "recommended daily allowances" (RDA) for a number of vitamins and minerals. Unfortunately, the concept on which the RDA is set for each micronutrient is based on the avoidance of deficiency diseases such as scurvy, pellagra, or beri-beri. While such data may still be important in parts of the world, it would seem that in the developed countries a newer concept needs to be established to set what we would call "optimal daily allowances" or ODA. These new values, to be established through further research efforts, would be related to the optimal intake of macronutrients and micronutrients needed to lower risk of chronic diseases: cancer, heart diseases, stroke, diabetes, and even forms of arthritis.

Other major nutritional elements uncovered by research in the last ten years concern fat and fiber and the desirable levels and type of each needed to minimize chronic disease risks. The type and amount of fat have been demonstrated to be important in the occurrence of atherosclerosis and coronary heart disease, and for major kinds of neoplasms such as colon, pancreas, breast, prostate, ovary, and endometrium. While in the past, these were diseases of the Western World, lifestyle changes everywhere, but particularly in Japan, indicate that changing nutritional traditions tend to decrease consumption of total, especially of saturated, fat in the Western World, but to increase it in countries such as Japan and the Soviet Union. Efforts by the U.S. Federal government and associations such as the American Heart Association and the American Cancer Society have altered public dietary habits with the beneficial effect of a lower rate of fatal heart disease. The average diet of the American people now is at the point where fat intake is about 30–35% of calories. However, in Japan, where the diet used to be 10–15% of calories as fat taken in mostly from fish and fish oils, current data show that fat accounts for about 22% of calories. In a Princess Takamatsu lecture in 1971 (*74*), we expressed the hope that worldwide nutritional customs would stabilize fat intake at the current Japanese levels, 20–25% of

TABLE 3. Long Term Goals for Chronic Disease Prevention

Action	Benefit
Control smoking—less harmful cigarettes	Coronary heart disease; cancers of the lung, kidney, bladder, pancreas
Lower total fat intake	Coronary heart disease; cancers of the colon, breast, prostate, ovary, endometrium
Lower salt Na^+ intake—balance $K^+ + Ca^{2+}/Na^+$ ratio	Hypertension, stroke, cardiovascular disease
Increase natural fiber, Ca^{2+} (?)	Colorectal cancer
Avoid pickled, smoked, highly salted foods	Cancer of the stomach, esophagus, nasopharynx, liver (?); hypertension, stroke
Avoid mycotoxins, senecio alkaloids, bracken fern	Cancer of the liver, stomach, bladder (?)
Increase and balance micronutrients, vitamins, minerals	Cardiovascular disease, several types of cancer
Lower intake of fried foods	Cancers of colon, breast, pancreas (?)
Practice sexual hygiene	Cancer of the cervix, penis

fat calories. While exact data are not yet available, this intake of fat may be optimal for prevention of major chronic diseases without handicapping too severely feasible nutritional traditions in the Western World.

Contemporary findings indicate that adequate intake of cereal bran fiber can serve effectively to avoid specific intestinal diseases, including colon cancer. Such a recommendation was made by Burkitt (*75*) many years ago, and current research findings are consonant with this action. Basically, the data show that intake of various types of fiber are desirable in such amounts where the daily stool bulk amounts to about 200 g.

Overall, medical research has collected an adequate data base for recommending a number of specific actions to the public (Table 3). Sugimura in his presidential address to the 3rd International Environmental Mutagen Association has reached a similar conclusion (*76*). We who are in research are not totally satisfied that the details of these recommendations will all be equally successful. Certainly much more effort is needed worldwide. The area of research concerned with food mutagens formed during processing or cooking is less than ten years old, and to fully satisfy Koch's postulates, more data are needed to demonstrate the relevance of these products in human disease causation, even though there is a strong suggestion that they do indeed play a role. Another area deals with the molecular biology of the neoplastic process. This area may provide future tools for new approaches to disease prevention, and more particularly, for early disease diagnosis leading to excellent secondary prevention. It is clear that were Prof. Nakahara alive today, he could look back on his generation of students and young associates and note that knowledge of the mechanisms of carcinogenesis and the causes of important types of cancer has accrued to the point where the prevention of death from many forms of cancer and, indeed, from hypertension and stroke, myocardial infarction, and other chronic diseases is in sight.

ACKNOWLEDGMENTS

Some of the research discussed herein was supported by USPHS grants CA-29602, CA-30658, and CA-24217 (Large Bowel Cancer Project) awarded by the National Cancer Institute (NIH). The author wishes to thank Mrs. Clara Horn for her editorial assistance as well as preparing the manuscript.

REFERENCES

1. Nakahara, W. (ed.) Chemical Tumor Problems, Japanese Society for the Promotion of Science, Tokyo, 1970.
2. Sugimura, T. Carcinogenicity of mutagenic heterocyclic amines formed during the cooking process. Mutat. Res., *150*: 33–41, 1985.
3. Miller, E. C. and Miller, J. A. Mechanisms of chemical carcinogenesis. Cancer, *47*: 1055–1064, 1981.
4. Williams, G. M. and Weisburger, J. H. Chemical carcinogens. *In;* J. Doull, C. D. Klaassen, and M. O. Amdur (eds.), Casarett and Doull's Toxicology. The Basic Science of Poisons, 3rd ed., Macmillan, New York, 1986 (in press).
5. Slaga, T. J. Overview of tumor promotion in animals. Environ. Health Perspect., *50*: 3–14, 1983.
6. Pitot, H. and Sirica, A. The stages of initiation and promotion in hepatocarcinogenesis. Biochim. Biophys. Acta, *605*: 191–195, 1980.
7. Highman, B., Norvell, M. J., and Shellenberger, T. E. Pathological changes in female C3H mice continuously fed diets containing diethylstilbestrol or 17β-estradiol. J. Environ. Pathol. Toxicol., *1*: 1–30, 1977.
8. Schottenfeld, D. and Fraumeni, J. F., Jr. Cancer Epidemiology and Prevention. W. B. Saunders Co, Philadelphia, 1982.
9. Mirvish, S. S. The etiology of gastric cancer. J. Natl. Cancer Inst., *71*: 631–647, 1983.
10. Henderson, B. E. (ed.) Third Symposium on Epidemiology and Cancer Registries in the Pacific Basin. NCI Monogr. 62, Natl. Cancer Inst., Bethesda, MD, 1982.
11. Waterhouse, J.A.H. Epidemiology of upper gastrointestinal tumors. *In;* M. Friedman, M. Ogawa, and D. Kisner (eds.), Diagnosis and Treatment of Upper Gastrointestinal Tumors, pp. 309–341. Excerpta Medica, Amsterdam-Oxford-Princeton, 1980.
12. Crespi, M., Muñoz, N., Grassi, A., Qiong, S., Jing, W. K., and Jien, L. J. Precursor lesions of oesophageal cancer in a low-risk population in China: comparison with high-risk populations. Int. J. Cancer, *34*: 599–602, 1984.
13. Day, N. E. The geographic pathology of cancer of the oesophagus. Br. Med. Bull., *40*: 329–334, 1984.
14. Joossens, J. V. and Geboers, J. Diet and environment in the etiology of gastric cancer. *In;* R. H. Riddell and B. Levin (eds.), Frontiers of Gastrointestinal Cancer, pp. 167–183, Elsevier Sci. Publ., New York, 1984.
15. Coggon, D. and Acheson, E. D. The geography of cancer of the stomach. Br. Med. Bull., *40*: 335–341, 1984.
16. Hirayama, T. Diet and cancer. Nutr. Cancer, *1*: 67–79, 1979.
17. Weisburger, J. H. and Horn, C. L. Human and laboratory studies on the causes and prevention of gastrointestinal cancer. Scand. J. Gastroenterol., *19* (Suppl. 104): 15–26, 1984.
18. Binkerd, E. F. and Kolari, O. E. The history and use of nitrate and nitrite in the curing of meat. Food Cosmet. Toxicol., *13*: 655–661, 1975.

19. Newmark, H. L. and Mergens, W. J. Block nitrosamine formation using ascorbic acid and alpha-tocopherol. Banbury Rep., *7*: 285–290, 1981.
20. Colditz, G. A., Branch, L. G., Lipnick, R. J., Willett, W. C., Rosner, B., Posner, B. M., and Hennekens, C. H. Increased green and yellow vegetable intake and lowered cancer deaths in an elderly population. Am. J. Clin. Nutr., *41*: 32–36, 1985.
21. Montes, G., Cuello, C., Correa, P., Haenszel, W., Zarama, G., and Gordillo, G. Mutagenic activity of nitrosated foods in an area with a high risk for stomach cancer. Nutr. Cancer, *6*: 171–175, 1984.
22. Yang, D., Tannenbaum, S. R., Büchi, G., and Lee, G.C.M. 4-Chloro-6-methoxyindole is the precursor of a potent mutagen (4-chloro-6-methoxy-2-hydroxy-1-nitroso-indolin-3-one oxime) that forms during nitrosation of the fava bean (*Vicia faba*). Carcinogenesis, *5*: 1219–1224, 1984.
23. Ochiai, M., Wakabayashi, K., Nagao, M., and Sugimura, T. Tyramine is a major mutagen precursor in soy sauce, being convertible to a mutagen by nitrite. Gann, *75*: 1–3, 1984.
24. Bruce, W. R., Varghese, A. J., Fuller, R., and Land, P. C. A mutagen in the feces of normal humans. *In;* H. H. Hiatt, J. D. Watson, and J. A. Winsted (eds.), Origins of Human Cancer, pp. 1641–1646, Cold Spring Harbor Laboratory, New York, 1977.
25. Wilkins, T. D., Lederman, M., Van Tassell, R. L., Kingston, D.G.I., and Henion, J. Characterization of a mutagenic bacterial product in human feces. Am. J. Clin. Nutr., *33*: 2513–2520, 1980.
26. Reddy, B. S., Ekelund, G., Bohe, M., Engle, A., and Domellöf, L. Metabolic epidemiology of colon cancer: Dietary pattern and fecal sterol concentrations of 3 populations. Nutr. Cancer, *5*: 34–40, 1983.
27. Gupta, I., Baptista, J., Bruce, W. R., Che, C. T., Furrer, R., Gingerich, J. S., Grey, A. A., Marai, L., Yates, P., and Krepinsky, J. J. Structures of fecapentaenes, the mutagens of bacterial origin isolated from human feces. Biochemistry, *22*: 241–245, 1983.
28. Hirai, N., Kingston, D.G.I, Van Tassell, R. L., and Wilkins, T. D. Structure elucidation of a potent mutagen from human feces. J. Am. Chem. Soc., *104*: 6149–6150, 1982.
29. Sugimura, T. and Sato, S. Mutagens-carcinogens in foods. Cancer Res., *43*: 2415s–2421s, 1983.
30. Barnes, W. S., Lovelette, C. A., Tong, C., Williams, G. M., and Weisburger, J. H. Genotoxicity of the food mutagen, 2-amino-3-methylimidazo[4,5-f]quinoline (IQ) and analogs. Carcinogenesis, *6*: 441–444, 1985.
31. Ohgaki, H., Kusama, K., Matsukura, N., Morino, K., Hasegawa, H., Sato, S., Sugimura, T., and Takayama, S. Carcinogenicity in mice of a mutagenic compound, 2-amino-3-methylimidazo[4,5-f]-quinoline, from broiled sardine, cooked beef, and beef extract. Carcinogenesis, *5*: 921–924, 1984.
32. Takayama, S., Nakatsuru, Y., Masuda, M., Ohgaki, H., Sato, S., and Sugimura, T. Demonstration of carcinogenicity in F344 rats of 2-amino-3-methylimidazo[4,5-f] quinoline (IQ) from broiled sardine, fried beef and beef extract. Gann, *75*: 467–470, 1984.
33. Tanaka, T., Barnes, W. S., Weisburger, J. H., and Williams, G. M. Multipotential carcinogenicity of the fried food mutagen 2-amino-3-methylimidazo[4,5-f]quinoline (IQ) in rats. Jpn. J. Cancer Res. (Gann), *76*: 570–576, 1985.
34. Reddy, B. S., Cohen, L. A., McCoy, G. D., Hill, P., Weisburger, J. H., and Wynder, E. L. Nutrition and its relationship to cancer. Adv. Cancer Res., *32*: 237–345, 1980.

35. National Academy of Sciences National Research Council Committee on Diet, Nutrition, and Cancer. Diet, Nutrition, and Cancer. National Academy Press, Washington, D.C., 1982.
36. Correa, P., Strong, J.P., Johnson, W. D., Pizzolato, P., and Haenszel, W. Atherosclerosis and polyps of the colon. Quantification of precursors of coronary heart disease and colon cancer. J. Chronic Dis., *35*: 313–320, 1982.
37. Akazaki, K. and Stemmermann, G. N. Comparative study of latent carcinoma of the prostate among Japanese in Japan and Hawaii. J. Natl. Cancer Inst., *50*: 1137–1142, 1973.
38. Wynder, E. L., Leveille, G. A., Weisburger, J. H., and Livingston, G. E. (eds.) Environmental Aspects of Cancer: The Role of Macro and Micro Components of Foods, Food and Nutrition Press, Westport, Connecticut, 1983.
39. Kalamegham, R. and Carroll, K. K. Reversal of the promotional effect of high-fat diet on mammary tumorigenesis by subsequent lowering of dietary fat. Nutr. Cancer, *6*: 22–31, 1984.
40. Carroll, K. K. and Braden, L. M. Dietary fat and mammary carcinogenesis. Nutr. Cancer, *6*: 254–259, 1985.
41. Autrup, H. and Williams, G. M. (eds.) Experimental colon carcinogenesis, CRC Press, Inc., Boca Raton, Florida, 1983.
42. Longnecker, D. S. and Morgan, R. G. Diet and cancer of the pancreas. Epidemiological and experimental evidence. *In;* B. S. Reddy and L. A. Cohen (eds.), Diet, Nutrition and Cancer: A Critical Evaluation, vol. 1, pp. 11–26, CRC Press, Boca Raton, Florida, 1985.
43. Birt, D. F., Salmasi, S., and Pour, P. M. Enhancement of experimental pancreatic cancer in Syrian golden hamsters by dietary fat. J. Natl. Cancer Inst., *67*: 1327–1332, 1981.
44. Cohen, L. A., Thompson, D. O., Maeura, Y., Choi, K., Blank, M. E., and Rose, D. P. Dietary fat and mammary cancer. 1. Promoting effects of different dietary fats on NMU-induced mammary tumorigenesis. J. Natl. Cancer Inst., *77*: 1986 (in press.)
45. Reddy, B. S. and Maeura, Y. Tumor promotion by dietary fat in azoxymethane-induced colon carcinogenesis in female F344 rats: influence of amount and source of dietary fat. J. Natl. Cancer Inst., *72*: 745–750, 1984.
46. Christakis, G., Fordyce, M. K., and Kurtz, C. S. The biological and medical aspects of olive oil. *In;* Proceedings of the IIIrd International Congress on the Biological Value of Olive Oil, pp. 85–120, Subtropical Plants and Olive Trees Institute of Chainia, Crete, Greece, 1980.
47. Manousos, O., Day, N. E., Trichopoulos, D., Gerovassilis, F., Tzonou, A., and Polychronopoulou, A. Diet and colorectal cancer: A case-control study in Greece. Int. J. Cancer, *32*: 1–5, 1983.
48. Talamini, R., LaVecchia, C., Decarli, A., Franceschi, S., Grattoni, E., Grigoletto, E., Liberati, A., and Tognoni, G. Social factors, diet and breast cancer in a northern Italian population. Br. J. Cancer, *49*: 723–729, 1984.
49. Ferro-Luzzi, A., Strazzullo, P., Scaccini, C., Siani, A., Sette, S., Mariani, M. A., Mastranzo, P., Dougherty, R. M., Iacono, J. M., and Mancini, N. Changing the Mediterranean diet: effects on blood lipids. Am. J. Clin. Nutr., *40*: 1027–1037, 1984.
50. Cohen, L. A., Thompson, D. O., Maeura, Y., and Weisburger, J. H. Influence of dietary medium-chain triglycerides on the development of N-methylnitrosourea-induced rat mammary tumors. Cancer Res., *44*: 5023–5028, 1984.
51. Gabor, H. A. Fish oils and cancer. N. Engl. J. Med., *313*: 823–824, 1985.

52. Jurkowski, J. J. and Cave, W. T., Jr. Dietary effects of menhaden oil on the growth and membrane lipid composition of rat mammary tumors. J. Natl. Cancer Inst., *74*: 1145–1150, 1985.
53. Reddy, B. S. and Maruyama, H. Effect of dietary fish oil on azoxymethane-induced colon carcinogenesis in male F344 rats. Cancer Res., *49*: 1986 (in press).
54. Welsch, C. W. Host factors affecting the growth of carcinogen-induced rat mammary carcinomas: a review and tribute to Charles Brenton Huggins. Cancer Res., *45*: 3415–3443, 1985.
55. Perkins, E. G. (ed.) Dietary Fats and Health. American Oil Chemists' Society, Champaign, Illinois, 1983.
56. DeRubertis, F. R., Craven, P. A., and Saito, R. Bile, eicosanoids, and colonic cell proliferation. J. Clin. Invest., *74*: 1614–1624, 1984.
57. Bird, R. P., Medline, A., Furrer, R., and Bruce, W. R. Toxicity of orally administered fat to the colonic epithelium of mice. Carcinogenesis, *6*: 1063–1066, 1985.
58. Bull, A. W., Nigro, N. D., Golembieski, W. A., Crissman, J. D., and Marnett, L. J. *In vivo* stimulation of DNA synthesis and induction of ornithine decarboxylase in rat colon by fatty acid hydroperoxides, autoxidation products of unsaturated fatty acids. Cancer Res., *44*: 4924–4928, 1984.
59. Sakaguchi, M., Hiramatsu, Y., Takada, H., Yamamura, M., Hioki, K., Saito, K., and Yamamoto, M. Effect of dietary unsaturated and saturated fats on azoxymethane-induced colon carcinogenesis in rats. Cancer Res., *44*: 1472–1477, 1984.
60. Aylsworth, C. F., Van Vugt, D. A., Sylvester, P. W., and Meites, J. Role of estrogen and prolactin in stimulation of carcinogen-induced mammary tumor development by a high-fat diet. Cancer Res., *44*: 2835–2840, 1984.
61. Ip, C., Carter, C. A., and Ip, M. M. Requirement of essential fatty acid for mammary tumorigenesis in the rat. Cancer Res., *45*: 1997–2001, 1985.
62. Manni, A. and Wright, C. Polyamines as mediators of the effect of prolactin and growth hormone on the growth of N-nitroso-N-methylurea-induced rat mammary tumor cultured *in vitro* in soft agar. J. Natl. Cancer Inst., *74*: 941–944, 1985.
63. Rose, D. P. (ed.) Endocrinology of Cancer, vol. I–III, CRC Press, Boca Raton, Florida, 1982.
64. Reddy, B. S. and Cohen, L. A. (eds.) Diet, Nutrition, and Cancer: A Critical Evaluation, vol. 1–2, CRC Press, Boca Raton, Florida, 1986.
65. Wetsel, W. C., Rogers, A. E., Rutledge, A., and Leavitt, W. W. Absence of an effect of dietary corn oil content on plasma prolactin, progesterone, and 17β-estradiol in female Sprague-Dawley rats. Cancer Res., *44*, 1420–1425, 1984.
66. Aylsworth, C. F., Van Vugt, D. A., Sylvester, P.O.W., and Meites, J. Failure of high dietary fat to influence serum prolactin levels during the estrous cycle in female Sprague-Dawley rats. Proc. Soc. Exp. Biol. Med., *175*: 25–29, 1984.
67. Aylsworth, C. F., Jone, C., Trosko, J. E., Meites, J., and Welsch, C. W. Promotion of 7,12-dimethylbenz[a]anthracene-induced mammary tumorigenesis by high dietary fat in the rat: possible role of intercellular communication. J. Natl. Cancer Inst., *72*: 637–645, 1984.
68. Oyaizu, N., Morii, S., Saito, K., Katsuda, Y., and Matsumoto, J. Mechanism of growth enhancement of 7,12-dimethylbenz[a]anthracene-induced mammary tumors in rats given high polyunsaturated fat diet. Jpn. J. Cancer Res. (Gann), *76*: 676–683, 1985.
69. Carroll, K. K. and Kalamegham, R. Lipid components and cancer. *In;* E. L. Wynder, G. A. Leveille, J. H. Weisburger, and G. E. Livingston (eds.), Environmental Aspects

of Cancer: The Role of Macro and Micro Components of Foods, pp. 101–107, Food and Nutrition Press, Westport, Connecticut, 1983.

70. King, M. M. and McCay, P. B. Modulation of tumor incidence and possible mechanisms of inhibition of mammary carcinogenesis by dietary antioxidants. Cancer Res., *43*: 2485s–2490s, 1983.

71. Newmark, H. L., Wargovich, M. J., and Bruce, W. R. Colon cancer and dietary fat, phosphate, and calcium: a hypothesis. J. Natl. Cancer Inst., *72*: 1323–1325, 1984.

72. Garland, C., Shekelle, R. B., Barrett-Connor, E., Criqui, M. H., Rossof, A. H., and Paul, O. Dietary vitamin D and calcium and risk of colorectal cancer: a 19 year prospective study in men. Lancet, *1*: 307–309, 1985.

73. Heilbrun, L. K., Nomura, A., Hankin, J. H., and Stemmermann, G. N. Dietary vitamin D and calcium and risk of colorectal cancer. Lancet, *1*: 925, 1985.

74. Weisburger, J. H. Model studies on the etiology of colon cancer. *In;* W. Nakahara, S. Takayama, T. Sugimura, and S. Odashima (eds.), Topics in Chemical Carcinogenesis, pp. 159–170. Japan. Sci. Soc. Press, Tokyo, 1972.

75. Burkitt, D. P. Non-infective disease of the large bowel. Br. Med. Bull., *40*: 387–389, 1984.

76. Sugimura, T. A view of a cancer researcher on environmental mutagens. *In;* T. Sugimura, S. Kondo, and H. Takebe (eds.), Environmental Mutagens and Carcinogens, pp. 1–20, Univ. of Tokyo Press, Tokyo, 1982.

EPIDEMIOLOGICAL STUDIES ON IMMIGRANTS

DIET, NUTRITION AND CANCER, Y. HAYASHI ET AL. (EDS.),
JAPAN SCI. SOC. PRESS, TOKYO/VNU SCI. PRESS, UTRECHT, PP. 29–40, 1986

Multiethnic Studies of Diet, Nutrition, and Cancer in Hawaii

Laurence N. KOLONEL, Jean H. HANKIN, and Abraham M. Y. NOMURA
Cancer Research Center, University of Hawaii, Honolulu, Hawaii 96813, U.S.A.

Abstract: Epidemiologic studies of diet and cancer have been facilitated in Hawaii by the multiethnic composition of its population and the consequent heterogeneity in dietary intakes. Studies of migrant populations, particularly the Japanese, have firmly supported the conclusions that environmental factors are of predominant etiologic significance for most major sites of cancer, and that these factors may exert their influences at particular periods of life. Recent observations on Filipino migrants reproduce most of the findings in the Japanese, although they do not show the same abrupt increase in colon cancer rates to the high levels found in Caucasians. Data on dietary intakes in these populations support several of the prevailing hypotheses regarding the etiology of certain gastrointestinal and hormone-dependent cancers.

Several case-control studies of diet and cancer have been completed or are ongoing in Hawaii. Some of these have included comparable studies in Japan, but the findings in Hawaii have generally not been reproduced in Japan. Weak associations with dietary fat have been found in Hawaii for breast cancer (particularly in Japanese women) and for prostate cancer (particularly in men ≥ 70 years of age). Vitamin A (especially carotene) has been shown to be inversely associated with lung cancer risk in men, but positively associated with prostate cancer risk in older men. Vitamin C may be inversely related to bladder cancer risk, but has shown no relationship to lung or prostate cancer risk. These and other findings are discussed in terms of future needs for epidemiologic research in this field.

Epidemiologic studies of diet, nutrition, and cancer are likely to be most productive when carried out in populations with maximum variability in intakes on the factors under study. In this regard, Hawaii is a particularly apt place for such research, because the local mixture of ethnicities and cultures provides considerable dietary heterogeneity in the population. In addition, the geographic isolation of the islands, and the presence of a high quality population-based cancer registry facilitate long-term follow-up on prospective cohorts. These advantages, together with a cooperative population, have resulted in a notable concentration of epidemiologic research programs in Hawaii.

Studies on Migrant Populations

1. Japanese

The relative recency of migration of many of the people in Hawaii has led to a particular research interest in the effects of migration on cancer risks. Most studied of the ethnic groups in the islands have been the Japanese. Comparisons of cancer incidence and mortality rates between Japan and Hawaii, and between first and second generation Japanese (Issei and Nisei) in Hawaii have provided important clues to cancer etiology and mechanisms. Figure 1, for example, shows incidence rates for four cancer sites in four comparison groups. In general, the rates in the migrants shift towards the levels prevalent in the U.S., the host country, an observation that has also been made in other parts of the U.S. and the world (*1–4*). The rapidity of these changes is striking, in that Issei rates show at least a 2-fold shift for all four sites. The results for colon cancer differ from the other two sites in that a gradient is lacking, the change from Japan to Hawaii being abrupt. For most other sites that have been examined in Hawaii and elsewhere, the change is the more gradual one seen for breast and stomach cancers.

From these results, one can conclude that: 1) environmental factors seem to be the primary determinants of risk for these cancer sites; and 2) the effects of early life exposures limit the extent to which risk can be modified in later life for some sites (*e.g.*, breast, stomach) but not others (*e.g.*, colon). The latter observation suggests a

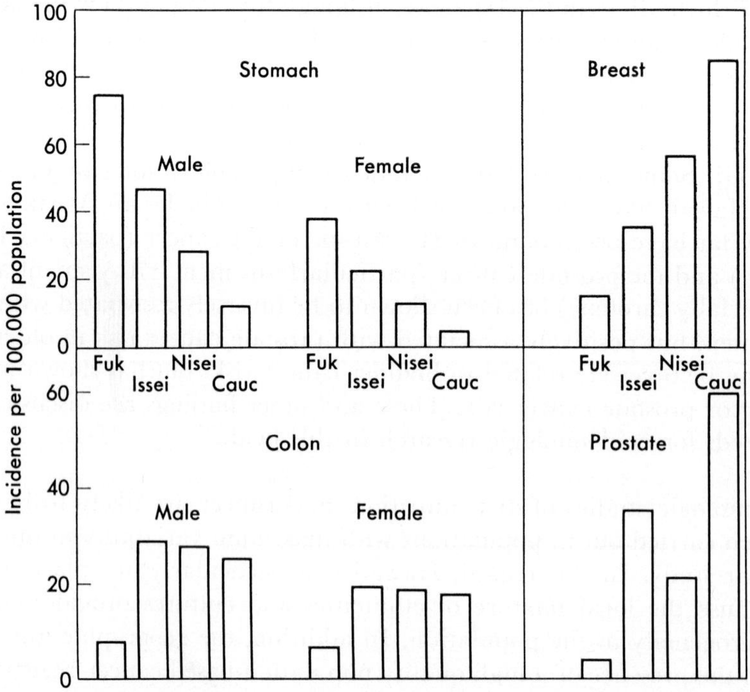

FIG. 1. Age-adjusted cancer incidence rates (World Population Standard) in Japanese migrants to Hawaii and indigenous populations, for the period 1973–77 (Hawaii) and 1974–75 (Japan). Fuk, Fukuoka Japanese; Cauc, Hawaii Caucasians.

TABLE 1. Mean Food and Nutrient Intakes of Issei and Nisei in Hawaii, 1977–1979[a]

Dietary component	Males		Females	
	Issei (n=98)	Nisei (n=772)	Issei (n=200)	Nisei (n=820)
Foods (weekly)				
Rice (g)	2,464	2,369	1,997	1,363[b]
Tofu (g)	219	169[b]	196	135[b]
Sashimi (g)	69	63	50	30[b]
Meat (g)	484	539[b]	360	378[b]
Fish: all (g)	226	224	179	160[b]
,, : dried/salted (g)	5	4	3	2
Coffee (cups)	14	16[b]	10	13[b]
Green tea (cups)	12	8[b]	17	8[b]
Nutrients (daily)				
Fat: total (g)	64.9	68.0[b]	50.8	53.6[b]
,, : animal (g)	41.6	43.2[b]	32.7	34.0
,, : cholesterol (mg)	356.3	326.7	273.3	246.2
Protein: total (g)	65.0	65.9[b]	52.1	51.3
,, : animal (g)	39.6	40.7[b]	29.7	30.8
Carbohydrate (g)	194.6	195.5	164.3	146.5[b]
Vitamin A[c] (I.U.)	6,811	5,568[b]	5,838	5,575
Vitamin C[c] (mg)	314	288	332	278

[a] Intakes are age-adjusted by analysis of covariance. [b] Differs significantly from Issei ($p<0.05$). [c] Includes supplement use.

multistage process for carcinogenesis, with determinants that may be active at different periods of life or biologic responses that differ with age. The initiation-promotion theory is consistent with these results, and would imply, for example, that initiators for colon cancer, acting early in life, may be equally prevalent in Japan and Hawaii, whereas promoters, acting later in life, are more prevalent in Hawaii. On this basis, one might expect that major risk factors for colon cancer in adults would be equally prevalent among Issei and Nisei in Hawaii, whereas corresponding adult risk factors for breast cancer would be more prevalent among Nisei than Issei women. This would lead one away from the hypothesis that dietary fat is a major promoter for both breast and colon cancers in women.

A number of studies on the relationship of exposure factors to cancer risk have been conducted in the Japanese population of Hawaii. Many of these studies have examined dietary factors as the likeliest agents to account for the varied cancer risks within this group. For example, we have obtained data on dietary intakes of representative Issei and Nisei men and women. Some of the findings are summarized in Table 1. As one would expect, Issei consume more "Japanese" items (rice, tofu, sashimi, salted fish, green tea) whereas Nisei consume more "Western" items (meat, coffee). This is reflected in nutrient intake differences as well. Fat consumption (both total and animal) is greater among Nisei, but, perhaps surprisingly, cholesterol consumption is greater among Issei. In contrast, protein intake is very similar in both groups. Thus, with regard to the observations on incidence patterns in Issei and Nisei, these results would support the hypothesis that dietary fat may be a promoter for breast cancer but not colon cancer. Indeed, the findings on protein intake are more consistent with the observations on colon cancer. However, as has been shown elsewhere (*5*), dietary fat and protein intake are highly correlated in the population of Hawaii, and associations with one of these nutrients cannot exclude the possibility that any true effect is due to the other (*6*).

Parallel case-control studies of diet in relation to stomach, large bowel, and breast cancers have been carried out among the Japanese populations of Hawaii and Japan. Haenszel and co-workers found an association between consumption of pickled vegetables and dried/salted fish and stomach cancer risk in Hawaii (*7*), but not in Japan, where an inverse association was found with consumption of lettuce and celery (*8*). The same workers found a direct association between beef consumption and large bowel cancer in Hawaii (*9*), but not in Japan, where an inverse association was found with consumption of cabbage (*10*). The failure to reproduce the Hawaii findings in Japan may be a reflection of greater dietary heterogeneity within the Japanese population in Hawaii than in Japan, particularly with regard to such food items as salted fish and beef. The inverse associations found in Japan could be indirect reflections of the same positive associations found in Hawaii, in that subjects consuming more fresh vegetables in Japan may actually consume less pickled and preserved items. A case-control study of breast cancer has also shown some difference between the results in Hawaii and those in Japan. Whereas there were some dietary differences between Japanese cases and controls in Hawaii (discussed below), there were no apparent differences in Japan (*11*).

A cohort of Japanese men in Hawaii, including both Nisei and Issei members, is being followed prospectively for cancer incidence in relation to dietary and other exposure factors, and many of the observations on this group have been published (*12*). A report on the findings with regard to fat and cancer in this cohort will be presented elsewhere in this Symposium (*13*).

2. *Other ethnic groups*

More recently, other migrant ethnic groups in Hawaii have been studied. A comparison of data among Filipinos in Hawaii and the Philippines, for example, showed that the usual patterns of change in migration occur in this population as well (Table 2). Although the incidence data from the Philippines (Manila area only) may be somewhat underreported, the overall patterns are likely to be real (*14*). Like the Japanese migrants to Hawaii, the Filipinos show a decline in stomach, liver, and cervical cancer incidence, and an increase in the incidence of cancers of the colon, thyroid, prostate, and breast. Unlike the Japanese, however, the Filipinos do not show the abrupt increase in colon cancer rates to those seen in the Caucasian population. In fact, a more detailed analysis of colon cancer incidence trends in Hawaii (Fig. 2) shows that the differential between Filipinos and Caucasians is persisting despite an increasing trend among both. This lack of convergence may be a result of continued adherence to traditional dietary patterns by Filipinos, and may also reflect in part the continued high immigration rate in this group.

Some dietary comparisons of Filipinos and Caucasians in Hawaii are shown in Table 3. Filipinos, like the Japanese, consume more of the traditional foods of South-

TABLE 2. Cancer Incidence Rates in the Philippines and among Filipinos and Caucasians in Hawaii[a]

Cancer site	Philippines	Hawaii		
	Manila 1977	Filipinos 1962–65	Filipinos 1978–81	Caucasians 1978–81
Males				
Stomach	10	7.1	6.6	14.8
Colon	5	14.3	18.8	33.6
Rectum	6	12.4	16.2	17.5
Liver	20	7.5	4.1	5.6
Lung	29	15.9	27.3	75.3
Thyroid	2	5.0	5.6	1.0[b]
Prostate	8	14.0	33.4	69.0
Females				
Colon	4	6.4[b]	11.6	26.7
Lung	9	17.3[b]	16.3	36.4
Thyroid	5	22.4	17.4	6.3
Breast	31	18.2	36.2	92.9
Cervix	16	16.6[b]	7.5	10.2

[a] Average annual incidence per 100,000 population, age-adjusted to the World Population Standard. [b] Rate is based on fewer than 10 cases.

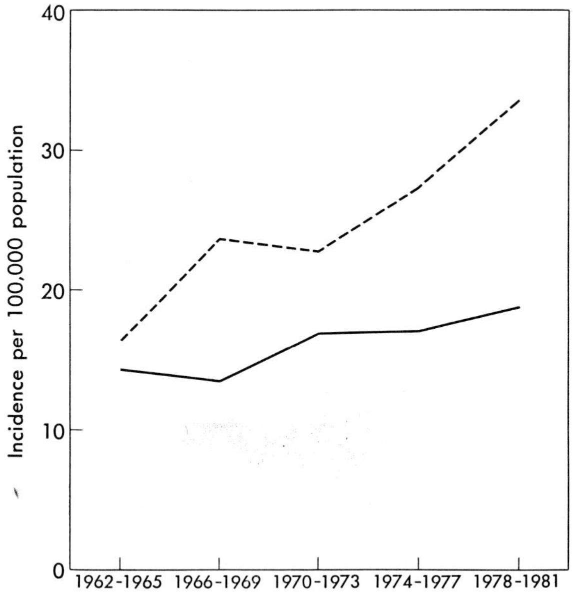

FIG. 2. Trends in age-adjusted colon cancer incidence (World Population Standard) among Filipino (——) and Caucasian (----) males in Hawaii, 1962–81.

east Asia (*e.g.*, rice and fish) than do the Caucasians. On the other hand, Filipino beef, fat, and vitamin C intakes are lower than in Caucasians. In terms of current etiologic hypotheses (*6*), the higher fat intake is consistent with the higher breast and prostate cancer rates among Caucasians.

TABLE 3. Comparison of Filipinos and Caucasians in Hawaii on Selected Daily Food and Nutrient Intakes, 1977–79

Dietary component	Males[a]		Females[a]	
	Filipino	Caucasian	Filipino	Caucasian
Foods				
Rice (g)	431.9	90.2	300.6	59.8
Fish (g)	32.7	23.0	25.8	19.1
Pork (g)	37.7	34.3	27.1	22.2
Beef(g)	34.6	51.4	24.2	37.0
Coffee (cups)	2.0	3.1	1.6	2.8
Nutrients				
Fat	57.9	81.9	48.1	63.0
Protein	61.8	69.6	51.2	54.4
Vitamin C[b]	136.1	305.6	208.1	349.9

[a] Intakes are for adults age 45 and older and are age-adjusted by analysis of covariance. [b] Includes supplements.

Interestingly, Caucasian incidence rates seem to differ by migrant status in Hawaii, as well. The only site yet examined in detail is the stomach, where incidence rates are clearly higher among those born in Hawaii than among those born on the U.S. mainland (*15*). Differential exposure to dietary factors etiologically related to this site, such as dried/salted fish, is one plausible explanation for this intriguing observation.

Multiethnic Studies

Much of the more recent research in Hawaii has been focused on interethnic comparisons among the five major racial groups: Caucasians, Japanese, Hawaiians, Filipinos, and Chinese. Early data on cancer incidence in these groups showed that their rates varied rather substantially. Some examples of these variations are shown in Fig. 3. Apart from cigarette smoking, no exposure factor of widespread distribution in the population of Hawaii other than diet showed sufficient variability to be a likely explanation for these differences. Initial correlations suggested that certain nutrients which varied substantially among the groups might be of significance, particularly dietary fat and protein (*5, 16*). On the other hand, we developed a particular interest in the role of vitamin A because of a lack of correlation, *i.e.*, the failure of smoking patterns to correspond with lung cancer rates (*17*).

Follow-up on these observations resulted in our conducting several case-control studies. A study of breast cancer included Japanese women in both Hawaii and Japan, as well as Caucasian women in Hawaii, and was mentioned earlier. This study examined the relationship of animal protein and dietary fat to disease risk. Among Japanese and Caucasian women in Hawaii, there were only small increases in fat intake by cases compared with controls and these were not statistically significant (Table 4). For the Japanese women, there was also increased animal protein consumption by the cases. In a similar study of prostate cancer, we also found higher

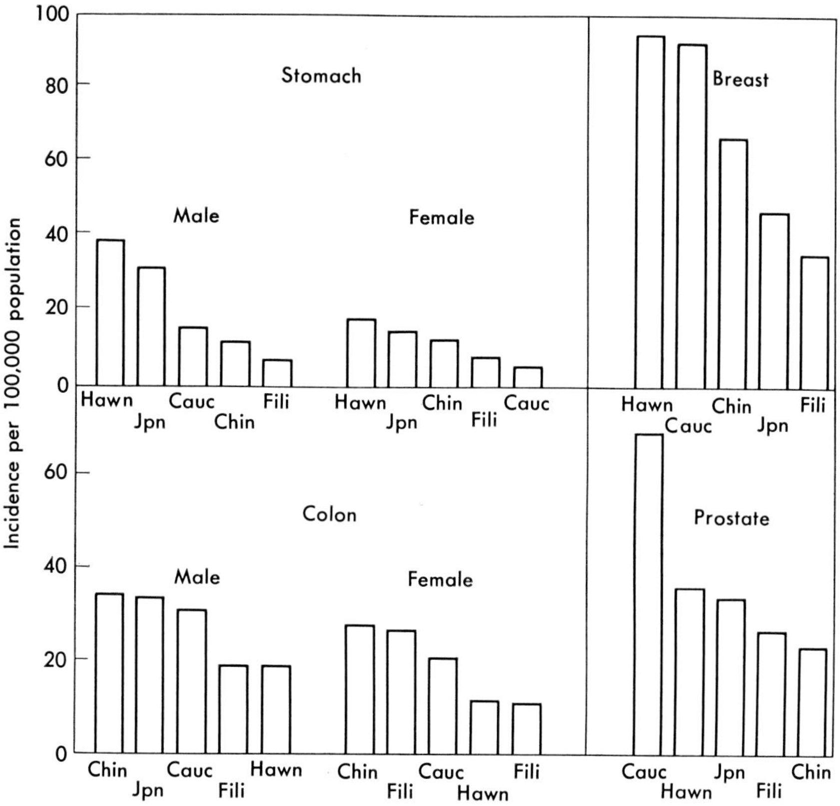

FIG. 3. Age-adjusted cancer incidence (World Population Standard) for five ethnic groups in Hawaii, 1978–81. Hawn, Hawaiian; Jpn, Japanese; Cau, Caucasian; Chin, Chinese; Fili, Filipino.

TABLE 4. Dietary Fat and Animal Protein Consumption by Breast and Prostate Cancer Subjects in Hawaii

Cancer site	Subgroup (No. of cases / No. of controls)	Nutrient component	Mean weekly intake (g): Cases	Controls	p-value for difference
Breast[a]	Japanese (183/183)	Total fat	365	338	0.12
		Saturated fat	121	111	0.12
		Animal protein	254	237	0.08
	Caucasian (161/161)	Total fat	434	405	0.12
		Saturated fat	152	140	0.09
		Animal protein	305	305	0.86
Prostate[b]	≥70 years (263/508)	Total fat	498	473	0.17
		Saturated fat	169	155	0.04
	<70 years (189/391)	Total fat	572	572	1.00
		Saturated fat	191	186	0.53

[a] Intakes simultaneously adjusted by multiple covariance analysis for current age, age at first birth, age at menopause, history of benign breast disease and family history of breast cancer. [b] Intakes simultaneously adjusted by multiple covariance analysis for age, race, and socioeconomic status.

fat consumption by cases than controls, but this was apparent only for the men 70 years of age and older, and was most evident with respect to saturated fats (Table 4). Other case-control studies of dietary fat in relation to these cancers have been carried out in North America, and, with the exception of one negative breast cancer study (*18*), all have reported positive associations of a generally modest degree as here (*19–26*). If fat is indeed acting as a promoter, one may only expect to see relatively small increases in risk in the absence of additional data on exposure to initiating factors and other modifying factors as well.

Our case-control studies on vitamins A and C have also been illuminating. Vitamin A intake was clearly and inversely related to lung cancer risk in men but not in women (Table 5). When carotenes and retinol were examined separately, it appeared that the effect was due to the former component. Several case-control and cohort investigations elsewhere have also examined this relationship (*27–37*). Most studies have found this inverse association, particularly in men. The findings in women have been mixed. Like us, Gregor *et al.* (*31*) did not find the association in women, whereas Wu *et al.* (*37*) and Kvale *et al.* (*33*) did.

In an ongoing case-control study of bladder cancer, preliminary data are also suggesting a possible protective effect of vitamin A (*38*). On the other hand, in the case-control study of prostate cancer mentioned earlier, we found a direct association for vitamin A in the older men (≥70 years); this effect was not apparent in the younger men, where there was a non-significant inverse linear trend (Table 5). These relationships were not significantly altered by additionally adjusting for fat intake in the analysis. There is little epidemiologic evidence yet in support of the animal work showing a protective effect of vitamin A on prostate cancer (*39*), although two studies have reported an inverse relationship with consumption of carrots and other vegetables (*26, 40*). More significant has been the evidence for a direct association, now reported in three studies including ours (*23, 41*). Although animal work supports this direct effect of vitamin A (*6, 42*), the significance of these findings in human populations is not yet clear.

Both the lung and prostate case-control studies also examined the relationship of

TABLE 5. Vitamin A Consumption and Risk for Lung and Prostate Cancers in Hawaii

Cancer site	Subgroup	Vitamin A component	Odds ratios[a] by intake quartile			
			1 (highest)	2	3	4 (lowest)
Lung[b]	Males	Total	1.0	1.2 (0.7–2.0)	1.3 (0.8–2.2)	1.8 (1.1–3.1)
		Carotenes	1.0	1.1 (0.4–2.4)	1.1 (0.3–1.7)	2.2 (0.3–1.6)
	Females	Total	1.0	1.0 (0.4–2.4)	0.7 (0.3–1.7)	0.7 (0.3–1.6)
		Carotenes	1.0	0.9 (0.4–2.0)	0.9 (0.4–2.1)	0.6 (0.2–1.5)
Prostate[c]	≥70 years	Total	1.9 (1.2–3.0)	1.3 (0.8–2.0)	1.4 (0.9–2.1)	1.0
		Carotenes	1.6 (1.0–2.5)	1.6 (1.0–2.5)	1.5 (1.0–2.3)	1.0
	<70 years	Total	0.8 (0.5–1.3)	1.0 (0.6–1.6)	1.3 (0.8–2.1)	1.0
		Carotenes	0.9 (0.5–1.4)	1.1 (0.7–1.8)	1.2 (0.7–1.9)	1.0

[a] Including 95% confidence intervals. [b] Odds ratios adjusted for age, ethnicity, cholesterol intake, occupational status, and cigarette smoking by multiple logistic regression. [c] Odds ratios adjusted for age and race by multiple logistic regression.

vitamin C in the diet to cancer. In neither study was any effect of this nutrient found. On the other hand, the preliminary data from our bladder cancer study suggested a possible protective effect for vitamin C, which is of interest because of the hypothesis that nitrosamine formation might be etiologically related to this cancer site (*6*, *43*).

Although we are following a large cohort of Hawaii residents in order to further examine some of these dietary relationships, sufficient incidence cases have not yet accrued to warrant any major analyses.

Comment

The several studies of diet and cancer in Hawaii are contributing to a growing understanding of the role of dietary factors in the risk for cancer in human populations. Despite much effort in Hawaii and elsewhere, however, little can be concluded definitively at present. Some discrepancies in findings among studies have been noted, and possible reasons for the small relative risks observed have been mentioned. For the future, certain needs can be identified to further research in this area. If better biologic markers of dietary exposure could be found, then some of the exposure misclassification which undoubtedly occurs in interview studies could be considerably reduced. Closer collaboration between epidemiologists and basic scientists can contribute to this effort. Indeed, the identification of meaningful markers of long-term dietary exposures by laboratory researchers could be of much greater public health significance than further elucidations of mechanisms.

The problems of multiple causes, including both positive and negative interactions among several factors and different critical periods of life for their effects, continue to plague this field of epidemiologic research. Although statistical attempts to control for confounders are routinely included in analyses, these problems can seldom be satisfactorily addressed. Perhaps closer collaboration between epidemiologists and laboratory researchers could result in the quantification of some of the relationships among these interrelated factors. For example, cadmium is carcinogenic in animals, but its effect can be blocked by zinc (*44*). If a counterpart exists in man, one might fail to observe the true association if exposures to cadmium alone are studied, since an individual with a higher than usual intake of cadmium but a correspondingly higher intake of zinc may be at no greater risk than a person with a low cadmium intake. Similarly, examining stomach cancer risks in persons with exposure to nitrosamine precursors, by the additional classification of subjects into high and low vitamin C groups, hardly seems an adequate means of dealing with this biochemical interaction. Because we are dealing with risk factors that are undoubtedly less strong than radiation or cigarette smoking, we probably will need to be more refined and specific in our future analytic approaches.

Despite the inherent difficulties in nutrition and cancer research, progress is being made. Dietary guidelines have already been proposed for populations in the U.S., Japan, and elsewhere, and though these are controversial because they are not based on definitive data, they do seem rational. As a result of these actions, some degree of change in the eating patterns of large segments of our populations will no doubt continue to occur. As has already happened with heart disease, mortality from some of the major diet-related cancers may thereby actually begin to decline before scientists have definitive data as to their etiology.

ACKNOWLEDGEMENTS

This work was supported in part by Research Grants CA15655, CA20897, CA25903, CA26515, CA28943, and CA33619, and by Contracts CP53511 and CB53884 from the U.S. National Cancer Institute, Department of Health and Human Services.

REFERENCES

1. Haenszel, W. Cancer mortality among the foreign-born in the United States. J. Natl. Cancer Inst., *26*: 37–132, 1961.
2. Kmet, J. The role of migrant population in studies of selected cancer sites: a review. J. Chronic Dis., *23*: 305–324, 1970.
3. Staszewski, J. and Haenszel, W. M. Cancer mortality among the Polish-born in the U.S. J. Natl. Cancer Inst., *35*: 291–297, 1965.
4. Tulchinsky, D. and Modan, B. Epidemiological aspects of cancer of the stomach in Israel. Cancer, *20*: 1311–1317, 1967.
5. Kolonel, L. N., Hankin, J. H., Lee, J., Chu, S. Y., Nomura, A.M.Y., and Hinds, M. W. Nutrient intakes in relation to cancer incidence in Hawaii. Br. J. Cancer, *44*: 332–339, 1981.
6. Committee on Diet, Nutrition and Cancer. Diet, Nutrition and Cancer. Washington, D.C., National Academy Press, 1982.
7. Haenszel, W. M., Kurihara, M., Segi, M., and Lee, R.K.C. Stomach cancer among Japanese in Hawaii. J. Natl. Cancer Inst., *49*: 969–988, 1972.
8. Haenszel, W. M., Kurihara, M., Locke, F. B., Shimizu, K., and Segi, M. Stomach cancer in Japan. J. Natl. Cancer Inst., *56*: 265–274, 1976.
9. Haenszel, W., Berg, J. W., Segi, M., Kurihara, M., and Locke, F. B. Large-bowel cancer in Hawaiian Japanese. J. Natl. Cancer Inst., *51*: 1765–1779, 1973.
10. Haenszel, W., Locke, F. B., and Segi, M. A case-control study of large bowel cancer in Japan. J. Natl. Cancer Inst., *64*: 17–22, 1980.
11. Hirohata, T., Shigematsu, T., Nomura, A.M.Y., Nomura, Y., Horie, A., and Hirohata, I. Occurrence of breast cancer in relation to diet and reproductive history: a case-control study in Fukuoka, Japan. Natl. Cancer Inst. Monogr., *69*: 187–190, 1985.
12. Nomura, A., Stemmermann, G. N., and Heilbrun, L. K. Gastric cancer among the Japanese in Hawaii: a review. Hawaii Med. J., *44*: 301–303, 1985.
13. Stemmermann, G. N., Nomura, A.M.Y., and Heilbrun, L. K. Cancer risk in relation to fat and energy intake among Hawaii Japanese: a prospective study. This volume, pp. 265–274, 1986.
14. Kolonel, L. N. Cancer incidence among Filipinos in Hawaii and the Philippines. Natl. Cancer Inst. Monogr., *69*: 93–98, 1985.
15. Kolonel, L. N., Nomura, A.M.Y., Hirohata, T., Hankin, J. H., and Hinds, M. W. Association of diet and place of birth with stomach cancer incidence in Hawaii Japanese and Caucasians. Am. J. Clin. Nutr., *34*: 2478–2485, 1981.
16. Kolonel, L. N., Hankin, J. H., Nomura, A.M.Y., and Chu, S. Y. Dietary fat intake and cancer incidence among five ethnic groups in Hawaii. Cancer Res., *41*: 3727–3728, 1981.
17. Kolonel, L. N. Smoking and drinking patterns among different ethnic groups in

Hawaii. Natl. Cancer Inst. Monogr., *53*: 81–87, 1979.
18. Graham, S., Marshall, J., Mettlin, C., Rzepka, T., Nemoto, T., and Byers, T. Diet in the epidemiology of breast cancer. Am. J. Epidemiol., *116*: 68–75, 1982.
19. Miller, A. B., Kelly, A., Choi, N. W., Matthews, V., Morgan, R. W., Munan, L., Burch, J. D., Feather, J., Howe, G. R., and Jain, M. A study of diet and breast cancer. Am. J. Epidemiol., *107*: 499–509, 1978.
20. Lubin, J. H., Burns, P. E., Blot, W. J., Ziegler, R. G., Lees, A. W., and Fraumeni, J. F., Jr. Dietary factors and breast cancer risk. Int. J. Cancer, *28*: 685–689, 1981.
21. Phillips, R. L. Role of life-style and dietary habits in risk of cancer among Seventh-Day Adventists. Cancer Res., *35*: 3513–3522, 1975.
22. Nomura, A., Henderson, B. E., and Lee, J. Breast cancer and diet among the Japanese in Hawaii. Am. J. Clin. Nutr., *31*: 2020–2025, 1978.
23. Heshmat, M. Y., Kaul, L., Kovi, J., Jackson, M. A., Jackson, A. G., Jones, G. W., Edson, M., Enterline, J. P., Worrell, R. G., and Perry, S. L. Nutrition and prostate cancer: a case-control study. Prostate, *6*: 7–17, 1985.
24. Snowdon, D. A., Phillips, R. L., and Choi, W. Diet, obesity and risk of fatal prostate cancer. Am. J. Epidemiol., *120*: 244–250, 1984.
25. Rotkin, I. D. Studies in the epidemiology of prostatic cancer: expanded sampling. Cancer Treat. Rep., *61*: 173–180, 1977.
26. Schuman, L. M., Mandel, J. S., Radke, A., Seal, U., and Halberg, F. Some selected features of the epidemiology of prostatic cancer: Minneapolis-St. Paul, Minnesota case control study, 1976–1979. *In;* K. Magnus (ed.), Trends in Cancer Incidence: Causes and Practical Implications, Hemisphere Publishing Corp., pp. 345–354, Washington, New York, and London, 1982.
27. Bjelke, E. Dietary vitamin A and human lung cancer. Int. J. Cancer, *15*: 561–565, 1975.
28. MacLennan, R., Da Costa, J., Day, N. E., Law, C. H., Ng, Y. K., and Shanmugaratnam, K. Risk factors for lung cancer in Singapore Chinese, a population with high female incidence rates. Int. J. Cancer, *20*: 854–860, 1977.
29. Hirayama, T. Diet and cancer. Nutr. Cancer, *1*: 67–81, 1979.
30. Mettlin, C., Graham, S., and Swanson, M. Vitamin A and lung cancer. J. Natl. Cancer Inst., *62*: 1435–1438, 1979.
31. Gregor, A., Lee, P. N., Roe, F.J.C., Wilson, M. J., and Melton, A. Comparison of dietary histories in lung cancer cases and controls with special reference to vitamin A. Nutr. Cancer, *2*: 93–97, 1980.
32. Shekelle, R. B., Lepper, M., Liu, S., Maliza, C., Raynor, W. J., Jr. Rossof, A. H., Paul, O., Shryock, A. M., and Stamler, J. Dietary vitamin A and risk of cancer in the Western Electric study. Lancet, *2*: 1185–1190, 1981.
33. Kvale, G., Bjelke, E., and Gart, J. J. Dietary habits and lung cancer risk. Int. J. Cancer, *31*: 397–405, 1983.
34. Ziegler, R. G., Mason, T. J., Stemhagen, A., Hoover, R., Schoenberg, J. B., Gridley, G., Virgo, P. S., Altman, R., and Fraumeni, J. F., Jr. Dietary carotene and vitamin A and risk of lung cancer among white men in New Jersey. J. Natl. Cancer Inst., *73*: 1429–1435, 1984.
35. Samet, J. M., Skipper, B. J., Humble, C. G., and Pathak, D. R. Lung cancer risk and vitamin A consumption in New Mexico. Am. Rev. Respir. Dis., *131*: 198–202, 1985.
36. Nomura, A.M.Y., Stemmermann, G. N., Heilbrun, L. K., Salkeld, R. M., and Vuel-

leumier, J. P. Serum vitamin levels and the risk of cancer of specific sites in men of Japanese ancestry in Hawaii. Cancer Res., *45*: 2369–2372, 1985.
37. Wu, A. H., Henderson, B. E., Pike, M. C., and Yu, M. C. Smoking and other risk factors for lung cancer in women. J. Natl. Cancer Inst., *74*: 747–751, 1985.
38. Kolonel, L. N., Hinds, M. W., Nomura, A.M.Y., Hankin, J. H., and Lee, J. Relationship of dietary vitamin A and ascorbic acid intake to the risk of cancers of the lung, bladder, and prostate in Hawaii. Natl. Cancer Inst. Monogr., *69*: 137–142, 1985.
39. Lasnitzki, I. and Goodman, D. S. Inhibition of the effects of methylcholanthrene on mouse prostate in organ culture by vitamin A and its analogs. Cancer Res., *34*: 1564–1571, 1974.
40. Hirayama, T. Epidemiology of prostate cancer with special reference to the role of diet. Natl. Cancer Inst. Monogr., *53*: 149–154, 1979.
41. Graham, S., Hughey, B., Marshall, J., Priore, R., Byers, T., Rzepka, T., Mettlin, C., and Pontes, J. E. Diet in the epidemiology of carcinoma of the prostate gland. J. Natl. Cancer Inst., *70*: 687–692, 1983.
42. Verma, A. K., Conrad, E. A., and Boutwell, R. K. Differential effects of retinoic acid and 7,8-benzoflavone in the induction of mouse skin tumors by the complete carcinogenesis process and by the initiation-promotion regimen. Cancer Res., 42: 3519–3525, 1982.
43. Mirvish, S. S., Wallcave, L., Eagen, M., and Shubik, P. Ascorbate-nitrite reaction: possible means of blocking the formation of carcinogenic N-nitroso-compounds. Science, *177*: 65–68, 1972.
44. Gunn, S. A., Gould, T. C., and Anderson, W.A.D. Effect of zinc on cancerogenesis by cadmium. Proc. Soc. Exp. Biol. Med., *115*: 653–657, 1964.

DIET, NUTRITION AND CANCER, Y. HAYASHI ET AL. (EDS.),
JAPAN SCI. SOC. PRESS, TOKYO/VNU SCI. PRESS, UTRECHT, PP. 41-53, 1986

A Large Scale Cohort Study on Cancer Risks by Diet—with Special Reference to the Risk Reducing Effects of Green-Yellow Vegetable Consumption

Takeshi HIRAYAMA
Institute of Preventive Oncology, Tokyo 162, Japan

Abstract: Using materials obtained in a large scale cohort study of 265,118 adults in Japan from 1966 to 1982, effects of diet and nutrition on cancer motality were reviewed.

Daily consumption of green-yellow vegetables (GYV) rich in beta-carotene, vitamin C, calcium, and dietary fiber was observed to lower risks for selected cancers such as lung, stomach, prostate, and cervix. The risk reducing effect appeared more striking in cigarette smokers.

Risks for cancer of the stomach in males and females and cancer of the breast in females were observed to be lower with the increase in frequency of soybean paste soup consumption which frequently contains GYV.

In daily meat consumers risks were higher for cancer of the lung in both sexes and for cancer of the breast in females. The habit of cigarette smoking was found to comfound the apparently elevated risk in daily meat consumers for lung cancer. For breast cancer daily smoking interacted with daily meat consumption in raising the risk. The extent of risk elevation by daily meat consumption was limited when GYV was taken daily. Those who do not consume GYV daily with habits of daily smoking, daily drinking and daily meat intake were found to carry the highest risks for cancer of all sites and for cancers of selected sites such as the mouth and pharynx, esophagus, stomach, liver, larynx, lung, and urinary bladder. When GYV were consumed daily, considerably lower risk was observed for each of these cancers, even if other habits remained unchanged.

Possible reasons for these phenomena and selected dietary guidelines derived from these findings were discussed in relation to programs for primary prevention of cancer.

Risks for cancer of selected sites were observed to be strongly influenced by life styles such as cigarette smoking, alcohol drinking, and diet in a large-scale 17 year cohort study of 265,118 Japanese adults aged 40 and above. For example, frequent consumers of green-yellow vegetables (GYV) showed reduced risks of cancer of all sites and of selected sites. In this aspect, the most recent follow-up results confirmed findings reported previously (*1–5*). Such risk lowering effects of daily GYV consumption were studied by sex, age, and by habits of smoking, drinking, and meat consumption, and the possible roles of promoter inhibitors included in GYV were considered. Other examples include significantly lowered mortality rates among daily consumers of soybean paste soup for cancer of stomach in males and females and cancer of breast in females, and significantly elevated mortality rates in daily meat consumers for cancer of breast in females. Possible reasons for these associations are discussed together with their potential values for primary prevention of cancer.

A large-scale population-based cohort study was conducted in Japan over a 17 year period (1966–82). Life styles including diet of 122,261 men and 142, 857 women aged 40 years and older residing in 29 Health Center Districts in 6 prefectures (Miyagi, Aichi, Osaka, Hyogo, Okayama, and Kagoshima) were surveyed and followed; this number represented 95% of the census population in October 1965. The study was conducted through home visits of participants made by public health nurses and midwives trained in the standardized method of the interview. The form itself was simple and straightforward, and only items easy for participants to answer were included. Examples of guidelines provided to interviewers were as follows: 1) on items regarding diet, information on current habits throughout the year was required and not seasonal peculiarities. 2) criteria for frequency of consumption were:

TABLE 1. Number of Deaths from Cancer by Site (Prospective Study, 1966–82)

Cancer site	Total (14,740)	Male (8,794)	Female (5,946)
Buccal cavity	91	59	32
Pharynx	39	28	11
Esophagus	585	438	147
Stomach	5,247	3,414	1,833
Intestine	574	256	318
Rectum	563	316	247
Bile duct and gallbladder	530	240	290
Liver	1,251	788	463
Pancreas	679	399	280
Nose, nasal cavities	115	71	44
Larynx	102	83	19
Lung	1,917	1,454	463
Breast	243	2	241
Cervix uteri	589	—	589
Ovary	106	—	106
Prostate	183	183	—
Kidney	92	58	34
Bladder	248	173	75
Skin	90	40	50
Brain	139	77	62
Thyroid gland	55	22	33
Lymphoma	388	231	157
Leukemia	206	107	99
Other	708	355	353

daily=four or five or more times a week; rare= one to three times a month; occasional=in-between first 2 criteria; none=up to two to three times a year (such as at a festival).

Ascertainment of dietary information such as GYV and meat consumption was satisfactory according to the results of comparisons of the 1965 and 1971 surveys on 7,507 individuals who were randomly selected from the original population. The rate of agreement by daily consumption was: 91.3% for smoking, 99.3% for GYV, and 96.3% for meat consumption. The rate of exact agreement (daily, occasionally, rarely, and never) for smoking, GYV, and meat was 84.8%, 65.7%, and 63.3%, respectively. Thus, information on diet was nearly as stable as that of smoking, especially in the daily or non-daily categories.

To enable the follow-up of study subjects, a system of record linkage was established between the original risk factor records and those of the annual census for residence check and death certificates.

During the 17 year follow-up period, 31,979 men and 23,544 women died, out of which 8,794 men and 5,946 women died of cancer. Number of deaths by each site of cancer is shown in Table 1. Total observed person-years were 1,709,273 for men and 2,140,364 for women.

Green-yellow Vegetables

In general, the more frequently GYV were consumed the lower the risk of these cancers, age standardized mortality ratios for daily, occasional, rare, and non-

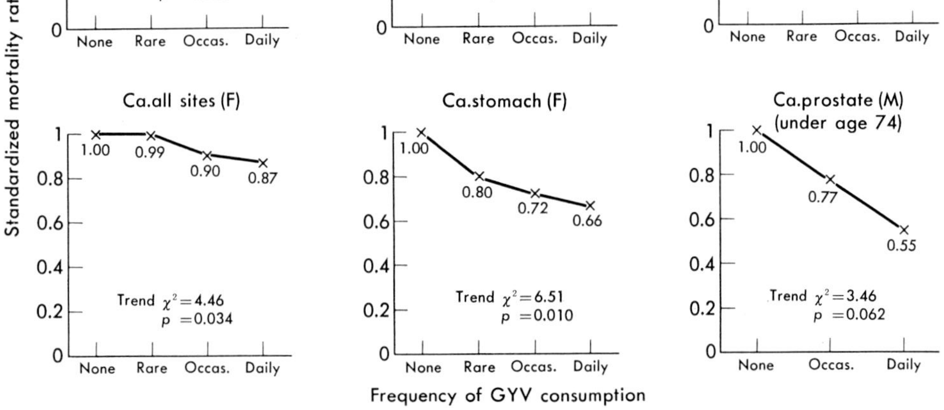

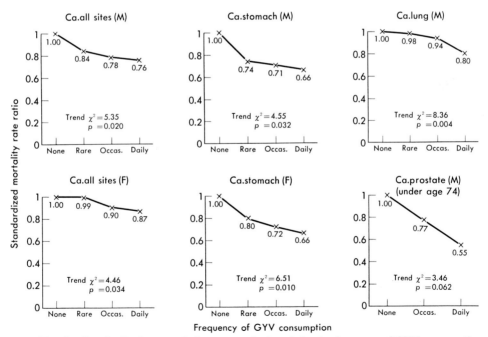

FIG. 1. Standardized rate ratio for cancer of selected sites by frequency of GYV consumption (prospective study, 1966–82, Japan).

consumers being 0.76, 0.78, 0.84, 1.00 for cancer of all sites, 0.66, 0.71, 0.74, 1.00 for stomach cancer, and 0.80, 0.94, 0.98, 1.00 for lung cancer. A similar trend was also observed for cancer of the prostate, age standardized mortality ratios under age

TABLE 2. Risk of Cancer of All Sites with Frequent GYV Consumption (Prospective Study, 1966–82, Japan)

Sex	Frequency of GYV consumption		40–44	45–49	50–54	55–59	60–64	65–69	70–74	75–	Standardized rate ratio	
M	None	OBS[a]	1	2	1	3	3	13	11	5	1.000	Mantel-extension chi
		PY[b]	146	435	669	933	1,001	970	724	506		
	Rare	OBS	1	0	7	26	34	52	49	47	0.840	
		PY	1,069	3,168	4,938	6,481	6,518	6,230	4,601	3,442		−2.312
	Occas.	OBS	7	44	101	232	399	489	498	492	0.782	
		PY	13,996	40,356	63,187	82,201	78,243	70,102	49,313	35,155		One-tail p value
	Daily	OBS	14	89	293	625	994	1,456	1,407	1,321	0.757	0.01039
		PY	36,307	110,872	177,148	230,883	221,619	199,161	141,036	99,912		
F	None	OBS	0	3	1	2	6	2	2	4	1.000	Mantel-extension chi
		PY	143	449	786	1,044	1,073	977	756	615		
	Rare	OBS	1	4	11	13	22	22	26	19	0.993	
		PY	1,018	3,392	5,498	7,095	6,826	5,894	4,185	3,001		−2.113
	Occas.	OBS	10	37	119	191	235	302	263	261	0.901	
		PY	15,523	48,891	77,926	98,933	89,778	75,657	52,214	41,622		One-tail p value
	Daily	OBS	32	130	270	509	748	878	875	906	0.867	0.01730
		PY	45,166	145,921	236,054	305,666	283,848	245,963	175,941	135,351		

[a] Observed number of deaths. [b] Person-years of observation.

TABLE 3. Risk of Stomach Cancer with Frequency of GYV Consumption (Prospective Study, 1966–82, Japan)

Sex	Frequency of GYV consumption		40–44	45–49	50–54	55–59	60–64	65–69	70–74	75–	Standardized rate ratio	
M	None	OBS	0	1	0	1	2	5	7	1	1.000	Mantel-extension chi
		PY	146	435	669	933	1,001	970	724	506		
	Rare	OBS	0	0	3	12	13	18	21	16	0.741	
		PY	1,069	3,168	4,938	6,481	6,518	6,230	4,601	3,442		−2.134
	Occas.	OBS	4	21	48	81	160	208	187	184	0.707	
		PY	13,996	40,356	63,187	82,201	78,243	70,102	49,313	35,155		One-tail p value
	Daily	OBS	9	43	120	252	422	579	502	457	0.664	0.01642
		PY	36,307	110,872	177,148	230,883	221,619	199,161	141,036	99,912		
F	None	OBS	0	2	0	0	3	1	0	2	1.000	Mantel-extension chi
		PY	143	449	786	1,044	1,073	977	756	615		
	Rare	OBS	0	1	6	2	10	3	10	6	0.799	
		PY	1,018	3,392	5,498	7,095	6,826	5,894	4,185	3,001		−2.551
	Occas.	OBS	5	12	45	66	76	103	62	89	0.720	
		PY	15,523	48,891	77,926	98,933	89,778	75,657	52,214	41,622		One-tail p value
	Daily	OBS	12	33	79	131	205	266	287	304	0.662	0.00537
		PY	45,166	145,921	236,054	305,666	283,848	245,963	175,941	135,351		

74 for daily, occasional, rare or none being 0.55, 0.77, and 1.00, respectively. A suggestive reverse trend was observed for age 75 or older prostate cancer cases.

Similar trends were also observed for selected cancers in women, the age-standardized mortality ratio for daily, occasional, rare, and non-consumers of GYV being 0.87, 0.90, 0.99, 1.00 for cancer of all sites, 0.66, 0.72, 0.80, and 1.00 for stomach cancer (Fig. 1, Tables 2, 3, 4, and 5). No significant trend was observed for lung cancer in females.

TABLE 4. Risk of Lung Cancer with Frequency of GYV Consumption (Prospective Study, 1966–82, Japan)

	Frequency of GYV consumption		Age in years								Standardized rate ratio	
			40–44	45–49	50–54	55–59	60–64	65–69	70–74	75–		
M	None	OBS	0	0	0	1	0	2	0	3	1.000	Mantel-extension chi
		PY	146	435	669	933	1,001	970	724	506		
	Rare	OBS	0	0	3	5	5	8	10	8	0.982	−2.892
		PY	1,069	3,168	4,938	6,481	6,518	6,230	4,601	3,442		
	Occas.	OBS	0	6	8	35	65	92	111	93	0.937	
		PY	13,996	40,356	63,187	82,201	78,243	70,102	49,313	35,155		One-tail *p* value
	Daily	OBS	0	5	31	88	130	233	267	237	0.798	0.00198
		PY	36,307	110,872	177,148	230,883	221,619	199,161	141,036	99,912		
F	None	OBS	0	0	0	1	0	0	0	0	1.000	Mantel-extension chi
		PY	143	449	786	1,044	1,073	977	756	615		
	Rare	OBS	0	0	0	0	0	0	2	0	0.361	−0.391
		PY	1,018	3,392	5,498	7,095	6,826	5,894	4,185	3,001		
	Occas.	OBS	0	2	4	16	16	35	32	19	1,615	
		PY	15,523	48,891	77,926	98,933	89,778	75,657	52,214	41,622		One-tail *p* value
	Daily	OBS	1	11	13	28	53	67	73	87	1.351	0.34790
		PY	45,166	145,921	236,054	305,666	283,848	245,963	175,941	135,351		

TABLE 5. Risk of Cancer of the Prostate by Frequency of GYV Consumption (Males, Prospective Study, 1966–82, Japan)

Frequency of GYV consumption		Age in years							
		40–44	45–49	50–54	55–59	60–64	65–69	70–74	75–
None	OBS	0	0	0	0	1	1	2	0
	PY	1,215	3,603	5,607	7,414	7,519	7,200	5,325	3,948
Occas.	OBS	0	1	0	2	3	12	12	19
	PY	13,996	40,356	63,187	82,201	78,243	70,102	49,313	35,155
Daily	OBS	0	0	1	3	7	21	28	69
	PY	36,307	110,872	177,148	230,883	221,619	199,161	141,036	99,912

Frequency of GYV consumption	Standardized rate ratio	
	All ages	Under age 74
None and rare	1.000	1.000
Occas.	1.305	0.772
Daily	1.227	0.545
Mantel-extension chi	−0.147	−1.860
one-tail *p* value	0.44157	0.03144

Cross-tabulation with smoking habits revealed a general tendency of greater risk lowering effect of GYV daily consumption in heavier smokers than non-smokers. Results of such cross-tabulation for cancer of all sites, cancers of lung and cervix are shown in Figs. 2, 3, and 4. In each case a clear-cut dose-response relationship was observed between amount of smoking and cancer risk. When compared to nonsmokers

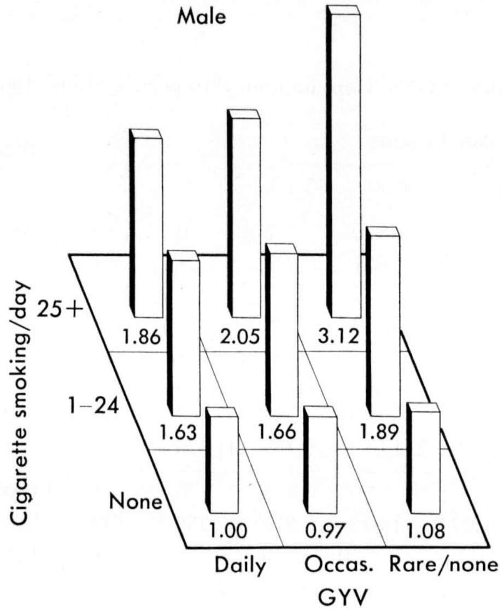

FIG. 2. Relative risk of cancer of all sites in relation to frequency of GYV and number of cigarettes smoked per day (prospective study, 1966–82, Japan).

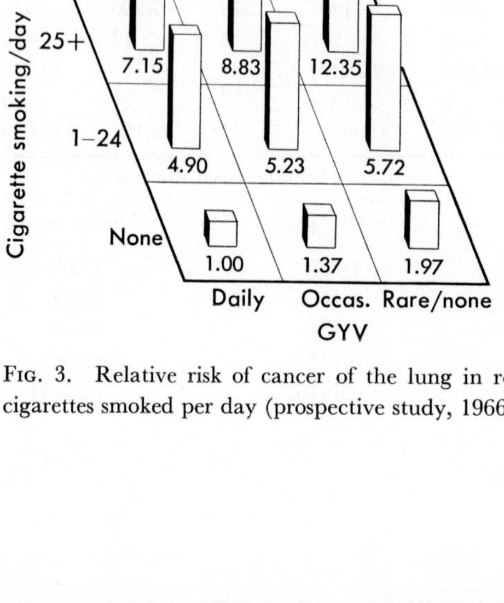

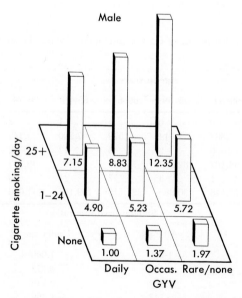

FIG. 3. Relative risk of cancer of the lung in relation to frequency of GYV and number of cigarettes smoked per day (prospective study, 1966–82, Japan).

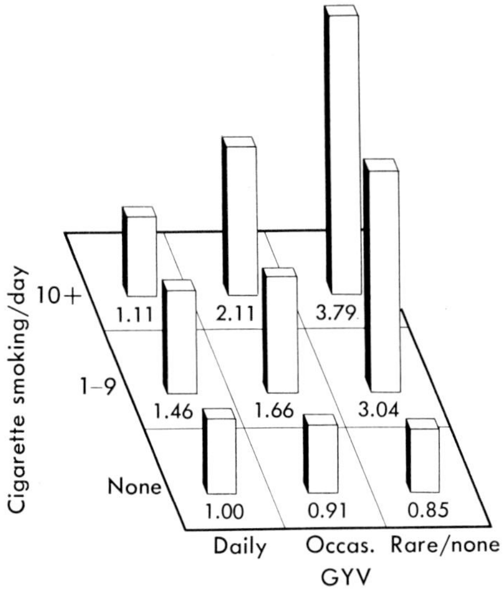

FIG. 4. Relative risk of cancer of the cervix uterine in relation to frequency of GYV and number of cigarettes smoked per day (prospective study, 1966–82, Japan).

consuming GYV daily, smokers of 25 or more cigarettes daily consuming GYV rarely or never were observed to carry 3.12 (confidence limits 2.43–4.07) times higher risk for cancer of all sites and 12.35 (8.25–19.90) times higher risk for cancer of lung. For cancer of cervix uteri, the risk was 3.79 (2.03–9.25) times higher in rare or non-consumers of GYV who smoked 10 or more cigarettes daily compared to non-smokers consuming GYV daily.

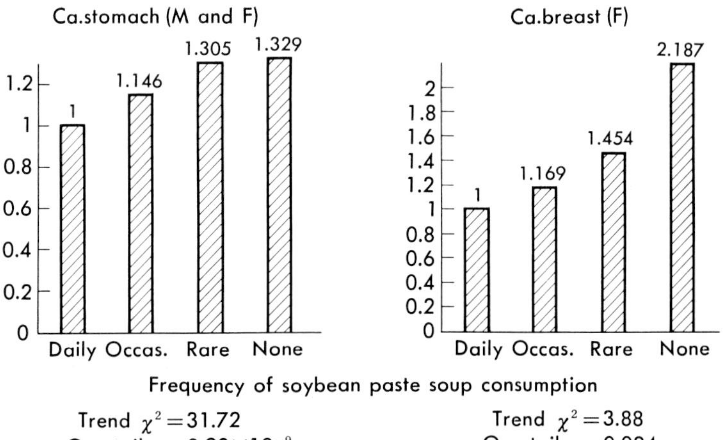

FIG. 5. Standardized rate ratio for cancer of stomach (both sexes) and cancer of breast (females, age 50 and up) by frequency of soybean paste soup consumption (prospective study, 1966–82, Japan).

Soybean Paste Soup

The standardized mortality rates for stomach cancer in males and females and for cancer of breast in females were found to be significantly lower with the increase in frequency of soybean paste soup consumption (Fig. 5). The association was found to be independent of socio-economic variations and other habits such as smoking (*7*). This effect might come either from soybean components or from nutritious foods such as GYV frequently included in the soup.

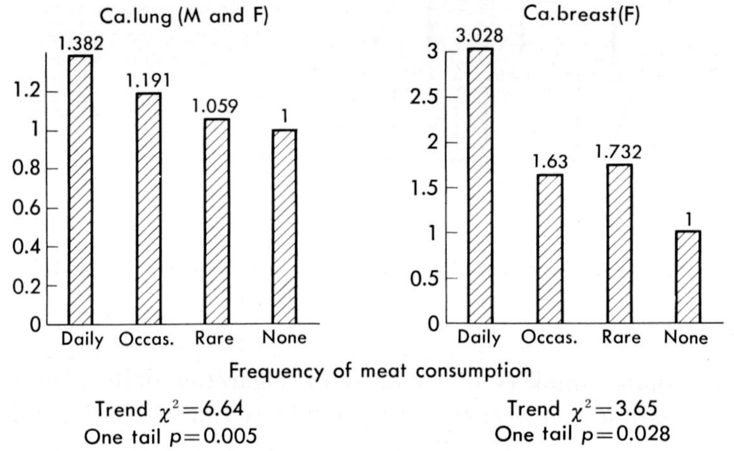

FIG. 6. Standardized rate ratio for cancer of lung (both sexes) and cancer of breast (females, age 50 and up) by frequency of meat consumption (prospective study, 1966–82, Japan).

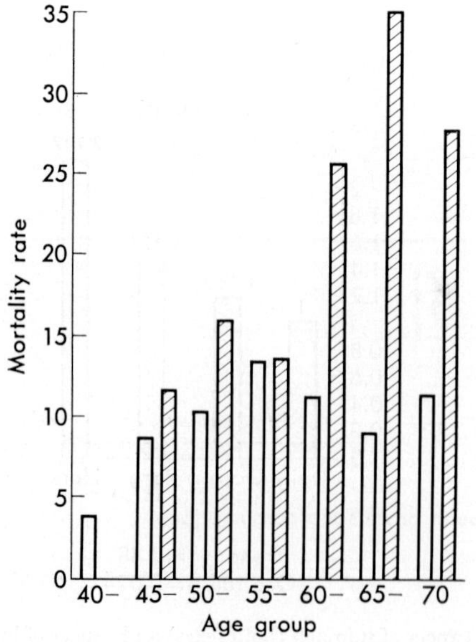

FIG. 7. Age specific mortality rate per 100,000 for breast cancer in daily meat consumers (shaded bar) and in non-daily meat consumers (blank bar) (prospective study, 1966–82, Japan).

Meat

The standardized mortality rates for lung cancer in males and females and breast cancer in females were observed to become significantly higher with the increase in the consumption of meat (Fig. 6). For lung cancer cross-tabulation with the smoking habit revealed that the observed association between lung cancer and meat consumption was entirely due to the comfounding with smoking, χ^2 for trend and one tail p being 0.014 and 0.453 respectively, and lung cancer risk for none, rare, occasional, and daily meat consumers being 1.000, 1.000, 1.073, and 1.034 respectively when the smoking habit was adjusted. On the other hand, association between breast cancer and meat consumption remained significant after cross-tabulation with the smoking habit, those women smoking 10 or more cigarettes daily and consuming meat daily being 4.9 (1.7–8.5) times higher risk for breast cancer than nonsmokers consuming meat rarely or not at all. When the age specific mortality rates for breast cancer were compared between daily meat consumers and others, the curve for daily meat consumers was observed to be similar to that of western countries, the

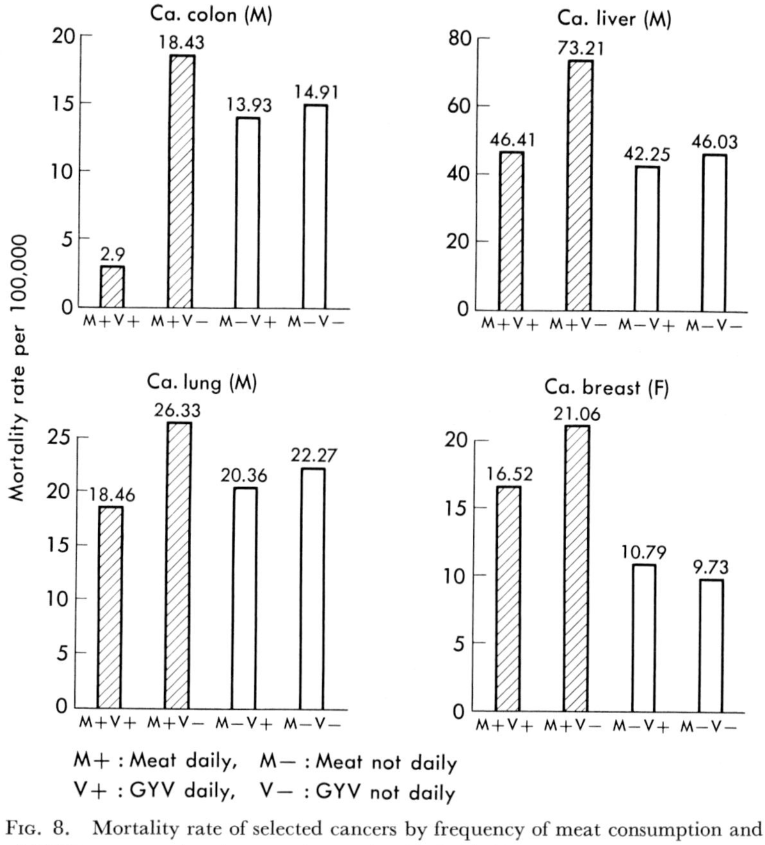

FIG. 8. Mortality rate of selected cancers by frequency of meat consumption and by frequency of GYV consumption (prospective study, 1966–81, Japan). M+, meat daily; M−, meat not daily; V+, GYV daily; V−, GYV not daily.

rates kept going up after menopause. In non-daily meat consumers, the curve was flat or slightly downward after menopause, fitting the so-called Japanese type breast cancer age curve (Fig. 7)

A common tendency at most cancer sites was that the risk enhancing effect of daily meat intake was observed to be less when GYV were eaten daily (Fig. 8). Significantly lower risk of colon cancer was observed in males who consumed both meat and GYV daily, while daily meat consumption was observed to increase colon cancer risk when GYV was not eaten daily (*8*). The beneficial effect of the daily consumption of GYV is particularly apparent in this case.

Lifestyles Carrying Highest Cancer Risk and Role of GYV Daily Consumption in Lowering Such Risk

Results of more detailed cross-tabulation of habits of smoking, drinking, meat and GYV consumption revealed clearly the highest and lowest cancer risk groups. Those smoking daily, drinking daily, consuming meat daily, and not consuming GYV daily were observed to carry the highest cancer risk, not only for cancer of all

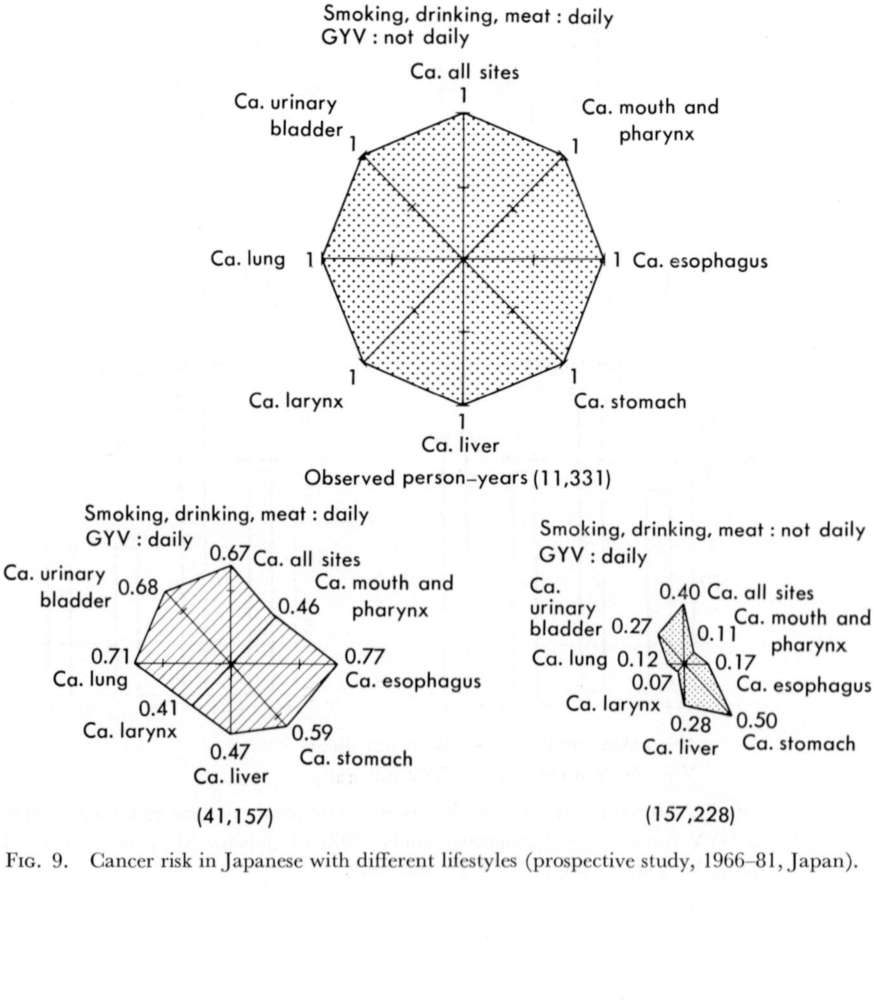

FIG. 9. Cancer risk in Japanese with different lifestyles (prospective study, 1966–81, Japan).

sites but also for selected sites such as mouth and pharynx, esophagus, stomach, liver, larynx, lung, and urinary bladder. The lowest cancer risk was noted for those with entirely opposite lifestyles, *i.e.*, consuming GYV daily and not smoking, drinking, and eating meat every day. The age standardized mortality rates of the latter group for cancer of all sites were 60% lower, for cancers of mouth and pharynx, larynx and lung about 90% lower, for cancer of esophagus, liver, and urinary bladder around 70% lower and for stomach cancer 50% lower than the highest risk group (Fig. 9). A noteworthy risk reducing effect of daily GYV consumption was observed in those with otherwise hazardous habits. The age standardized mortality rate for those smoking daily, drinking daily, with daily meat intake and consuming GYV daily was between the highest and lowest risk groups described above. Compared to the highest risk group, the mortality rate for cancer of all sites was over 30% lower with daily ingestion of GYV (Fig. 9). Cancers of other selected sites showed a similar pattern. These observations further underline the potential importance of encouraging people to consume GYV more frequently in order to reduce the risk of cancers of selected sites, even though those individuals may have hazardous habits.

DISCUSSION

A large-scale, 17 year cohort study of 265,118 adults revealed reduced risks of cancer of all sites and of cancers of stomach and other selected sites in frequent consumers of GYV, confirming our previous reports on shorter follow-up periods (*1–5*). The current observation showed that the lung cancer mortality rates became significantly lower with the increase in frequency of GYV consumption in males but not in females. The prostate cancer mortality rates were also lowered in frequent consumers of GYV under age 74 but not in those aged 75 and above, suggesting a longer lantency period if GYV is consumed in higher frequency. These observations are in line with those of a cohort study of Japanese in Hawaii (*9*).

GYV are defined as vegetables containing over 600 μg of carotene per 100 g, such as carrots, spinach, pumpkin, green lettuce, and green asparagus. Biological mechanisms of risk lowering effects of daily consumption of GYV observed by epidemiological studies in the literature, including ours, are currently being studied taking into consideration the possible effect of beta-carotene/vitamin A, vitamin C, minerals, such as calcium, and dietary fibers plus other substances rich in such vegetables.

The risk lowering effect of daily GYV consumption on lung cancer was observed throughout different socio-economic strata (*3*). Among ex-smokers, reduction in lung cancer risk with the lapse of years after smoking cessation was greater among daily consumers of GYV than others (*3*). Such risk lowering effect of daily GYV consumption was nearly absent in those who began smoking in their teen years (*3*). The risk lowering effect of daily GYV consumption on lung cancer was also observed in non-smoking wives with smoking husbands, suggesting the beneficial effect of GYV consumption in reducing risk due to passive smoking as well that due to active smoking (*6*). These results plus current observations of more striking risk reducing effect of daily GYV consumption in heavier smokers strongly suggest the possible roles played by promotor inhibitors or inhibitor-like substances which are included in GYV.

For those who increased GYV consumption within five years after the start of the current cohort study a noteworthy reduction of stomach cancer risk was observed in the subsequent follow-up period (*5*). Therefore, increased consumption of GYV appeared to be promising in reducing stomach cancer risk even after reaching adult age.

As no side effects are conceivable at this moment it is advisable to teach the public to consume GYV more frequently, particularly smokers, along with guidance on smoking cessation, as both apparently contribute to reducing risks of cancers of lung and other selected sites.

The risk lowering effects of soybean paste soup consumption on stomach cancer and breast cancer are interpreted as partly due to its contents such as GYV which are frequently used. In addition, reported results of inhibitory effects of soybean components themselves on cancer promoters (protease inhibitors(*9–12*), lignan (*13*) *etc.*) should also be considered as contributory factors.

Increased meat consumption was observed to raise the risk of cancer of breast in females. Meat is a major source of animal fat in Japan. Excess consumption of fat should be avoided in order not to follow bitter examples of western countries. It is noteworthy that the extent of risk elevation by increased meat consumption becomes much lower when GYV is taken daily. Dietary fibers and calcium included in GYV, in addition to other selected nutrients, might explain the reasons for such risk reduction.

The observation of effects of various combinations of lifestyle components revealed that non-daily consumers of GYV with combined habits of daily smoking, daily drinking, and daily meat intake carried the highest risks for cancer of all sites and also of selected sites as mouth and pharynx, esophagus, stomach, liver, larynx, lung, and urinary bladder. It is noteworthy that for those consuming GYV daily, even though other habits were unchanged, considerably lower mortality for cancer of each site was found. This observation further fortifies the advisability of daily GYV consumption as a means of cancer prevention.

Our large scale cohort study in Japan thus clearly underlined the importance of diet and nutrition as potential risk modifiers of cancer in man. Such observations are believed to be of value not only in studying the etiology of human cancer but also in formulating the guidelines for lifestyle modification aiming at the primary prevention of cancer in man.

ACKNOWLEDGMENT

This work was supported by Grants-in-Aid for Cancer Research from the Ministry of Health and Welfare, Japan.

REFERENCES

1. Hirayama, T. A population prospective study on cancer epidemiology in Japan, Proc. Jpn. Cancer Assoc., 36th annual meeting, p. 271, 1977.
2. Hirayama, T. Epidemiology of prostatic cancer with special reference to the risk lowering effect of green-yellow vegetable intake, Proc. Jpn. Cancer Assoc., 37th annual meeting, p. 280, 1978.
3. Hirayama, T. Diet and cancer. Nutr. Cancer, *1* (3): 67–81, 1979.
4. Hirayama, T. Epidemiology of prostate cancer with special reference to the role of diet, Second Symp. on Epidemiology and Cancer Registries in the Pacific Basin. Natl. Cancer Inst. Monogr., *53*: 149–153, 1979.
5. Hirayama, T. Does daily intake of green-yellow vegetable lower down the risk of cancer in man? An example of application of epidemiological methods to identify individuals at low risk. *In;* H. Bartsch, B. Armstrong, and W. Davis (eds.), Proc. Symp. Host Factors in Human Carcinogenesis, IARC Scientific Pub. 39, pp. 531–540, IARC, Lyon, 1982.
6. Hirayama, T. Lung cancer in Japan. Effects of nutrition and passive smoking. *In;* M. Mizell and P. Correa (eds.), Lung Cancer, Causes and Prevention, pp. 175–195, Verlag Chemie International Inc., New York, 1984.
7. Hirayama, T. Relationship of soybean paste soup intake to gastric cancer risk. Nutr. Cancer, *3* (4): 223–233, 1982.
8. Hirayama, T. A cohort study on cancer in Japan. *In;* W. J. Blot, T. Hirayama, and D. G. Hoel (eds.), Statistical Method in Cancer Epidemiology, pp. 73–91, Radiation Effects Research Foundation, Hiroshima, 1985.
9. Kolonel, L. N., Hinds, M. W., Nomura, A.M.Y., Hankin, J. H., and Lee, J. Relationship of dietary vitamin A and ascorbic acid intake to the risk of cancers of the lung, bladder, and prostate in Hawaii. Natl. Cancer Inst. Monogr., *70* (in press).
10. Troll, W., Wiesner, R., Belman, S., and Shellabarger, C. J. Inhibition of carcinogenesis by feeding diets containing soybeans. Proc. Am. Assoc. Cancer Res. 20, p. 265, 1979.
11. Troll, W., Wiesner, R., Shellabarger, C. J., Holtzman, S., and Stone, J. P. Soybean diet lowers breast tumor incidence in irradiated rats. Carcinogenesis, *1*: 469–472, 1980.
12. Troll, W. Tumor promotion by protease inhibitors. *In;* P. N. Magee, S. Takayama, T. Sugimura, and T. Matsushima (eds.), Fundamentals in Cancer Prevention, pp. 41–53, Japan Sci. Soc. Press, Tokyo, 1976.
13. Horwitz, C. and Walker, A.R.P. Lignans—additional benefits from fiber? Nutr. Cancer, *6* (2): 73–76, 1984.

MUTAGENS/CARCINOGENS IN FOODS

DIET, NUTRITION AND CANCER, Y. HAYASHI ET AL. (EDS.),
JAPAN SCI. SOC. PRESS, TOKYO/VNU SCI. PRESS, UTRECHT, PP. 57-65, 1986

Cancer Risks Posed by Aflatoxin M_1

D.P.H. HSIEH, J. M. CULLEN,*[1] L. S. HSIEH, Y. SHAO,*[2] and B. H. RUEBNER

Department of Environmental Toxicology and Department of Pathology, University of California, Davis, California 95616, U.S.A.

Abstract: The suspect milk-borne carcinogen, aflatoxin M_1 (AFM), was produced and isolated from the rice culture of the fungus *Aspergillus flavus* NRRL3251 for confirmation and determination of the potency of its carcinogenicity in the male adult Fischer rat. The carcinogen was mixed into an agar-based, semisynthetic diet at 0, 0.5, 5, and 50 ppb (μg/kg) and was fed to groups of animals continuously for 19–21 months. Aflatoxin B_1 (AFB), of which AFM is a metabolite, at 50 ppb was used as a positive control. Hepatocarcinogenicity of AFM was detected at 50 ppb, but not at 5 or 0.5 ppb, with a potency of 2–10% that of AFB. A low incidence of intestinal adenocarcinomas was found in the AFM 50 ppb group, but not in any other groups. At 0.5 ppb, the action level enforced by the U.S.A. Food and Drug Administration, AFM induced no liver lesions in the rats but stimulated the animals' growth. On the average, the rats in the 0.5 ppb group weighed 11% ($p<0.001$) more than those in the control group. This increased growth was associated with increased feed intake. Based on the biological activity of AFM at the relevant low doses and the estimated level of human exposure to AFM through consumption of milk, the cancer risk posed by this contaminant for human adults is assessed to be very low. For infants, further studies are warranted because milk constitutes the major ingredient of the infant diet and because infant animals have been shown to be more sensitive to the carcinogenicity of AFB than adult animals.

In 1978, a shocking episode occurred in the Southwestern U.S.A. where commercial milk was labelled as "poison" by panic-struck consumers because samples of the milk were found to contain aflatoxin M_1 (AFM) at levels greater than 0.5 μg/kg (ppb), the regulatory action level enforced by the U.S.A. Food and Drug Administration (*1*). Hundreds of gallons of the contaminated milk were thrown away (*2*). Val-

*[1] Present address: School of Veterinary Medicine, North Carolina State University, Raleigh, NC 27606, U.S.A.
*[2] Present address: School of Public Health, Shanghai Medical University, Shanghai, People's Republic of China.

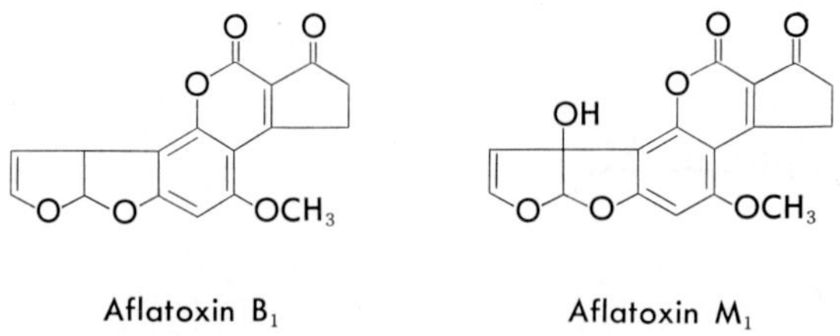

FIG. 1. Structures of AFB and AFM.

uable food was thus destroyed and local streams were polluted by a flood of organic matter.

AFM is a hydroxylated metabolite of aflatoxin B_1 (AFB) (Fig. 1). The latter is an established potent carcinogen produced by the fungi *Aspergillus flavus* and *Aspergillus parasiticus* which can infest a variety of agricultural commodities such as corn, peanuts, and cottonseeds (*3*). In the rat and rainbow trout, AFB is the most potent hepatocarcinogen known. When feed containing AFB-contaminated ingredients is ingested by dairy cattle, as much as 2% of the AFB may appear in the milk as AFM (*4–6*). Food surveillance has revealed that more than 50% of milk samples were contaminated by AFM at sub-ppb (less than 1 μg/kg) levels from many sampling sites (*7*). The occurrence of AFM in commercial milk and dairy products, which are directly consumed by humans and especially by young humans, has aroused a worldwide concern about the potential cancer hazard posed by AFM because of its structural similarity to AFB.

In fact, AFM was found to be almost as toxic as AFB to one-day-old ducklings (*8*) and adult male Fischer rats (*9*). It is also genotoxic to *Salmonella typhimurium* (*10, 11*) and rat liver cells (*12, 13*), and is hepatocarcinogenic to the rat (*14*) and rainbow trout (*15, 16*), though considerably less potent than AFB. These toxicity evaluations for AFM were done on a rather small scale due to a limited supply of the pure compound. It is not feasible to routinely isolate AFM from contaminated milk because the concentrations are at ppb levels, necessitating an extremely tedious purification procedure. The supply problem, however, was overcome by a fermentation method for producing AFM, utilizing a rice culture of a special toxigenic strain of *A. flavus*. Sufficient quantities of AFM have been purified for further evaluation of the biological activity of the compound.

Production and Purification of AFM

The mutant strain of *A. flavus* NRRL3251, which produced a relatively large quantity of AFM (*17*), was used to obtain the compound. The strain was cultivated on 50 g portions of long grain rice supplemented with yeast extract (2%), $ZnSO_4$ (26 μg/g), and water (40%), shaken continuously in a 500 ml conical flask at 28 °C in subdued lighting for 10 days. The AFM produced in the fermented rice was extracted with dichloromethane and cleaned up on a silica gel column by elution with a series of solvents of increasing polarity. Further purification of AFM was achieved by normal-phase column chromatography using a prepacked silica gel 60 column

(LoBar, E. Merck), with dichloromethane: hexane: methanol (75+25+2) as the mobile phase, followed by reverse-phase column chromatography using a prepacked RP-8 column (LoBar), and eluting AFM with acetonitrile: water (25+75) (*18*).

The purity and identity of AFM were confirmed by its co-chromatography with a reference compound prepared by transformation of AFB by microsomal enzymes prepared from the rat liver (*19*), and also by ultraviolet, mass, and proton nuclear magnetic resonance spectra. The levoratary form of the compound, as the natural AFM found in milk, was confirmed by its optical rotation.

The AFM so produced was mutagenic to a frame-shift mutation tester strain of *S. typhimurium* TA98 (*20*) at 1/100 the potency of AFB. The mutagenic potency found for this product, compared with that of AFB, was somewhat lower than the values reported previously, *e.g.*, 1/33 and 1/63 as activated by the liver S-9 preparations of rat and rainbow trout, respectively (*10, 11*). This reduced potency is possibly attributed to the greater purity achieved by the use of reverse-phase chromatography, which eliminates the possibility of AFB contamination, since AFM is eluted from the column before AFB. We have found this product to be approximately 1/10 and 2/7 as acutely toxic (lethal) as AFB to 7-day old male Fischer rats and chicken embryos, respectively.

Carcinogenicity

The AFM isolated and purified from the *A. flavus* culture was used in a recently completed long-term feeding experiment to determine the carcinogenicity of AFM in the male Fischer rat, a strain known to be very sensitive to the hepatocarcinogenic effect of AFB (*21*). The AFM was mixed into an agar-based, semisynthetic diet, similar to that used by Wogan *et al.* (*22*), at levels of 0, 0.5, 5, and 50 ppb. These experimental diets, prepared weekly, were fed continuously to four groups of 62 animals. The feeding experiment started when the animals were 7 weeks old, weighed 140–170 g, and lasted 21 months. A positive-control group of 42 animals was maintained on a similar diet containing 50 ppb of AFB. A negative-control group of 23 animals was maintained on a regular laboratory rat chow (Purina 5002).

Body weights and feed consumption were monitored, first weekly and then monthly, throughout the study. Animals from each group were killed after appropriate periods of continuous treatment for pathological studies to examine the progress of lesions. At each necropsy, the number, size, site, and appearance of lesions in the liver, lung, kidney, and intestine were assessed grossly, and representative samples of any lesions found were studied histologically and histochemically. Particular attention was given to the liver which is the primary target organ of AFB.

Animals were scored on the presence or absence of preneoplastic foci, neoplastic nodules, and hepatocellular carcinomas in two sections from the right, left, and median liver lobes. Foci were defined as those lesions which were not visible to the naked eye, usually smaller than a lobule, not compressing the adjacent parenchyma, and having hepatic plates which generally blended smoothly with surrounding hepatic plates. Neoplastic nodules were defined as larger than a lobule, usually visible by gross inspection and at least 1 mm in diameter. They compressed adjacent paren-

TABLE 1. Incidence and Time of Appearance of Lesions in Male Fischer Rats Treated with AFB and AFM

Treatment	Duration of experiment (months)	Time of appearance of earliest lesions, months (incidence)			Incidence of lesions at last sacrifice		
		F	N	C	F	N	C
AFB 50 ppb	16–17	10 (3/7)	16 (9/9)	16 (6/9)	29/29	28/29	25/29[a]
AFM 50 ppb	19–21	16 (2/6)	17 (1/6)	21 (2/18)	28/37	6/37	2/37[b]
AFM 5 ppb	19–21	17 (1/3)	ND	ND	3/47	0/47	0/47
AFM 0.5 ppb	19–21	16 (1/5)	ND	ND	1/32	0/32	0/32
Control (agar)	19–21	21 (1/21)	ND	ND	1/31	0/31	0/31
Control (chow)	19–21	21 (8/10)	ND	ND	8/16	0/16	0/16

[a] Each of these lesions represents multiple carcinomas. [b] Each of these lesions represents a single carcinoma. F, foci; N, nodules; C, carcinoma; ND, none detected.

chyma and their plates generally did not blend with surrounding normal plates, often intersecting at right angles. Hepatocellular carcinomas were distinguished by a prominent disruption of normal architecture. Detailed criteria for these lesions were derived from guidelines published by the U.S. National Cancer Institute (*23*).

The development of liver lesions in animals under different treatments is shown in Table 1. Depending upon the health condition of the animals, the feeding experiment lasted 16 to 17 months for the AFB group and 19 to 21 months for the other groups. No hepatic lesions of any type were found in any group of animals before month 10, indicating that AFB and AFM at the dose used showed negligible acute liver toxicity. At month 10, histologically detectable foci were evident in rats fed AFB. These foci were not detected in the AFM treated groups until after month 16.

At month 16, all nine animals taken from the AFB group had multiple liver neoplastic nodules varying in size from 2 to 20 mm in diameter. Six of the nine also had hepatocellular carcinomas. A single neoplastic nodule was first detected in one of the six animals from the AFM 50-ppb group at month 17. Hepatocellular carcinomas were first detected in the AFM 50-ppb group in month 21, when single hepatocarcinomas were found in two of the eighteen animals sacrificed. Low incidences of foci were detected in all the other groups, including the controls.

No neoplastic nodules or carcinomas were found in the livers of any animals from the AFM 5-ppb, the AFM 0.5-ppb, and the control groups at any time. The low incidence of spontaneous liver tumours in animals maintained on the agar-based diet used in this experiment was also observed by Wogan *et al.* (*22*).

Based on the incidence of liver tumours found at the last necropsy (19–21 months), where six of the 37 animals of the AFM 50-ppb group had single neoplastic nodules and two had single hepatocellular carcinomas, it is evident that AFM is hepatocarcinogenic to the male Fischer rat. The carcinogenic potency of AFM, however, is considerably lower than that of AFB, as shown by comparing the animal responses to 50 ppb of each of the two carcinogens. While multiple hepatocellular carcinomas developed in 95% (19/20) of the AFB-treated animals within 17 months, only single hepatocellular carcinomas were found in 2/37 of the animals fed 50 ppb of AFM at 19–21 months of treatment. The high incidence of liver tumours and the relatively short time before their appearance in our AFB-treated group are in agreement with observations reported by Wogan *et al.* (*22*), who used the same strain of rat and a similar experimental protocol. The incidence of hepatocellular carcinomas induced by 50 ppb of AFM in our experiment (2/37 at 19–21 months) is comparable to that induced by 1–5 ppb AFB in the experiment of Wogan *et al.* (*22*). Thus, AFM appears to have 2–10% of the hepatocarcinogenic potency of AFB under our experimental conditions.

None of the rats from the AFM 5-ppb and the AFM 0.5-ppb groups developed neoplastic nodules or carcinomas. Preneoplastic foci were detected in 3/47, 1/32, and 1/31 of the animals sacrificed at month 19–21 in the 5-ppb, the 0.5-ppb, and the control agar-diet groups, respectively. Statistically, there is no significant difference in the incidence of the foci in the three groups. The relatively high incidence of foci in the control lab-chow group (8/16) may be due to some carcinogenic ingredients of the chow although neither AFB nor AFM was detected in the chow.

Intestinal adenocarcinomas were found in three animals exposed to 50 ppb of AFM. Two neoplasms involved the small intestine: one occurred at month 9, and the other at month 18. The third occurred in the colon at month 19. No intestinal neoplasms were found in the other AFM groups, the AFB group, or either control group. There was no evidence of metastasis by any of the neoplasms described above. The detection of the intestinal adenocarcinomas suggests that the possible carcinogenicity of AFM in the intestine warrants further investigations.

Biological Activity of AFM at 0.5 ppb

Animals fed AFM at the Food and Drug Administration action level of 0.5 ppb grew significantly faster than the control group during the first 12 months of the experiment. This effect was not seen in animals fed AFM at 5 ppb or at 50 ppb. At the end of 12 months, the animals fed 0.5 ppb of AFM weighed on the average 10.6% ($p<0.001$) more than the animals maintained on the two control diets free of AFM. This increased growth was associated with an increased feed intake. On the average, the animals fed 0.5 ppb of AFM consumed 7.7% ($p<0.001$) more feed than the animals fed the control diet.

Our findings suggest that there seems to be an "apparent threshold" for the carcinogenicity of AFM in the rat and that AFM appears to possess significant biological activity even below that threshold. Moreover, the biological activity of AFM at the lowest level tested is totally unrelated to that observed at a dose one-hundred times greater. Our results indicate that prediction of the carcinogenicity at a low dose, such as may occur in a natural environment, by "no threshold" extrapolation of the carcinogenicity observed at experimentally convenient, but high, doses is not valid.

Risk Assessment

Food surveys have indicated that the major sources of AFM ingested by humans are milk and dairy products. Using the 1981 U.S.A. data as an example, the average per capita consumption rate of milk is 242 g/capita/day (*24*). The frequency of contamination, based on California milk samples analyzed during the summer of 1979, was found to be 50% (*25*). The AFM concentrations in the samples ranged from 0.1 to 0.9 ppb (at the quantitation limit of 0.1 ppb). The worst case of daily intake of AFM, in which all (100%) the milk samples are assumed to contain 0.9 μg/kg of AFM, was calculated to be 218 ng/capita/day. Further, assuming that per capita food consumption rate is 2 kg of moist solid food per day (*26*), the average dietary concentration of AFM would be approximately 0.11 ppb (218/2000 ppb). At this level, AFM is not carcinogenic to even the aflatoxin-sensitive male Fischer rat.

Humans appear more resistant than the rat to the hepatocarcinogenicity of AFM, based on three observations. First, rhesus monkeys are much more resistant than the rat to the hepatocarcinogenicity of AFB (*27*), and the postmitochondrial preparation of the human liver metabolized AFB in a manner very similar to that of the monkey liver, but not that of the rat liver (*28*, *29*). Second, the ability to activate AFB is found to be greater in the enzyme preparation produced from the rat liver than that from the human liver, while third, the opposite is true with respect to the ability to detoxify AFB (*29*, *30*).

The very low cancer risk posed by AFM as analyzed herein is substantiated by the result of a recent study of the probability of AFB as a cause of primary hepatocellular carcinoma (PHC) in the U.S.A. conducted by Stoloff (*31*). In this study, the life-time risk of PHC for rural, white males ingesting AFB at 13–197 ng/kg body weight/day was found to exceed that of a similar U.S. population ingesting AFB at 0.2–0.3 ng/kg body weight/day by merely 10%. This dose-response relationship is much weaker than what would have been anticipated from experiments with rats and from prior epidemiological studies in Africa and Asia. The author (*32*) further pointed out a number of methodological flaws in these epidemiological studies and questioned the strong correlation between the incidence of PHC and the intake of dietary AFB, as indicated by these studies (*26*, *33*, *34*). If the correlation between PHC and dietary AFB is tenuous, the correlation between the cancer and dietary AFM must be even weaker. As given above, the worst case level of exposure to AFM is 218 ng/capita/day, or 4.4 ng/kg body weight/day, assuming an average body weight being 50 kg per capita. The level of exposure to AFM is considerably lower than the level of exposure to AFB for the populations at higher risk (13–197 ng/kg body weight/day). Consequently, the contribution of ingestion of AFM to the cancer risk of these populations is most likely negligible in view of the low potency of AFM relative to AFB.

It should be noted that the risk assessment presented above is based on data obtained from either adult animals or adult humans. The significance of the risk attributable to ingestion of AFM by infant human populations needs to be evaluated separately. This is because milk represents the major constituent of the infant diet and because infant animals have been found to be considerably more susceptible to AFB carcinogenicity than adult animals (*35*).

Epilogue

Current costly regulatory control of chemical carcinogens in foods is based on the carcinogenicity of suspect chemicals in laboratory animals at experimentally feasible high doses. Once a chemical is found to be carcinogenic in this manner, it is not allowed to exceed an experimentally unattainable low tolerance level (in many cases zero tolerance). This is based on the assumption that at such a prescribed low level, the carcinogenicity would be a fraction of that observed at higher levels. This no-threshold assumption is questioned by our observation on the carcinogenicity of AFM in the range of doses used in our experiment.

ACKNOWLEDGMENTS

The authors gratefully acknowledge the technical assistance provided by Linda Beltran, Patricia Dunn, Deborah Fasulo, and Dr. Dallas Hyde in the feeding and morphometry studies and the helpful discussion offered by Mr. Leonard Stoloff. The authors' research in this area has been supported in part by USPHS Grant CA 27426, the USDA Western Regional Research Project W122, and the Dairy Council of California.

REFERENCES

1. United States Food and Drug Administration. Criteria and procedures for evaluating assays for carcinogenic residues in edible products of animals. Fed. Regist., *42*: 61630; Guideline 7406.06, 1977.
2. Kobbe, B., Hsieh, D.P.H., and Dunkley, W. L. Aflatoxin in milk. What is it, how does it get there? The Dairyman, July: A6–A8, 1979.
3. Busby, W. F. and Wogan, G. N. Aflatoxins. *In;* C. E. Searle (ed.), Chemical Carcinogens, pp. 948–965, American Chemical Society, Washington, D. C., 1984.
4. Patterson, D.S.P., Glancy, E. M., and Roberts, B. A. The "carry over" of aflatoxin M_1 into the milk of cows fed rations containing a low concentration of aflatoxin B_1. Food Cosmet. Toxicol., *18*: 35–37, 1980.
5. Hsieh, D.P.H. Metabolism and transmission of mycotoxins. *In;* Proceedings of the International Symposium on Mycotoxins, pp. 151–166, National Research Center, Cairo, 1983.
6. Goto, T. and Hsieh, D.P.H. Fractionation of radioactivity in the milk of goats administered ^{14}C-aflatoxin B_1. J. Assoc. Off. Anal. Chem., *68*: 456–458, 1985.
7. Stoloff, L. Aflatoxin M_1 in perspective. J. Food Prot., *43*: 226–230, 1980.
8. Purchase, I.F.H. Acute toxicity of aflatoxins M_1 and M_2 in one-day-old ducklings. Food Cosmet. Toxicol., *5*: 339–342, 1967.
9. Pong, R. S. and Wogan, G. N. Toxicity and biochemical and fine structural effects of synthetic aflatoxins M_1 and B_1 in rat liver. J. Natl. Cancer Inst., *47*: 585–592, 1971.
10. Wong, J. J. and Hsieh, D.P.H. Mutagenicity of aflatoxins related to their metabolism and carcinogenic potential. Proc. Natl. Acad. Sci. U.S.A., *73*: 2241–2244, 1976.
11. Coulombe, R. A., Shelton, D. W., Sinnhuber, R. O., and Nixon, J. E. Comparative mutagenicity of aflatoxins using a *Salmonella*/trout hepatic activation system. Carcinogenesis, *3*: 1261–1267, 1982.
12. Lutz, W. K., Jaggi, W., Luthy, J., Sagelsdorff, P., and Schlatter, C. *In vivo* covalent binding of aflatoxin B_1 and aflatoxin M_1 to liver DNA of rat, mouse, and pig. Chem.

Biol. Interactions, *32*: 249–256, 1980.

13. Green, C. E., Rice, D. W., Hsieh, D.P.H., and Byard, J. L. The comparative metabolism and toxic potency of aflatoxin B_1 and aflatoxin M_1 in primary cultures of adult-rat hepatocytes. Food Chem. Toxicol., *20*: 53–60, 1982.
14. Wogan, G. N. and Palialunga, S. Carcinogenicity of synthetic aflatoxin M_1 in rats. Food Cosmet. Toxicol., *12*: 381–384, 1974.
15. Sinnhuber, R. O., Lee, D. J., Wales, J. H., Landers, M. K., and Keyl, A. C. Hepatic carcinogenesis of aflatoxin M_1 in rainbow trout (*Salmo gairdneri*) and its enhancement by cyclopropene fatty acids. J. Natl. Cancer Inst., *53*: 1285–1288, 1974.
16. Canton, J. H., Kroes, R., van Logten, M. J., van Schothorst, M., Stavenutter, J.F.C., and Verhuledonk, C.A.H. The carcinogenicity of aflatoxin M_1 in rainbow trout. Food Cosmet. Toxicol., *13*: 441–443, 1975.
17. Stubblefield, R. D., Shannon, G. M., and Shotwell, O. L. Aflatoxins M_1 and M_2: preparation and purification. J. Am. Oil Chem. Soc., *47*: 389–390, 1970.
18. Hsieh, D.P.H., Beltran, L., Fukayama, M. Y., Rice, D. W., and Wong, J. J. Production and isolation of aflatoxin M_1 for toxicological studies. J. Assoc. Off. Anal. Chem. (in press).
19. Rice, D. W. and Hsieh, D.P.H. Aflatoxin M_1: *in vitro* preparation and comparative *in vitro* metabolism *versus* aflatoxin B_1 in the rat and mouse. Res. Commun. Chem. Pathol. Pharmacol., *35*: 467–490, 1982.
20. Ames, B. N., McCann, J., and Yamasaki, E. Methods for detecting carcinogens and mutagens with the *Salmonella*/mammalian-microsome mutagenicity test. Mutat. Res., *31*: 347–364, 1975.
21. Wogan, G. N. Aflatoxin carcinogenesis. *In;* H. Busch (ed.), Methods in Cancer Research, vol. 7, pp. 309–319, Academic Press, New York, 1973.
22. Wogan, G. N., Paglialunga, S., and Newberne, P. M. Carcinogenic effects of low dietary levels of aflatoxin B_1 in rats. Food Cosmet. Toxicol., *12*: 681–685, 1974.
23. Stewart, H. L., Williams, G. M., Keysser, C. H., Lombard, L. S., and Montali, R. J. Histologic typing of liver tumours of the rat. J. Natl. Cancer Inst., *64*: 179–206, 1980.
24. United States Department of Agriculture. Food consumption, prices and expeditures. 1960–1980. *In*; Economic Research Service. Statistical Bulletin No. 672, pp. 62 and 75, USDA, 1981.
25. Fukayama, M. Y. The production, analysis, occurrence, and toxicity of aflatoxin M_1. M.S. Thesis, pp. 60–70, University of California, Davis, 1979.
26. Carlborg, F. W. Cancer, mathematical models and aflatoxin. Food Cosmet. Toxicol., *17*: 159–166, 1979.
27. Seiber, S. M., Correa, P., Dalgard, D. W., and Adamson, R. H. Induction of osteogenic sarcomas and tumors of the hepatobiliary system in non-human primates with aflatoxin B_1. Cancer Res., *39*: 4545–4554, 1979.
28. Buchi, G. H., Muller, P. M., Roebuck, B. D., and Wogan, G. N. Aflatoxin Q_1: a major metabolite of aflatoxin B_1 produced by human liver. Res. Commun. Chem. Pathol. Pharmacol., *8*: 585–591, 1974.
29. Hsieh, D.P.H., Wong, Z. A., Wong, J. J., Michas, C., and Ruebner, B. H. Comparative metabolism of aflatoxin. *In;* J. V. Rodrick, C. W. Hesseltine, and M. A. Mehlman (eds.), Mycotoxins in Human and Animal Health, pp. 37–50, Pathotox, Illinois, 1977.
30. Hsieh, D.P.H., Wong, J. J., Wong, Z. A., Michas, C., and Ruebner, B. H. Hepatic transformation of aflatoxin and its carcinogenicity. *In;* H. H. Hiat, J. D. Watson, and J. A. Winsten (eds.), Origins of Human Cancer, pp. 697–707, Cold Spring Harbor Laboratory, New York, 1977.

31. Stoloff, L. Aflatoxin as a cause of primary liver-cell cancer in the United States: a probability study. Nutr. Cancer, *5*: 165–186, 1983.
32. Stoloff, L. A rationale for the control of aflatoxin in human foods. *In;* 6th IUPAC, Symposium of Mycotoxins and Phytoalexins, Pretoria, South Africa, 1985.
33. Van Rensburg, S. J. Role of epidemiology in the elucidation of mycotoxin health risks. *In;* J. V. Rodrick, C. W. Hesseltine, and M. A. Mehlman (eds.), Mycotoxins in Human and Animal Health, pp. 699–711, Pathotox, Illinois, 1977.
34. Van Rensburg, S. J. Role of mycotoxins in endemic liver and oesophageal cancer. *In;* 6th IUPAC, Symposium of Mycotoxins and Phytoalexins, Pretoria, South Africa, 1985.
35. Vesselinovitch, S. D., Mihailovich, N., Wogan, G. N., Lombard, L. S., and Rao, K.V.N. Aflatoxin B_1, a hepatocarcinogen in the infant mouse, Cancer Res., *32*: 2289–2291, 1972.

DIET, NUTRITION AND CANCER, Y. HAYASHI ET AL. (EDS.),
JAPAN SCI. SOC. PRESS, TOKYO/VNU SCI. PRESS, UTRECHT, PP. 67–75, 1986

Diet and Exposure to N-Nitroso Compounds

Steven R. TANNENBAUM
Department of Applied Biological Sciences, Massachusetts Institute of Technology, Cambridge, Massachusetts 02139, U.S.A.

Abstract: The hypothesis linking nitrate and increased risk of cancer rests on the proposition that nitrate is endogenously reduced to nitrite by bacteria and that carcinogenic N-nitroso compounds are formed. A large number of foods and biological material have been examined for their ability to generate mutagens or carcinogens under simulated gastric conditions in the presence of nitrite. Only a limited number of foods qualify under these conditions for consideration as potential sources of genotoxic agents. Foods that have generated mutagens following nitrosation include beans, salt-preserved fishery products, fermented soy products, and certain moldy foods. In each case there appears to be a potential link between formation of the nitroso compound and epidemiological evidence of increased risk for specific cancers. The present state of knowledge is reviewed and the chemistry of the nitrosation of specific chemicals of interest is discussed. A major problem for the future will be to demonstrate that these N-nitroso compounds form in the population at risk and react with cellular nucleophiles to produce genetic damage.

Exposure to N-nitroso compounds is now known to occur through endogenous processes which can be modulated by dietary factors. Although the original occurrence of this process was first demonstrated in 1967 (*1*) it was not until 1981 that a quantitative method was developed for measuring the extent of this process (*2*). In fact, a model I proposed in 1980 which estimated endogenous dimethylnitrosamine formation to be on the order of tens of μg per day (*3*) was met with a great deal of skepticism. Since then, the work of several groups of investigators, using the N-nitrosoproline (NPro) method developed by Ohshima and Bartsch (*2*) have shown that endogenous exposure to individual nitrosamines may in fact be many tens of μg per day. In this paper I will review recent work in my laboratory and that of other selected studies relative to an understanding of some of the roles that diet may play in modulating endogenous formation of nitrosamines. A comprehensive review of the literature is outside the scope of this paper, and the reader is additionally referred to two recent reviews which complement this paper (*4*, *5*).

A Balance of Risk Factors

There are many dietary and nondietary factors that contribute to the risk balance for any individual for a given disease. In the case of gastric cancer we proposed a risk model which required a certain combination of events to occur in sequence (*6*). The absence of one of these factors would break the chain of events essential to causation of the disease. In the case of gastric cancer the sequence required a condition of the gastric mucosa which afforded permanent colonization by microorganisms capable of nitrate reduction (*e.g.*, chronic atrophic gastritis and/or intestinal metaplasia). The quantity of intragastric nitrite formed would then depend upon the flow of nitrate to the stomach. The quantity of carcinogen formed would depend upon the nature and quantity of an appropriate nitrogen compound present at the same time as the nitrite. Other factors could down-modulate this process, including dietary components such as ascorbic acid, α-tocopherol, or some phenolic substances. There are also factors which might enhance nitrosation, such as gastric juice thiocyanate. Finally, the condition of the tissue itself, *i.e.*, the rate of cell replication or regeneration would determine the probability of permanently fixing any genetic lesions. An enhanced rate of cell replication could be the result of exposure to excessive concentrations of dietary salt (*7*, *8*).

A generalization of the approach just described is given in Fig. 1. The risk of cancer is not an all-or-none situation, and as it will be seen later almost every member of the population is exposed to endogenously synthesized N-nitroso compounds. Since only a minor percentage of the population ever gets cancer of any type, we should be looking for those individuals who have the highest exposure to N-nitroso compounds to determine whether this is due to high exposure to precursors, or low exposure to inhibitors, or both. In other words, we have not yet done enough epidemiological studies in conjunction with measurement of biochemical parameters to know what the limiting factors are in terms of risk to a population.

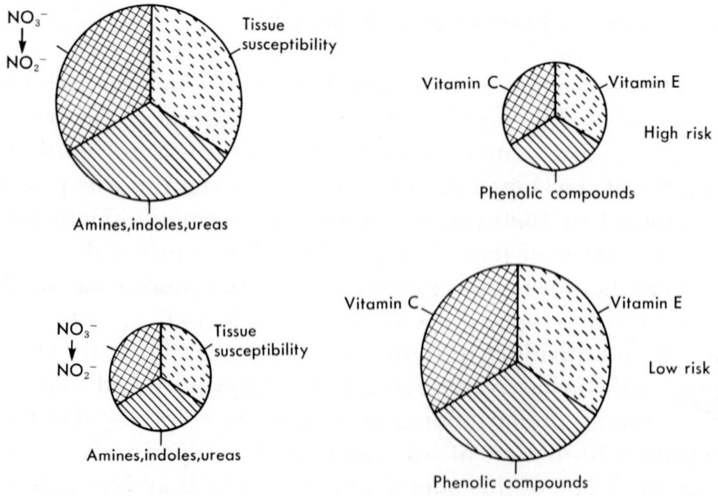

FIG. 1. Balance of risk factors for carcinogenic risk from N-nitroso compounds.

Nitrosoproline (NPro) and Nitrosothiazolidine Carboxylic Acid (NTCA)

Since NPro is excreted quantitatively into the urine we can estimate the total amount synthesized in an individual per day. This has now been done for several small groups in a variety of countries (*9*). However, in order to compare values between high and low risk populations, we will need to conduct studies on larger groups, and also on individuals over a longer period of time, in order to have a good estimate of the population mean and variance, and the individual mean and variance. To test the hypothesis that there is a significant difference between two populations, how many individuals should be tested in each population and how frequently should each individual be tested? Bartsch and his colleagues have begun to make such measurements in several high risk populations (H. Bartsch, personal communication), and my group has been making multiple measurements on individuals over a period of time up to six weeks. Undoubtedly there are numerous other studies by other investigators in progress at this time.

Our studies (*10*) have demonstrated that large doses of ascorbic acid (2 g per day) can inhibit the intragastric synthesis of NPro induced by nitrate and proline intake. To effectively draw this conclusion it was necessary to use $^{15}NO_3^-$ and measure incorporation into ^{15}N-NPro, because there is a background level of excretion that is not related to nitrate intake. Also as part of this study, we gave 10 individuals equivalent doses of nitrate and proline and measured their NPro excretion. There was a large interindividual variation (Table 1) with a greater than six-fold difference between the highest and the lowest values. We are interested in the source of this variability and believe it may lie in a combination of factors related to conversion of nitrate to nitrite and to the reactivity of nitrite in the gastric lumen. We have seen similar variability in experiments with dogs (*11*).

More recently, we have been conducting a large study in collaboration with Allan Conney and William Garland of Hoffmann-LaRoche in Nutley, New Jersey. The Roche investigators collected urines from over 20 individuals over a period of more than one month without additional intake of proline or nitrate. Urines were

TABLE 1. Nitrosoproline Synthesis Induced by Nitrate and Proline[a]

Subject	NPro synthesis (μg/day)
1	7.2
2	7.9
3	8.5
4	8.6
5	11.4
6	13.2
7	13.7
8	15.6
9	16.6
10	45.8
Mean	14.8

[a] Subjects were administered 3.5 mmol of sodium nitrate followed 1 hr later with 4.3 mmol of L-proline.

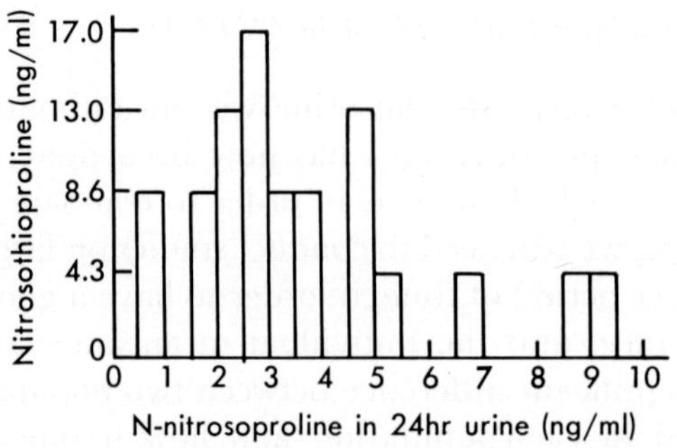

FIG. 2. Urinary NPRo for 23 individuals (means of 20 daily collections).

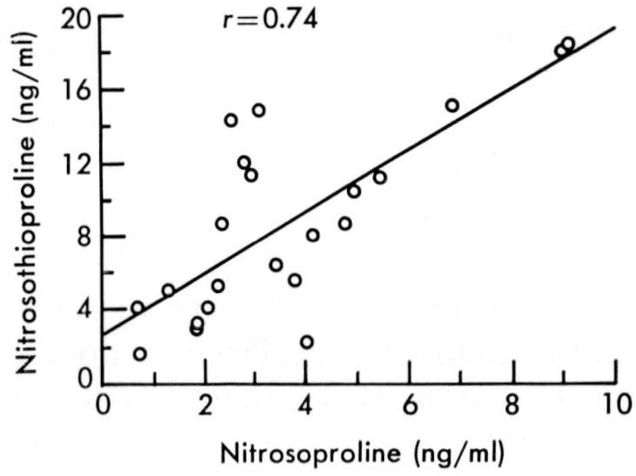

FIG. 3. Correlation of urinary NPro with urinary NCTA.

analyzed for NPro and dimethylnitrosamine (NDMA) by high resolution mass spectrometry at Roche; and for NPro and nitrosothiazolidine carboxylic acid (NTCA) (nitrosothioproline) by gas chromatography-thermal energy analysis (GC-TEA) at MIT. The result for NPro are shown in Fig. 2, plotted as a histogram of individual means. Note that these results represent the actual amount of NPro synthesized endogenously plus that consumed in the diet. Therefore this is the background exposure for each individual. Again there is great variability in this group, excretion ranging from undetectable (less than 1 ng/ml) to as much as 12 ng/ml.

Similar population variation was seen in the NTCA excretion, from undetectable up to 18 ng/ml. What is most interesting however, is the good correlation between NPro and NTCA excretion (Fig. 3). This figure is a plot of the means of the excretion of each compound for each individual. Although the relationship is highly significant, the value of the correlation coefficient suggests that multiple measurements are required to estimate the status of an individual. Nevertheless, the results also suggest that it may be possible to estimate differences in endogenous nitrosamine formation between individuals and populations *via* measurement of urinary NPro, and further that it may be possible to use NPro as a guide to the development of dietary intervention programs.

Two further questions can now be addressed:

a) What are the biologically important compounds produced by endogenous nitrosation?

b) How much exposure is there to these other compounds?

Mutagens, Carcinogens, and Toxic Compounds from Nitrosated Foods

If endogenously formed N-nitroso compounds cause human cancer, then precursor nitrogen compounds will have to be identified which yield nitroso carcinogens upon exposure to nitrite in the stomach or other part of the body. These compounds may be of endogenous or dietary origin or come from specific habits such as tobacco usage or betel quid chewing. To date, in fact, the only unique carcinogens identified are from nondietary (*e.g.*, tobacco, ref. *12*) or endogenous origin (*e.g.*, conjugated bile acids, ref. *13*). In addition there are many examples of precursors of simple nitrosamines, particularly NDMA from plant alkaloids, pharmaceuticals, agricultural, and industrial products, *etc.*

One group of compounds which we have characterized are the conjugated bile acids, N-nitrosoglycocholic acid (NOGC), and N-nitrosotaurocholic acid (NOTC), which yield the respective N-nitrosocarboxamides. Both NOGC and NOTC are mutagens for bacteria and diploid human lymphoblasts (*13*), and have also been shown to cause hepatocellular carcinomas and gastric tumors in the rat (*14*). Therefore, there is adequate characterization of the biological effects of these compounds to implicate them as a potential cause of cancer in humans if they were found to be formed endogenously. Although this has not yet been accomplished, we have shown that intragastric nitrosation of the precursors can occur at pH 3, and it is known that gram quantities of the precursors are refluxed from bile into the gastric lumen. Demonstration of the actual formation of the N-nitroso compounds or of their reaction products with DNA or protein is a high priority goal of our current research.

In contrast to NOGC and NOTC or to the simple dialkyl or heterocyclic nitrosamines which are known carcinogens, some compounds have been isolated from nitrosated foods which have been shown to be mutagens, but not yet shown to be carcinogens. In particular, experiments on the nitrosation of foods using bacterial mutation bioassays have disclosed some very interesting compounds derived from non-mutagenic precursors, examples of which are shown in Table 2. These particular

TABLE 2. Food Sources of Nitrite-derived Mutagens

Fava beans
4-Chloro-6-methoxyindole ⟶ 4-chloro-6-methoxy-2-hydroxy-1-nitrosoindolin-3-one oxime
Soy sauce
Tyramine ⟶ 4-(2-aminoethyl)-6-diazo-2,4-cyclohexadienone
1-methyl-1,2,3,4-tetrahydro-β-carboline-3-carboxylic acid ⟶ dinitroso derivative
Chinese cabbage
Indole-3-acetonitrile ⟶ 1-nitroso derivative
Salted fish and seafood pastes
Unknown precursors ⟶ direct acting mutagens
Corn
Unknown precursors ⟶ nitrohexane, 3-nonenyl-nitrolic acid

mutagens have been isolated and characterized from foods consumed in quantity by populations at high-risk for gastric cancer. The fava bean mutagen (trivially named Nitrosofavine) is among the most potent known bacterial mutagens (*15*). Some of the other foods are discussed in another paper in this Symposium (*16*). However, although these compounds may be potent mutagens they have not yet been tested for carcinogenicity, nor have they been demonstrated to form *in vivo*. Further research must be conducted in this direction, particularly on the biological properties of the products derived from nitrosation of indoles, which appear to derive from a wide variety of fermented and nonfermented foods.

There are also examples of foods which yield toxic but not mutagenic compounds upon nitrosation. For instance corn, a staple of the Colombian Andean diet, yields a nitroalkane and an unsaturated nitrolic acid (Table 2). These compounds are not mutagens, but nitroalkanes are known to be strong irritants to the gastrointestinal tract (*17*). To what extent have we neglected the influence of toxic substances on tissue susceptibility in favor of the exclusive study of genotoxic effects?

Clearly there is a great deal more of research to be done in this area. The biological importance of the above mutagens must be further evaluated, and other foods, such as salted fish and seafood paste must be further characterized (*18*), to enable the evaluation of the nitrosation hypothesis.

How Much Exposure to Other Nitrosamines?

Even in the cases where we know the structure of the precursor and the product (*e.g.*, dimethylamine and NDMA) it is exceedingly difficult to estimate the amount of exposure *via* endogenous formation, because the compounds are rapidly metabolized. One can approach this problem either by measurement of some metabolic endproduct or by calculation based upon known physico-chemical parameters and concentrations. This latter approach may be useful only for a very rough estimate of exposure, but even order of magnitude estimates can be useful. Therefore an approach is presented below for the case of NDMA formation from dimethylamine under a specific set of conditions.

The approach assumes that the quantity of any N-nitroso compound formed is a function of the intrinsic rate constants for nitrosation, the pK_a of the amine, and the ratio of the concentration of the amine to that of proline. It also assumes that pH, catalysts, and inhibitors are the same for an individual

The ratio of the third order rate constants (k) for nitrosation of dimethylamine (2×10^5 M^{-2} sec^{-1}, ref. *19*) and proline (1.4×10^5 M^{-2} sec^{-1}, ref. *20*) is 1.4.

The ratio of reacting species at pH 3 can be calculated from the pK_a of 10.7 for dimethylamine to obtain the concentration of the free base and the calculation of the concentration of neutral proline species (calculation at pH 3 from the microscopic equilibrium constants is given in ref. *20*). This ratio is 0.05.

The concentrations in gastric juice are 12 mM for dimethylamine (*21*) and 0.3 mM for proline (*22*). This ratio is 40.

The expected ratio of NDMA/NPro is calculated from the product of the above ratios since the

$$\text{Rate of nitrosamine formation} = k[\text{free amine concentration}]$$

other factors such as pH, nitrite concentration, temperature, *etc.* being equal. Thus, at pH 3

$$\text{NDMA} = (1.4)(0.05)(40)\text{NPro} = 2.8 \text{ NPro (molar basis)}$$

It is not important whether or not the above calculations are exact. Rather, it is important that they suggest exposure of some individuals (those with highest NPro) to perhaps tens of μg of NDMA per day. If appropriate data were available other nitrosamines might be estimated in similar fashion.

Prospectus

This paper has covered some aspects of human exposure to endogenously formed N-nitroso compounds. Many questions have been raised related to our current state of knowledge on which compounds are formed, and under what circumstances. We are close to, but not yet at, an understanding of the differences between individuals and populations for this type of carcinogenic exposure. The tools are now available to move forward and conduct appropriate epidemiological studies on high-risk populations.

Some comment should also be made on mechanisms of endogenous nitrosation other than nitrosation in gastric juice. We have recently demonstrated mammalian nitrate biosynthesis induced by infection and inflammation (*23*). Subsequently, it was shown that a major source of this biosynthetic process was induced macrophages (*24*). Most recently we have begun to conduct experiments on the capacity of these macrophages to effect nitrosamine synthesis, with early positive results (Miwa, Wishnok, and Tannenbaum, preliminary results). Great interest has also recently been shown in the bacterial catalysis of nitrosamine formation (*25*). It would be especially interesting if selected bacterial strains resident in injured or infected tissues were also competent in catalyzing nitrosation reactions.

New approaches must also be taken to more directly estimate exposure to metabolically labile nitrosamines. My own research has begun to focus on methods for the study of protein alkylation while others have continued to explore DNA adducts which are either excreted in urine or remain unrepaired in tissue DNA. We are only at the beginning of this approach for simple compounds of known structure, but this direction holds great promise for compounds of more complex structure once the structure of the adducts is known.

Finally, we must deal with the issue of the biological significance of this now established exposure to N-nitroso compounds. Essentially, I agree with the conclusions of Bartsch and Montesano (*5*) as follows:

"The evidence accumulated to date, indicates that the levels of nitrosamines found in man's environment may be involved in the causation of human cancers. However, it is perhaps difficult or impossible, to demonstrate in the general population a cause / effect relationship between exposure to low levels of nitrosamines and the incidence of certain cancers, due to the insensitivity of the epidemiological instruments available today and to the lack of truly unexposed populations that could be used as controls."

I believe, however, that further advances along the lines I have already discussed will lift the cloak of impossibility and allow a more definitive conclusion on the biological relevance of this interesting process of endogenous nitrosation first demonstrated only twenty years ago.

ACKNOWLEDGMENTS

This investigation was supported by Public Health Service Grant Number 1-POl-CA 26731-06, awarded by the National Cancer Institute, Department of Health and Human Services; and by grants from Hoffmann-LaRoche, Nutley, NJ and from the American Meat Institute.

REFERENCES

1. Sander, J. Kann nitrit in der menschlichen nahrung ursache einer krebsenstenhung durch nitrosaminblidung sein? Arch. Hyg. Bakt., *151*: 22–28, 1967.
2. Ohshima, H. and Bartsch, H. Quantitative estimation of endogenous nitrosation in humans by monitoring N-nitrosoproline excreted in the urine. Cancer Res., *41*: 3658–3662, 1981.
3. Tannenbaum, S. R. A model for estimation of human exposure to endogenous N-nitrosodimethylamine. Oncology, *37*: 232–235, 1980.
4. Tannenbaum, S. R. N-Nitroso compounds: a perspective on human exposure. Lancet, March 19: 629–632, 1983.
5. Bartsch, H. and Montesano, R. Relevance of nitrosamines to human cancer. Carcinogenesis, *5*: 1381–1389, 1984.
6. Correa, P., Haenszel, W., Cuello, C., Tannenbaum, S. R., and Archer, M. C. A model for gastric cancer epidemiology. Lancet, July: 58–62, 1975.
7. Charnley, G. and Tannenbaum, S. R. A flow cytometric analysis of the effect of sodium chloride on gastric cancer risk in the rat. Cancer Res., *45*: 5608–5616, 1985.
8. Takahashi, M. and Hasegawa, R. Enhancing effect of dietary salt on both initiation and promotion stages of rat gastric carcinogenesis. This volume, pp. 169–182, 1986.
9. Preussmann, R. Occurrence and exposure to N-nitroso compounds and precursors. *In;* I. K. O'Neill, R. C. Von Borstel, C. T. Miller, J. Long, and H. Bartsch (eds.), N-Nitroso Compounds: Occurrence, Biological Effects and Relevance to Human Cancer, IARC Sci. Pub. No. 57, pp. 3–15, International Agency for Research on Cancer, Lyon, 1984.
10. Wagner, D. A., Shuker, D.E.G., Bilmazes, C., Obiedzinski, M., Baker, I., Young, V. R., and Tannenbaum, S. R. Effect of vitamins C and E on endogenous synthesis of N-nitrosoamino acids in humans: precursor-product studies with ^{15}N-nitrate. Cancer Res. (in press).
11. Lintas, C., Fox, J., Tannenbaum, S. R., and Newberne, P. M. *In vivo* stability of nitrite and nitrosamine formation in the dog stomach. Effect of nitrite and amine concentration and of ascorbic acid. Carcinogenesis, *3*: 161–165, 1982.
12. Hoffmann, D. and Hecht, S. S. Nicotine-derived N-nitrosamines and tobacco-related cancer: current status and future directions. Cancer Res., *45*: 935–944, 1985.
13. Song, P., Shuker, D.E.G., Bishop, W. W., Falchuk, K. K., Tannenbaum, S. R., and Thilly, W. G. Mutagenicity of N-nitroso bile acid conjugates in *S. typhimurium* and

diploid human lymphoblasts. Cancer Res., *42*: 2601–2604, 1982.
14. Busby, W. F., Jr., Shuker, D.E.G., Charnley, G., Newberne, P. M., Tannenbaum, S. R., and Wogan, G. N. Carcinogenicity of the nitrosated bile acid conjugates, N-nitrosoglycocholic acid and N-nitrosotaurocholic acid. Cancer Res., *45*: 1367–1371, 1985.
15. Yang, D., Tannenbaum, S. R., Buchi, G., and Lee, G.C.M. 4-Chloro-6-methoxyindole is the precursor of a potent mutagen (4-chloro-6-methoxy-2-hydroxy-1-nitroso-indoline-3-one oxime) that forms during nitrosation of the fava bean (*Vicia faba*). Carcinogenesis, *5*: 1219–1224, 1984.
16. Nagao, M., Wakabayashi, K., Fujita, Y., Tahira, T., Ochiai, M., Takayama, S., and Sugimura, T. Nitrosatable precursors of mutagens in vegetables and soy sauce. This volume, pp. 77–86, 1986.
17. Hansen, T. J., Tannenbaum, S. R., and Archer, M. C. Identification of a nonenylnitrolic acid in corn treated with nitrous acid. J. Agric. Food Chem., *29*: 1008–1011, 1981.
18. Tannenbaum, S. R., Bishop, W., Yu, M. C., and Henderson, B. E. Attempts to isolate N-nitroso compounds from Chinese-style salted fish. J. Natl. Cancer Inst. Monogr. (in press).
19. Fan, T. Y. and Tannenbaum, S. R. Natural inhibitors of nitrosation reactions; the concept of available nitrite. J. Food Sci., *38*: 1067–1069, 1973.
20. Mirvish, S. S., Sams, J., Fan, T. Y., and Tannenbaum, S. R. Kinetics of nitrosation of the amino acids proline, hydroxyproline, and sarcosine. J. Natl. Cancer Inst., *51*: 967–969, 1973.
21. Zeisel, S. H., DaCosta, K.-A., and Fox, J. G. Endogenous formation of dimethylamine. Biochem. J., *232*: 403–408, 1985.
22. Muting, D. Über den Aminosaurengehalt des menschlichen Magen und Duodenalsaftes. Naturwissenschaften, *41*: 580, 1954.
23. Wagner, D. A., Young, V. R., and Tannenbaum, S. R. Mammalian nitrate biosynthesis: incorporation of [^{15}N]-ammonium into nitrate is enhanced by endotoxin treatment. Proc. Natl. Acad. Sci. U.S.A., *80*: 4518–4521, 1983.
24. Steuhr, D. J. and Marletta, M. A. Mammalian nitrate biosynthesis: mouse macrophages produce nitrite and nitrate in response to *Escherichia coli* lipopolysaccharide. Proc. Natl. Acad. Sci. U.S. A. (in press).
25. Calmels, S., Ohshima, H., Vincent, P., Gounot, A.-M., and Bartsch, H. Screening of microorganisms for nitrosation catalysis at pH 7 and kinetic studies on nitrosamine formation from secondary amines by *E. coli* strains. Carcinogenesis, *6*: 911–915, 1985.

DIET, NUTRITION AND CANCER, Y. HAYASHI ET AL. (EDS.),
JAPAN SCI. SOC. PRESS, TOKYO/VNU SCI. PRESS, UTRECHT, PP. 77–86, 1986

Nitrosatable Precursors of Mutagens in Vegetables and Soy Sauce

Minako NAGAO, Keiji WAKABAYASHI, Yuki FUJITA, Tomoko TAHIRA, Masako OCHIAI, Shozo TAKAYAMA, and Takashi SUGIMURA

Carcinogenesis Division, National Cancer Center Research Institute, Tokyo 104, Japan

Abstract: Nitrosatable precursors of mutagens that show mutagenicity to *Salmonella typhimurium* TA100 without S9 mix after treatment with nitrite at pH 3 were found in various foods. From Chinese cabbage, three indole compounds, indole-3-acetonitrile, 4-methoxyindole-3-acetonitrile, and 4-methoxyindole-3-aldehyde, were identified as mutagen precursors. 1-Methylindole and 2-methylindole, which are present in cigarette smoke showed strong mutagen precursor activity. *Escherichia coli* WP2 *uvrA*/pKM101 is more sensitive than *S. typhimurium* TA100 to nitrosatable precursors in soy sauce after treatment with 1–3 mM nitrite. The mutagenicity of soy sauce towards *E. coli* WP2 *uvrA*/pKM101 is partly explained by 1-methyl-1,2,3,4-tetrahydro-β-carboline-3-carboxylic acid (MTCA) and tyramine reported previously. Oral administration of soy sauce and nitrite to male Fischer 344 rats for 2 years induced basal cell proliferation of the forestomach and intestinal metaplasia of the glandular stomach, but did not induce cancers in any organ. 3-Diazotyramine, a mutagenic nitrosation product of tyramine that is present at high concentrations in various foods induced squamous cell carcinomas of the oral cavity of rats when given in their drinking water. The carcinogenesis by N-benzylmethylamine, a nitrosatable precursor and nitrite was prevented by thioproline.

The *Salmonella* mutagenesis assay is a rapid and reliable method for the primary screening of carcinogens. By use of this test many mutagenic heterocyclic amines have been isolated from ordinary cooked foods and shown to be carcinogenic (*1*).

Nitrosatable precursors of mutagens that show mutagenicity after nitrite treatment have also been found in various foods. Vegetables are a major source of nitrate and microflora in the gut reduce nitrate to nitrite (*2*). Nitrate and nitrite are also produced endogenously in the body from ammonia, amino acids, and peptides (*2–4*). Marked increases in the endogenous formations of nitrate and nitrite were observed under conditions in which macrophages were activated (*5*). A close correlation was found between nitrate intake and the death rate from stomach cancer (*6*). Since nitrite itself is not carcinogenic to experimental animals (*7*), nitrosatable precursors, such as secondary amines and alkylamides are suspected as causes of human cancers. In addition, new nitrosatable precursors of mutagens of various types have been reported. 4-Chloro-6-methoxyindole in fava beans (*8*), 1-methyl-1,2,3,4-tetrahydo-β-carboline-3-carboxylic acid (MTCA), and tyramine in soy sauce showed mutagenicity after nitrite treatment. Tyramine is present at high concentration in various foods (*9–12*).

This paper reports studies on the mutagenicities of indole compounds isolated from Chinese cabbage. The appearance of mutagenicity on treatment of soy sauce with a physiologically feasible concentration of nitrite is also reported. In addition experiments on the carcinogenesis of soy sauce and nitrite and the carcinogenicity of 3-diazotyramine(4-(2-aminoethyl)-6-diazo-2,4-cyclohexadienone), a nitrosation product of tyramine, are described. Production of mutagens/carcinogens by the reaction of precursors with nitrite should be prevented. Thioproline, a nitrite scavenger, was found to suppress the carcinogenicity of N-benzylmethylamine plus nitrite.

Nitrosatable Precursors of Mutagens in Chinese Cabbage

Chinese cabbage is a popular vegetable in Japan and its average consumption is 45 g/day/capita. Water extracts of fresh Chinese cabbage and Chinese cabbage pickled with salt were not mutagenic to *S. typhimurium* TA100. However, after treatment with nitrite in acidic conditions, they became mutagenic to *S. typhimurium* TA100 without S9 mix.

A water extract of fresh Chinese cabbage was fractionated by chromatographies on columns of XAD-7, Lobar Lichroprep RP-8 and Si60 (Merck, Darmstadt), and HPLC ODS-120A (Toyo Soda, Tokyo), and three indole compounds, indole-3-acetonitrile, 4-methoxyindole-3-acetonitrile, and 4-methoxyindole-3-aldehyde, were isolated as nitrosatable precursors of mutagens. Their specific mutagenic activities after treatment with 50 mM nitrite at pH 3 for 1 hr were 17,400, 31,800, and 156,900 revertants/mg, respectively, on *S. typhimurium* TA100 without S9 mix (*13, 14*). These three compounds accounted for about 20% of the total mutagenicity of a water extract of Chinese cabbage after nitrite treatment. 1-Nitrosoindole-3-acetonitrile was identified as direct-acting nitrosated form of indole-3-acetonitrile. The specific activities of 1-nitrosoindole-3-acetonitrile on *S. typhimurium* TA100 and TA98 were 45 and 30 revertants/μg, respectively (*15*). Indole compounds are ubiquitous in plants, and some of them serve as plant growth hormones (*16–18*).

Mutagenic Precursor Activities of Indole Compounds

We tested 30 indole compounds that are available commercially for mutagen precursor activity. After nitrite treatment 22 of these compounds showed mutagenicity to *S. typhimurium* TA100 without S9 mix (*19*). Of these, 1-methylindole, 2-methylindole, 1-methyl-D,L-tryptophan, harmaline, and (−)-(1*S*,3*S*)-1,2-dimethyl-1,2,3,4-tetrahydro-β-carboline-3-carboxylic acid ((−)-(1*S*,3*S*)-DiMTCA) were the most mutagenic, as shown in Table 1. 1-Methylindole and 2-methylindole are present in cigarette smoke (*20*). Harmaline is present in plants (*21*). Of the 22 mutagen

TABLE 1. Mutagenicities of Indole-compounds after Nitrite Treatment

Compound	Structure	Precursor activity[a] (Revertants/mg)
Indole	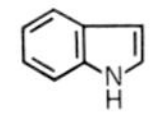	8,800
1-Methylindole	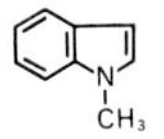	615,000
2-Methylindole	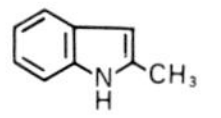	129,000
3-Methylindole	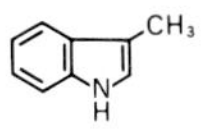	17,900
L-Tryptophan	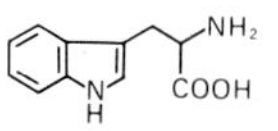	4,740
D-Tryptophan		4,800
1-Methyl-DL-tryptophan	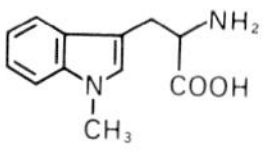	184,000
L-Abrine	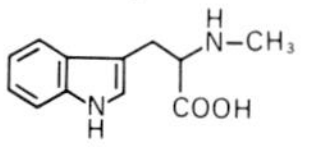	0 (100)[b]
Norharman	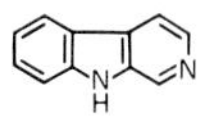	0 (1,000)
Harman	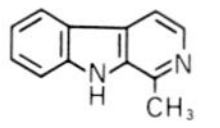	0 (500)
Harmaline	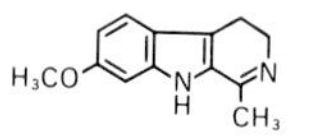	103,000
(−)-(1*S*, 3*S*)-MTCA	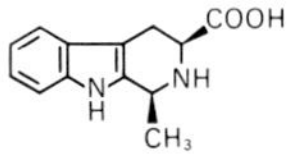	17,400
(−)-(1*R*, 3*S*)-MTCA	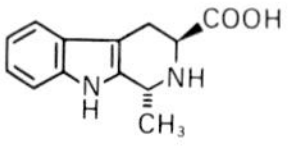	13,000
(−)-(1*S*, 3*S*)-DiMTCA	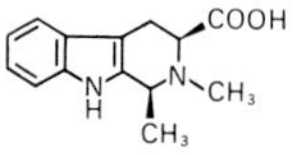	197,000
(−)-(1*S*, 3*S*)-Acetyl-MTCA	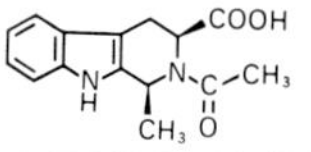	900

[a] Mutagenic precursor activity was tested using *S. typhimurium* TA100 without S9 mix. [b] Figures in parentheses show maximum dose in μg/plate used for tests with nitrite treatment.

precursors 11 are naturally occurring. The kinetics of nitrosation of mutagen precursors and the biological effects of nitrosated products on animals require further study.

Mutagenicity of Soy Sauce Treated with a Physiologically Feasible Concentration of Nitrite

We previously reported that soy sauce treated with 50 mM nitrite at pH 3 was mutagenic to *S. typhimurium* TA100 without metabolic activation, and we identified three mutagen precursors, (−)-(1*S*,3*S*)-MTCA, (−)-(1*R*,3*S*)-MTCA (*22*), and tyramine (*12*). These three compounds accounted for almost all the mutagenic activity of soy sauce, that appeared after its treatment with 50 mM nitrite at pH 3.

The average nitrite concentrations in human saliva and gastric juice are much less than 50 mM, although they may change by various conditions and with various kinds of foods (*2–5*, *23–26*). The mixture obtained by treating 50% soy sauce with 3 mM nitrite at pH 3 for 1 hr, was not mutagenic to *S. typhimurium* TA100, and only slightly mutagenic to *E. coli* WP2 *uvrA*/pKM101.

Since soy sauce is diluted during preparation of foods and is also diluted by gastric juice in the stomach, we tested the effect of the soy sauce concentration on the mutagenicity of a mixture of nitrite and soy sauce on *E. coli* WP2 *uvrA*/pKM101. When soy sauce at a concentration of 5% was incubated with 1 to 3 mM sodium nitrite at pH 3 for 1 hr, dose-dependent increases of mutants ($trp^- \rightarrow trp^+$) was observed (*27*). The more efficient formation of mutagens in dilute soy sauce may be due to the presence of nitrite scavengers that react with nitrite more rapidly than the precursor(s) of mutagen in soy sauce. Another possibility is that soy sauce may contain compounds such as phenolic compounds, that catalyze the nitrosation reaction more efficiently at low concentration than at high concentration (*28*, *29*).

A fraction that did not contain MTCAs or tyramine was obtained by chromatography of desalted soy sauce on XAD-7 and CM-Sephadex columns. This fraction showed preferential mutagenicity to *E. coli* after treatment with 1 to 3 mM nitrite. The structure(s) of the mutagen precursor in this fraction is being studied.

Long-term Feeding Experiment with Soy Sauce and Nitrite

The effects of feeding of soy sauce and nitrite on male Fischer 344 rats were studied. Each experimental group consisted of about 40 rats. Rats in Group I were given diet containing soy sauce and drinking water containing 0.15% sodium nitrite. The rats in Group II were given diet containing soy sauce and deionized water. Group III received basal diet and deionized water. Diet containing soy sauce was prepared by mixing soy sauce and basal diet CE-2 (CLEA Japan, Inc.; Tokyo) in a ratio of 1 ml/3 g and drying the mixture at 70°C for 4 hr to form pellets. Sodium nitrite solution in deionized water was prepared in light-proof containers. The experiment was continued for 122 weeks. No tumors other than spontaneous ones developed in any organs, and the survivals of the three groups were similar. The stomach, liver, kidney, spleen, adrenal gland, thyroid gland, and pituitary gland were examined histologically.

Table 2 summarizes findings in the stomach. Tumors of the forestomach were observed in 3 rats of Group I. One was a squamous cell carcinoma and 2 were papillomas. One squamous cell carcinoma of the forestomach was also found in a control rat (Group III). The incidences of tumor in these two groups were not significantly

TABLE 2. Effect of Soy Sauce and Nitrite on Rat Stomach

	Effective number of rats	Forestomach		Glandular stomach	
		Basal cell proliferation	Tumor	Hyperplasia	Intestinal metaplasia
Group I (soy sauce+nitrite)	42	30*	3	9	34*
Group II (soy sauce)	35	19*	0	0	19
Group III (none)	37	5	1	3	12

Soy sauce was mixed with the diet in a ratio of 1 ml/3 g. Nitrite was given in the drinking water at 0.15%. The experimental period was 122 weeks. * Significantly different from the value for group III at $p<0.01$.

different. Basal cell proliferation in the forestomach was observed in 30 of 42 rats in Group I, 19 of 35 rats in Group II, and 5 of 37 rats in Group III, and its incidences in Group I and Group II were significantly higher than that in the control group, although its incidences in Groups I and II were not significantly different. No tumors of the glandular stomach were observed in any group. Interestingly, however, intestinal metaplasia was observed in 34 rats in Group I, 19 in Group II, and 12 in Group III, and the incidence in Group I was significantly higher than that in Group III. Basal cell proliferation and intestinal metaplasia were observed in rats killed 100 weeks after the start of the experiment. A relation between intestinal metaplasia and stomach cancer is suspected. Really high incidences of both intestinal metaplasia and stomach cancer are observed in Japanese and Columbian (*30, 31*).

Other organs examined showed no marked histological differences in the three groups. This experiment showed that soy sauce alone and soy sauce plus nitrite affected the stomach mucosa of rats. There is a report that soy sauce alone induced mucus loss and altered nuclear chromatin pattern of cells of surface epithelium and gastric pit of rats (*32*).

Carcinogenicity of 3-Diazotyramine

3-Diazotyramine, a nitrosation product of tyramine showed direct-acting mutagenicity on *S. typhimurium* TA100 and TA98 (*12*), and Chinese hamster lung cells, inducing diphtheria toxin resistance.

3-Diazotyramine hydrochloride synthesized by Nard Institute was given to 28 male Fischer 344 rats as 0.1% solution in their drinking water for 116 weeks. In week 58, one rat with a tumor of the oral cavity was observed. Figure 1 shows a typical example of a huge tumor that developed in the oral cavity. Histologically, this tumor was a squamous cell carcinoma. Similar tumors developed in 19 of 28 rats in the experimental group, but not in control rats. The cumulative intake of 3-diazotyramine per rat at the time of appearance of the first tumor was 7.4 g. All seven tumors tested were all successfully transplanted into the subcutis of the same strain of rats.

There are few reports on induction of tumors of the oral cavity in experimental animals. Topical application of 4-nitroquinoline 1-oxide and 7,12-dimethylbenz[a]-anthracene induced squamous cell carcinomas of the oral cavity (*33, 34*). Oral ad-

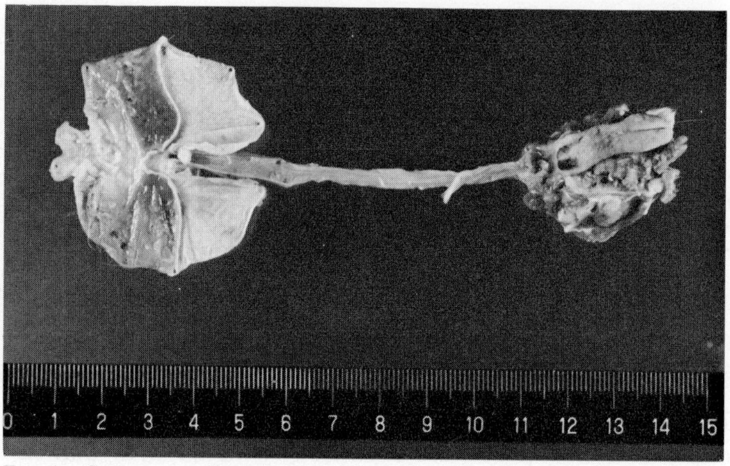

FIG. 1. Squamous cell carcinoma of the oral cavity induced by 3-diazotyramine.

ministration of solution of N-propyl- and N-butyl-N-nitrosourethan induced tumors of the upper parts of the digestive tract, including the oral cavity of rats (*35, 36*). But 3-diazotyramine is unique in specifically inducing squamous cell carcinomas of the oral cavity.

Inhibitory Effect of Thioproline on Carcinogenesis by N-Benzylmethylamine and Nitrite

Recently, nitrosothioproline was found as a major nitroso compound in urine (*37, 38*). Thioproline is known to be formed non-enzymatically from cysteine and formaldehyde, and its formation *in vitro* on addition of cysteine to a rat liver homogenate has been reported (*39*). The excretion of 20–150 μg/day of thioproline in human urine has also been reported (*40*).

We determined the kinetics of nitrosation of thioproline. The rate of the reaction increased with decrease in the pH value, and was proportional to the concentrations of total thioproline and nitrite at pH 2. The initial reaction rate followed the equation:

$$\text{rate} = k_4 \times [\text{thioproline}] \times [\text{NaNO}_2] \times [\text{H}^+]$$

The rate constant was found to be 49.4 $\text{M}^{-2} \cdot \text{sec}^{-1}$ at 37°C (*41*). Judging from this rate constant, thioproline may act as a nitrite scavenger *in vitro* and *in vivo*. As expected, thioproline inhibited the formation of N-nitroso-N-benzylmethylamine from N-benzylmethylamine and nitrite *in vitro*. Incubation of 20 mM N-benzylmethylamine and 20 mM sodium nitrite at pH 3 for 20 min resulted in formation of 0.74 mM N-nitroso-N-benzylmethylamine. Addition of 20 mM thioproline to this reaction mixture caused about 96% inhibition of the formation of N-nitroso-N-benzylmethylamine (*41*).

We tested the effect of thioproline on *in vivo* carcinogenesis by N-benzylmethylamine and nitrite. Eight male F344 rats in Group I were given powdered diet containing 0.25% N-benzylmethylamine and deionized water containing 0.2% sodium nitrite. Ten male F344 rats in Group II were given diet containing 0.25% N-benzyl-

TABLE 3. Inhibitory Effect of Thioproline on Carcinogenesis by N-Benzylmethylamine and Nitrite

	Number of rats		Carcinomas in forestomach	
	Initial	Effective	Invasive	*In situ*
Group I (N-benzylmethylamine+nitrite)	8	7	6	1
Group II (N-benzylmethylamine+thioproline+nitrite)	10	9	0	2

methylamine and 0.5% thioproline, and deionized water containing 0.2% sodium nitrite. The animals had free access to diet and water. The average intakes of diet and water per rat in Group I and Group II were almost the same and the body weights of rats in these groups were also the same during the 717 day experiment. The effective numbers of rats in Group I and Group II were 7 and 9, respectively.

The first tumor of the forestomach was found in a rat in Group I on day 633. Tumors of the forestomach developed in all 7 rats in Group I (Table 3). Six of these tumors were histologically squamous cell carcinomas invading the submucosa or serosa. A carcinoma *in situ* of the forestomach developed in the other rat. In Group II, carcinomas *in situ* developed in only 2 of 9 rats, and no carcinomas invaded the submucosa or serosa. No carinomas was found in any other organs of rats in Group I or II. These data show that thioproline prevents carcinogenesis by N-benzylmethylamine and nitrite. Under our experimental conditions, thioproline and nitrosothioproline had no toxic effects on the experimental animals.

DISCUSSION

Nitroso compounds are a group of carcinogenic compounds that includes N-methyl-N′-nitro-N-nitrosoguanidine, a carcinogen of the glandular stomach. We surveyed various foods for nitrosatable precursors of mutagens, and found that various vegetables and soy sauce showed mutagenicity after nitrite treatment. Structural determination of mutagen precursors in various foods revealed that many indole compounds have precursor activity (*8*, *9*, *13–15*, *19*, *22*). On the other hands, indole compounds such as indole-3-acetonitrile, indole-3-carbinol, and 3,3′-diindolylmethane were shown to have anti-carcinogenesis activity (*42*). For estimation of risks of indole compounds to human, studies on the metabolisms of precursor compounds in the body and the nitrosation kinetics are important. Elucidation of mechanisms of the anti-carcinogenic effect of indole compounds as well as information on the carcinogenicities of mutagens formed by nitrite treatment are also important.

We found that soy sauce contained a hitherto unknown mutagen precursor(s) which show mutagenicity preferentially to *E. coli* WP2 *uvrA*/pKM101 after treatment with a physiologically feasible low concentration of nitrite. Oral administration of soy sauce and nitrite to rats resulted in basal cell proliferation in the forestomach and intestinal metaplasia of the glandular stomach. Since soy sauce alone did not give significant increase of the incidence of intestinal metaplasia, soy sauce plus nitrite was suggested to damage more efficiently the epithelium, leading to abnormal differentiation of the glandular stomach mucosa. Japanese people have high incidences in both stomach cancer and intestinal metaplasia (*31*), and it is important to clarify what component in a mixture of soy sauce and nitrite induce intestinal metaplasia.

3-Diazotyramine, a nitrosation product of tyramine, specifically induced cancers of the oral cavity when given to rats in their drinking water. Tyramine is present at high concentration in various foods: Cheddar cheese, dried sausage, and soy sauce contain tyramine at more than 1 mg per g. The incidence of human cancer of the oral cavity is high in Sri Lanka and southern parts of India (*43*). Betel tobacco is considered to be one cause (*44*), but compounds such as tyramine and nitrite may also be causes.

Since nitrosatable precursors are now known to be present in various foods, various types of nitrite scavenger should be useful for preventing formation of nitrosated mutagens/carcinogens. The present study showed that thioproline prevented carcinogenesis induced by N-benzylmethylamine plus nitrite.

REFERENCES

1. Sugimura, T. Carcinogenicity of mutagenic heterocyclic amines formed during the cooking process. Mutat. Res., *150*: 33–41, 1985.
2. Tannenbaum, S. R. Endogenous formation of nitrite and N-nitroso compounds. *In;* E. C. Miller, I. Hirono, T. Sugimura, and S. Takayama (eds.), Naturally Occurring Carcinogens-mutagens and Modulators of Carcinogenesis, pp. 211–220, Japan Sci. Soc. Press, Tokyo and Univ. Park Press, Baltimore, 1979.
3. Wagner, D. A. Metabolic fate of an oral dose of ^{15}N-labeled nitrate in humans: effect of diet supplementation with ascorbic acid. Cancer Res., *43*: 1921–1925, 1983.
4. Dull, B. J. and Hotchkiss, J. H. Activated oxygen and mammalian nitrate biosynthesis. Carcinogenesis, *5*: 1161–1164, 1984.
5. Stuehr, D. J. and Marletta, M. A. Mammalian nitrate biosynthesis: mouse macrophages produce nitrite and nitrate in response to *Escherichia coli* lipopolysaccharide. Proc. Natl. Acad. Sci. U.S.A., *82*: 7738–7742, 1985.
6. Fine, D. H., Challis, B. C., Hartman, P., and Ryzin, J. V. Endogenous synthesis of volatile nitrosamine: model calculations and risk assessment. *In;* H. Bartsch, I. K. O'Neill, M. Castegnaro, M. Okada, and W. Davis (eds.), N-Nitroso Compounds: Occurrence and Biological Effects, IARC Sci. Publ. No. 41, pp. 379–396, IARC, Lyon, 1982.
7. Maekawa, A., Ogiu, T., Onodera, H., Furuta, K., Matsuoka, C., Ohno, Y., and Odashima, S. Carcinogenicity studies of sodium nitrite and sodium nitrate in F344 rats. Food Cosmet. Toxicol., *20*: 25–33, 1982.
8. Yang, D., Tannenbaum, S. R., Büchi, G., and Lee, G.C.M. 4-Chloro-6-methoxyindole is the precursor of a potent mutagen (4-chloro-6-methoxy-2-hydroxy-1-nitroso-indolin-3-one oxime) that forms during nitrosation of the fava beans (*Vicia faba*). Carcinogenesis, *5*: 1219–1224, 1984.
9. Wakabayashi, K., Nagao, M., Ochiai, M., Tsuda, M., Yamaizumi, Z., Saito, H., and Sugimura, T. Presence of 1-methyl-1,2,3,4-tetrahydro-β-carboline-3-carboxylic acids and tyramine as precursors of mutagens in soy sauce after nitrite treatment. *In;* I. K. O'Neill, R. C. von Borstel, C. T. Miller, J. Long, and H. Bartsch (eds.), N-Nitroso Compounds: Occurrence, Biological Effects and Relevance to Human Cancer, IARC Sci. Publ. No. 57, pp. 17–24, IARC, Lyon, 1984.
10. Smith, T. A. Amines in food. Food Chem., *6*: 169–200, 1981.

11. Wakabayashi, K., Nagao, M., Chung, T. H., Yin, M., Karai, I., Ochiai, M., Tahira, T., and Sugimura, T. Appearance of direct-acting mutagenicity of various foodstuffs produced in Japan and Southeast Asia on nitrite treatment. Mutat. Res., *158*: 119–124, 1985.
12. Ochiai, M., Wakabayashi, K., Nagao, M., and Sugimura, T. Tyramine is a major mutagen precursor in soy sauce, being convertible to a mutagen by nitrite. Gann, *75*: 1–3, 1984.
13. Wakabayashi, K., Nagao, M., Ochiai, M., Tahira, T., Yamaizumi, Z., and Sugimura, T. A mutagen precursor in Chinese cabbage, indole-3-acetonitrile, which becomes mutagenic on nitrite treatment. Mutat. Res., *143*: 17–21, 1985.
14. Nagao, M., Wakabayashi, K., Fujita, Y., Tahira, T., Ochiai, M., and Sugimura, T. Mutagenic compounds in soy sauce, Chinese cabbage, coffee and herbal teas. *In;* I. Knudsen (ed.), Genetic Toxicology of the Diet, Prog. Clin. Biol. Res., vol. 206., pp. 55–62, Alan R. Liss, Inc., New York, 1986.
15. Wakabayashi, K., Nagao, M., Tahira, T., Saito, H., Katayama, M., Marumo, S., and Sugimura, T. 1-Nitrosoindole-3-acetonitrile, a mutagen produced by nitrite treatment of indole-3-acetonitrile. Proc. Jpn. Acad., *61B*: 190–192, 1985.
16. Henbest., H. B., Johes, E.R.H., and Smith, G. F. Isolation of a new plant-growth hormone, 3-indolylacetonitrile. J. Chem. Soc., 3796–3801, 1953.
17. Nomoto, M. and Tamura, S. Isolation and identification of indole derivatives in clubroots of Chinese cabbage. Agric. Biol. Chem., *34*: 1590–1592, 1970.
18. Okamoto, T., Isogai, Y., Koizumi, T., Fujishiro, H., and Sato, Y. Studies on plant growth regulators, III. Isolation of indole-3-acetonitrile and methyl indole-3-acetate from the neutral fraction of the Moyashi extract. Chem. Pharm. Bull., *15*: 163–168, 1967.
19. Ochiai, M., Wakabayashi, K., Sugimura, T., and Nagao, M. Mutagenicity of indole and 30 derivatives after nitrite treatment. Submitted for *Mutat. Res.*
20. Grob, K. and Voellmin, J. A. GC-MS analysis of the "semi-volatiles" of cigarette smoke. J. Chromatogr. Sci., *8*: 218–220, 1970.
21. Allen, J.R.F. and Holmstedt, B. R. The simple β-carboline alkaloids. Phytochemistry, *19*: 1573–1582, 1980.
22. Wakabayashi, K., Ochiai, M., Saito, H., Tsuda, M., Suwa, Y., Nagao, M., and Sugimura, T. Presence of 1-methyl-1,2,3,4-tetrahydro-β-carboline-3-carboxylic acid, a precursor of a mutagenic nitroso compound, in soy sauce. Proc. Natl. Acad. Sci. U.S.A., *80*: 2912–2916, 1983.
23. Spiegelhalder, B., Eisenbrand., G., and Preussman., R. Influence of dietary nitrate on nitrite content of human saliva: possible relevance to *in vivo* formation of N-nitroso compounds. Food Cosmet. Toxicol., *14*: 545–548, 1976.
24. Tannenbaum, S. R., Weisman, M., and Fett, D. The effect of nitrate intake on nitrite formation in human saliva. Food Cosmet. Toxicol., *14*: 549–552, 1976.
25. Correa, P., Cuello, C., Gordillo, G., Zarama, G., Lopez, J., Haenszel, W., and Tannenbaum, S. The gastric microenvironment in population at high risk to stomach cancer. Natl. Cancer Inst. Monogr., *53*: 167–170, 1979.
26. Mirvish, S. S. The etiology of gastric cancer: intragastric nitrosamide formation and other theories. J. Natl. Cancer Inst., *71*: 631–647, 1983.
27. Tahira, T., Fujita, Y., Ochiai, M., Wakabayashi, K., Nagao, M., and Sugimura, T. Mutagenicity of soy sauce treated with a physiologically feasible concentration of nitrite. Mutat. Res. (in press).
28. Nakamura, M. and Kawabata, T. Effect of Japanese green tea on nitrosamine forma-

tion *in vitro*. J. Food Sci., *46*: 306–307, 1981.
29. Pignatelli, B., Bereziat, J. C., O'Neill, I. K., and Bartsch, H. Catalytic role of some phenolic substances in endogenous formation of N-nitroso compounds. *In;* H. Bartsch, I. K. O'Neill, M. Castegnaro, M. Okada, and W. Davis (eds.), N-Nitroso Compounds: Occurrence and Biological Effects, IARC Sci. Publ. No. 41, pp. 413–423, IARC, Lyon, 1982.
30. Cuello, C., Correa, P., Haenszel, W., Gordillo, G., Brown, C., Areher, M., and Tannenbaum, S. Gastric cancer in Colombia. I. Cancer risk and suspect environmental agents. J. Natl. Cancer Inst., *57*: 1015–1020, 1976.
31. Imai, T., Kudo, T., and Watanabe, H. Chronic gastritis in Japanese with reference to high incidence of gastric carcinoma. J. Natl. Cancer Res., *47*: 179–195, 1971.
32. MacDonald, W. C. and Dueck, J. W. Long-term effect of shoyu (Japanese soy sauce) of the gastric mucosa of the rat. J. Natl. Cancer Inst., *56*: 1143–1147, 1976.
33. Fujino, H., Chino, T., and Iwai, T. Experimental production of labial and lingual carcinoma by local application of 4-nitroquinoline N-oxide. J. Natl. Cancer Inst., *35*: 907–918, 1965.
34. Fujita, K., Kaku, T., Sasaki, M., and Onoe, T. Experimental production of lingual carcinoma in hamsters by local application of 9,10-dimethyl-1,2-benzanthracene. J. Dent. Res., *52*: 327–332, 1973.
35. Takeuchi, M., Kamiya, S., and Odashima, S. Induction of tumors of the forestomach, esophagus, pharynx, and oral cavity of the Donryu rat given N-butyl-N-nitrosourethan in the drinking water. Gann, *65*: 227–236, 1974.
36. Maekawa, A., Kamiya, S., and Odashima, S. Tumors of the upper digestive tract of ACI/N rats given N-propyl-N-nitrosourethan in the drinking water. Gann, *67*: 549–559, 1976.
37. Ohshima, H., Friesen, M., O'Neill, I., and Bartsch, H. Presence in human urine of a new N-nitroso compound, nitrosothiazolidine 4-carboxylic acid. Cancer Lett., *20*: 183–190, 1983.
38. Tsuda, M., Hirayama, T., and Sugimura, T. Presence of N-nitroso-L-thioproline and N-nitroso-L-methylthioprolines in human urine as major N-nitroso compounds. Gann, *74*: 331–333, 1983.
39. Cavallini, D., De Marco, C., Mondovi, B., and Trasanti, F. Studies of the metabolism of thiazolidine carboxylic acid by rat liver homogenate. Biochim. Biophys. Acta, *22*: 558–564, 1956.
40. Tsuda, M., Niitsuma, J., Sato, S., Hirayama, T., Kakizoe, T., and Sugimura, T. Increase in the levels of N-nitrosoproline, N-nitrosothioproline and N-nitroso-2-methylthioproline in human urine by cigarette smoking. Cancer Lett., *30*: 117–124, 1986.
41. Tahira, T., Tsuda, M., Wakabayashi, K., Nagao, M., and Sugimura, T. Kinetics of nitrosation of thioproline, the precursor of a major nitroso compound in human urine, and its role as a nitrite scavenger. Gann, *75*: 889–894, 1984.
42. Wattenberg, L. W. and Loub, W. D. Inhibition of polycyclic aromatic hydrocarbon-induced neoplasia by naturally occurring indoles. Cancer Res., *38*: 1410–1413, 1978.
43. Doll, R. Introduction, *In;* H. H. Hiatt, J. D. Watson, and J. A. Winsten (eds.), Origins of Human Cancer, pp. 1–12., Cold Spring Harbor Laboratory, Cold Spring Harbor, New York, 1977.
44. Cooke, R. R. Cancer of the lower alveolus., GANN Monogr. Cancer Res., *18*: 37–46, 1976.

DIET, NUTRITION AND CANCER, Y. HAYASHI ET AL. (EDS.),
JAPAN SCI. SOC. PRESS, TOKYO/VNU SCI. PRESS, UTRECHT, PP. 87–96, 1986

Effects of Meat Composition and Cooking Conditions on the Formation of Mutagenic Imidazoquinoxalines (MeIQx and Its Methyl Derivatives)

Margaretha JÄGERSTAD,*[1] Anita Laser REUTERSWÄRD,*[2] Spiros GRIVAS,*[3]
Kjell OLSSON,*[3] Chie NEGISHI,*[4] and Shigeaki SATO*[4]

Department of Applied Nutrition, University of Lund, Lund, Sweden,[1] *Swedish Meat Research Institute, Kävlinge, Sweden,*[2] *Department of Chemistry and Molecular Biology, Swedish University of Agricultural Sciences, Uppsala, Sweden,*[3] *and National Cancer Center Research Institute, Tokyo 104, Japan*[4]

Abstract: In recent years it has been shown that certain methyl derivatives of 3*H*-imidazo[4,5-*f*]quinoxaline-2-amine are responsible for a major part of the mutagenicity formed during frying, broiling or baking of meat, and also formed in the preparation of meat extracts. The present study describes the precursors of these compounds and their formation with participation of Maillard or nonenzymatic browning reactions. The formation of these IQ-type mutagens was shown to occur when model systems of creatin(in)e, reducing monosaccharides, and certain amino acids were heated at 128°C for 2 hr.

In meat experiments, the mutagenicity was found to be significantly correlated with the presence of creatin(in)e in the meat samples. The same conditions that are favorable for Maillard reactions, such as supply of starting materials, high temperature, and a suitable water concentration, also increased the yield of mutagenicity during cooking. Fats seemed to act as regulators of the amount of heat transferred into the product rather than as reactants in the formation of mutagenicity.

Although it has been known for many years that cooking may give rise to carcinogens, the significance of the latter in the development of human cancer is still unclear. Fortunately, however, we are making substantial progress today towards elucidating the role of these substances in human carcinogenesis. During the last decade alone three new categories of mutagenic heterocyclic amines have been isolated and identified in cooked meat, fish, and beef extracts using a short-term assay (Ames test). Several of these new mutagens have also been demonstrated to have tumor-inducing capacity in mice and rats in long-term studies carried out at the National Cancer Center and the Cancer Institute in Tokyo (see ref. *1*).

The three groups of new mutagenic heterocyclic amines are shown in Table 1. The first one comprises pyrolysates of amino acids and proteins, such as 3-amino-1,4-dimethyl-5*H*-pyrido[4,3-*b*] indole (Trp-P-1), 3-amino-1-methyl-5*H*-pyrido[4,3-*b*] indole (Trp-P-2), 2-amino-6-methyldipyrido[1,2-*a*: 3′,2′-*d*] imidazole(Glu-P-1), 2-aminodipyrido[1,2-*a*: 3′,4′-*d*] imidazole (Glu-P-2), 2-amino-9*H*-pyrido[2,3-*b*] indole

TABLE 1. Mutagens Identified in Cooked Protein-rich Foods. Mutagenic Activities (rev/μg) towards *Salmonella typhimurium* TA98 after S9 Activation

Mutagen	Rev/μg	Ref.
Amino acid pyrolysates		
Trp-P-1	39,000	*1*, *2*
Trp-P-2	104,000	*1*, *2*
Glu-P-1	49,000	*1*, *2*
Glu-P-2	1,900	*1*, *2*
Protein pyrolysates		
AαC	300	*1*, *2*
MeAαC	200	*1*, *2*
Imidazoquinolines		
IQ	433,000	*1*, *2*
MeIQ	661,000	*1*, *2*
Imidazoquinoxalines		
MeIQx	145,000	*1*, *2*
4,8-DiMeIQx	183,000	*1*, *2*
7,8-DiMeIQx	163,000	*1*, *2*
Imidazopyridines		
TMIP	100,000	*3*
PhIP	4,000	*3*

(AαC), and 2-amino-3-methyl-9*H*-pyrido[2,3-*b*] indole (MeAαC). These mutagens are sensitive to nitrite, which has made it possible to estimate that they constitute about 25% of the mutagenicity formed in fried beef (*2*). The second group is resistant to nitrite and is estimated to account for 75% of the mutagenicity produced in fried beef. Chemically, the compounds in this second group are methyl derivatives of 3*H*-imidazo[4,5-*f*]quinolin-2-amines (2-amino-3-methyl-3*H*-imidazo[4,5-*f*]quinoline (IQ), (2-amino-3,4-dimethyl-3*H*-imidazo[4,5-*f*]quinoline (MeIQ)) and 3*H*-imidazo [4,5-*f*]quinoxalin-2-amines(2-amino-3,8-dimethyl-3*H*-imidazo[4,5-*f*]quinoxaline (MeIQx), 2-amino-3,4,8-trimethyl-3*H*-imidazo[4,5-*f*]quinoxaline (4,8-DiMeIQx), 2-amino-3,7,8-trimethyl-3*H*-imidazo[4,5-*f*]quinoxaline (7,8-DiMeIQx)).

Recently, a third group of mutagenic heterocyclic amines has been isolated and identified in fried hamburger by Felton and coworkers (*3*). They found that the IQ-type mutagens, mainly MeIQx and 4,8-DiMeIQx, constituted 50% of the mutagenicity, whereas 50% originated from two new mutagens, 2-amino-n,n,n-trimethylimidazopyridine (TMIP) and 2-amino-N-methyl-5-phenylimidazopyridine (PhIP). Interestingly, although PhIP was produced in amounts ten times the concentration of the IQ-type mutagens its mutagenicity in the Ames test was very low (Table 1).

Finally, the present paper is focused on the IQ-type mutagens, their precursors and formation both in model systems and in meat experiments. Factors affecting their yield have been identified in order to be able to control their formation during the cooking of meat in the future.

IQ-type Mutagens—Precursors and Formation

Three years ago, we presented a possible pathway for the formation of the IQ-type mutagens (*4*). Considering the molecular structure of these compounds, a pathway based on Maillard reactions was proposed (*4, 12*). Three naturally occurring substances present in meat were assumed to participate: *viz.*, a hexose (glucose), an amino acid, and creatine (Fig. 1). The imidazole part of the molecule was thought to originate from creatine (probably *via* creatinine) and the pyridine or pyrazine part from Strecker degradation products formed in Maillard reactions between the hexose and the amino acid. Aldol condensations were suggested to link the two parts together *via* a Strecker aldehyde (or related Schiff base).

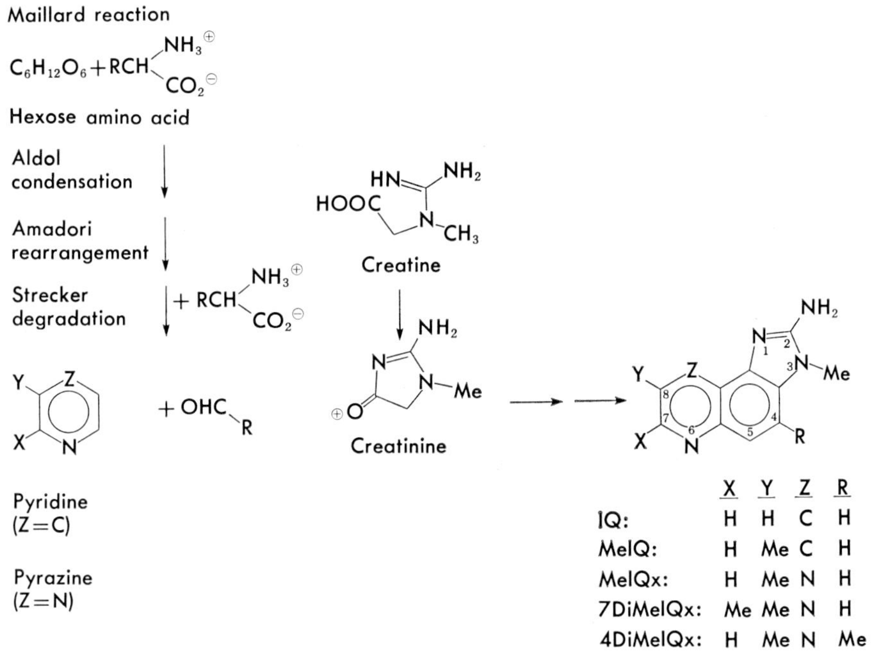

FIG. 1. Suggested pathway for the formation of the IQ-type mutagens (*4, 5*).

Model Systems Producing IQ, MeIQ, MeIQx, 4,8- and 7,8-DiMeIQx

A model system was developed to check the suggested pathway. By heating glucose, amino acid, and creatine (or creatinine) under reflux in a solution of diethylene glycol, containing 14% water, a reaction temperature of 128°C was maintained. Diethylene glycol was selected as a chemically inert solvent, with a boiling temperature that reflects the conditions occurring on the surface of meat during frying or baking. The reaction temperature and water concentration were thus favorable for Maillard reactions. Generally, the heating was interrupted after 2 hr (*4*). The yield of mutagenic activity was shown to vary with the amino acid used, being highest for threonine followed by glycine, lysine, alanine, *etc.* (*6*). One explana-

TABLE 2. Isolation and Identification of Mutagens Produced in Model Systems of Amino Acid, Monosaccharide, and Creatin(in)e

Amino acid	Monosaccharide	Isolated compounds	Ref.
Glycine	Glucose	MeIQx	*8*
Glycine	Glucose	7,8-DiMeIQx	*9*
Glycine	Fructose	IQ	*12*
Threonine	Glucose	MeIQx	*10*
Threonine	Glucose	4,8-DiMeIQx	*10*
Alanine	Fructose or ribose	4,8-DiMeIQx	*11, 13*
Alanine	Ribose	MeIQx	*11*
Alanine	Fructose	(MeIQ)[a]	*13*

[a] Due to a limited amount of MeIQ, its structure could not be verified by NMR or MS.

tion of these variations could be that amino acids differ in their capacity to produce pyrazine or pyridines. In a study by Wang and Odell (*7*), who heated single amino acids as dry substances at 180°C for 4 hr, threonine was the amino acid producing the largest amount of pyrazines.

The mutagenic substances were isolated and their identity was confirmed upon comparing the spectral similarities (NMR, UV, MS) to corresponding synthetic mutagens. The results obtained hitherto from these model systems are shown in Table 2. Generally, each model system produced two mutagens—MeIQx and one of its methyl derivatives (4,8- or 7,8-DiMeIQx). However, the proportion between MeIQx and its methyl derivatives and the position of the additional methyl group varied with the amino acid used.

Upon heating glycine (70 mmol), creatinine (70 mmol), and glucose (35 mmol) about 60 μg of MeIQx was produced, 10% of the mutagenicity produced being accounted for by 7,8-DiMeIQx (*8, 9*). In a similar model system where fructose was used instead of glucose, Grivas *et al.* (12) isolated 40–50 μg of MeIQx and 7 μg of IQ after heating glycine (35 mmol), creatinine (35 mmol), and fructose (17.5 mmol). Threonine as the amino acid resulted in a six-fold increase in mutagenicity compared with glycine, where 25% of the mutagenicity was ascribed to MeIQx and 75% to 4,8-DiMeIQx. Using alanine as the amino acid in this system, Muramatsu and Matsushima isolated both MeIQx and 4,8-DiMeIQx (*13*). Grivas *et al.* (*11*), by heathing alanine (35 mmol), creatinine (35 mmol), and fructose (17.5 mmol), isolated 15–20 μg of 4,8-DiMeIQx. A small amount (about 0.5 μg) of MeIQ was also indicated, but the limited amount did not permit an unequivocal identification.

Only very small amounts of IQ and possibly MeIQ were produced in two of the model systems (*11, 12*). In most of the model systems imidazoquinoxalines were produced, suggesting that pyrazines are more common than pyridines as Maillard reaction products at this temperature (128°C). Actually, recent reviews (*14, 15*) have reported that larger numbers of pyrazines than pyridines are also formed in cooked meat. Additionally, quantitative determinations of IQ-type mutagens occurring in cooked meat and extracts have shown that MeIQx and DiMeIQx clearly predominate over IQ and MeIQ (*3, 16–21*).

The two DiMeIQx compounds isolated from these model systems were previously not known, but one of them (4,8-DiMeIQx) has now been identified in both fried beef (*3*) and commercial beef extracts (*21*).

The mechanisms by which IQ, MeIQ, MeIQx, 4,8- and 7,8-DiMeIQx are formed need further substantiation. The model systems used in our investigations only indicate the precursors of these compounds. The very low yield in which these substances are produced in our systems is a serious limitation to establishing the reaction mechanisms. The possibility of alternative routes cannot be ruled out. For instance, Yoshida *et al.* (*22*) have reported that heating dry material of creatine and proline at 180°C for 1 hr produced IQ. Another "model system," introduced by Taylor *et al.* (*23*) has also been reported to produce IQ. A supernatant obtained by homogenization of lean round steak was boiled for *30 hr* at pH 4.0. IQ was isolated from this system and its yield was increased by prior addition of creatine phosphate.

Effect of Meat Composition and Cooking Conditions on Mutagenicity

The average concentrations of glucose, free amino acids, and creatine in beef are shown in Table 3. Glucose and its isomer fructose occur both free and bound to phosphate. In meat, no creatine phosphate occurs when assayed 24 hr after slaughter. Creatine is instead in free form and the creatinine concentration is almost zero. During cooking, however, a considerable amount of creatine is converted to creatinine (*4* and unpublished data). Data on free amino acids in muscle are few and not very detailed (*25*). The total amount of free amino acids is 0.1–0.3% (wet weight). Up to half of this is alanine, glutamic acid, and taurine, each at a level of 0.01–0.05%.

In a previous investigation (*4*), we reported that mutagenicity in fried beef as obtained with the Ames test varied with the glucose concentration as well as that of glucose and creatin(in)e concentrations. Similar findings were recently reported by Miller (*26*). He found in a limited survey of various heated muscle foods that the mutagenicity varied directly with the creatin(in)e level.

In a study now in progress in Lund, 12 muscle and organ samples have been screened for mutagenicity with the Ames test. Meat samples from ox (meat, heart, liver, and kidney) have been ground and formed into flat patties consisting of either pure samples or mixtures of meat and heart, meat and liver, or meat and kidney in proportions of 90: 10, 75: 25, and 60: 40, respectively. The patties were fried in

TABLE 3. Naturally Occurring Concentrations of Mutagen Precursors in Beef (% of Wet Weight)

Precursors	Concentration (%)	Ref.
Glucose	0.12	*24*
Glucose-6-phosphate	0.30	*24*
Glucose-1-phosphate	0.008	*24*
Fructose-6-phosphate	0.06	*24*
Creatine phosphate	0	*24*
Creatine	0.4	*4*
Creatinine	0.01	*4*
Free amino acids, total	0.1–0.3	*25*
Alanine	0.01–0.05	*25*

TABLE 4. Creatin(in)e and Hexose (Free and Bound) Content (μmol/g Dry Weight) in Raw Material of Meat, Heart, Liver, and Kidney from Ox. The Mutagenicity (rev/g, Crust TA98+S9) of These Products Was Assayed with the Ames Test after Frying on a Thermostatic Double-sided Teflon-coated Plate (200°C for 3 min)

	Meat	Heart	Liver	Kidney
Water, %	74.3	80.7	72.5	80.6
Creatine	119.7	12.7	5.9	9.4
Creatinine	7.9	9.8	1.8	2.4
Glucose	31.9	8.4	644	2.3
Glucose-6-Phosphate	11.7	1.3	1.1	1.4
Fructose-6-Phosphate	2.9	0.4	0	0
Glucose-1-Phosphate	0.6	0.4	0	0
Mutagenicity	1512±312	880±116	47±25	40±22

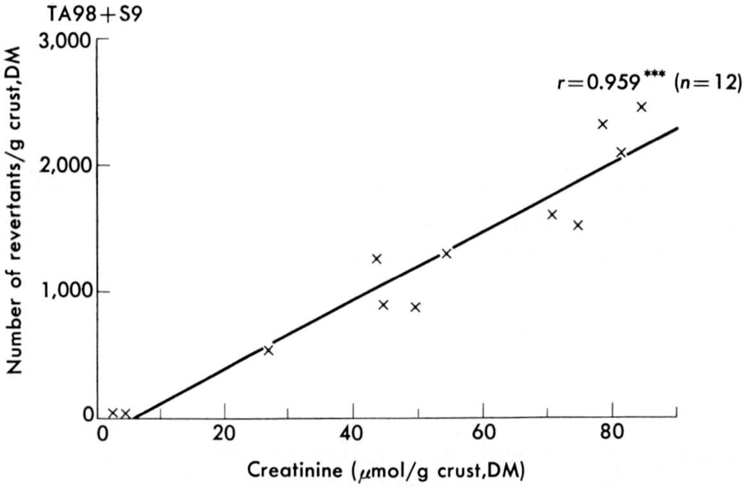

FIG. 2. Relationship between creatin(in)e levels of crust and mutagenicity in samples of meat, heart, liver, and kidney from ox, fried as ground patties at 200°C for 3 min without addition of fat on a thermostatic double-sided Teflon-coated plate. ***, $p<0.001$.

a thermostatic double-sided Teflon-coated plate at 200°C for 3 min. The crust was peeled off and assayed for mutagenicity according to an extraction procedure developed by Felton *et al.* (*19*). The pure muscle and organ samples varied markedly in their creatin(in)e and monosaccharide levels (Table 4). Liver and kidney contained very low levels of creatin(in)e and produced negligible amounts of mutagenicity compared with the meat and heart samples. As seen in Fig. 2, the mutagenicity was significantly ($p<0.001$) correlated with the creatin(in)e level of crusts.

In addition, four bouillon powders investigated for mutagenicity were all low, ranging from 30 to 250 rev/g. These values were much lower than those of the meat samples of muscle and heart (Table 4). The creatin(in)e content of the four bouillon samples was lower, too, ranging from 0 and to 40 μmol/g. In contrast, one commercial food-greade beef extract and one bacteriological grade beef extract (Difco) showed high mutagenicity, around 8,000 and 3,000 rev/g, respectively. Recently, Takahashi *et al.* (*20*) found 41.6 and 58.7 ng/g of IQ and MeIQx, respectively, in Difco beef extract, but only 3.1 ng/g of MeIQx in food-grade beef extract.

Fat, Acting as a Vehicle for Heat Transfer?

The mutagenicity formed during the cooking of meat and meat products has also been shown to increase with the intensity of the cooking conditions (time and temperature) (*27–31*).

The role of fat in this respect is interesting. In a study by Holtz *et al.* (*31*), meat loaves were prepared from ground beef mixed with various amounts of fat (2.8–16.6%). In one series of experiments meat loaves were baked in a convection oven until the surface temperature had reached 140°C. Although there was a large range in the amounts of fat used, the mutagenicity decreased with increasing fat content or remained almost constant if corrections for the dilution by the added fat were made (Table 5). The supplied energy (kJ) was similar for the first three fat levels, but lower for the highest fat level.When the amount of heat transferred to the meat loaves was fixed, this resulted in decreasing baking time with increasing fat content of the meat loaves. In a second experimental series the fat content was fixed but baking time gradually increased, resulting in increasing final surface temperature and mutagenicity (Table 5). Analysis of both experimental series gave no evidence of a chemical participation by fat. The variation in mutagenicity in these two series of experiments could mainly be attributed to the amount of brown color developed (Maillard reactions). However, further studies are needed to evaluate a mutual relationship.

This study and a study by Bjeldanes *et al.* (*30*) indicate that fat affects the amount of heat transferred into the product. The more fat that is present in the meat loaves, the less water in the product to be evaporated.

TABLE 5. Baking of Meat-loaves of Varying Fat Content Ranging between 2.8 and 16.6%. Effect on Baking Time, Final Surface Temperature and Mutagenicity (TA98+S9, rev/g Fat-free Dry Crust)

Experiment no.	Content (%)		Final surface temp. (°C)	Baking time (min)	Mutagenicity (rev/g)
	Fat	Water			
1	2.8	74.8	140	56	1,494
2	8.1	72.3	140	50	1,460
3	13.9	67.8	140	47	1,360
4	16.6	63.8	140	39	820
5	13.9	67.8	129	39	1,327
3	13.9	67.8	140	47	1,360
6	13.9	67.8	143	50	1,549
7	13.9	67.8	150	59	2,059
8	13.9	67.8	170	97	2,014

CONCLUSION

In a study on the importance of various structural features for the mutagenicity of the IQ-type mutagens, we found the imidazole part, especially its 2-amino group, to be crucial (*32*). Creatin(in)e present in meat was suggested as being the natural occurring precursor of the imidazole part. We found that the creation(in)e levels in samples from meat, heart, liver, and kidney have a great impact on the mutagenicity produced during cooking.

Our results also indicate that Maillard reactions are involved in the formation of the IQ-type mutagens. This could give us a tool for controlling the amount of mutagenicity produced during cooking; for example, we could reduce the amount of mutagenicity by keeping the cooking temperature low or by performing the cooking in such a way that the product is prevented from drying out excessively.

As regards the role of these mutagens/carcinogens in human cancer, more data are needed concerning our exposure to these compounds. Also important to study are factors interfering with their absorption and metabolism. From balance studies in rats, it is known that the IQ-type mutagens are easily absorbed and that they are excreted mainly in nonmutagenic form (*33*). The possibility that dietary fiber might interfere with their uptake has to date only been investigated *in vitro* (*34*, *35*). Such studies show that pectinlike fibers have a very low binding capacity for the IQ-type mutagens, whereas dietary fibers rich in Klason lignin, such as cereal fiber, have a high binding capacity for these mutagens.

ACKNOWLEDGEMENTS

Financial support from the Swedish Cancer Foundation (1824-B86-04XB), from the Foundation for Promotion of Cancer Research, Tokyo, and from the Japanese Ministry of Health and Welfare are gratefully acknowledged.

REFERENCES

1. Sugimura, T. Carcinogenicity of mutagenic heterocyclic amines formed during the cooking process. Mutat. Res., *150*: 33–41, 1985.
2. Sugimura, T. and Sato, S. Bacterial mutagenicity of natural materials, pyrolysis product and additives in foodstuffs and their association with genotoxic effects in mammals. *In;* A. W. Hayes, R. C. Schnell, and T. S. Miya (eds.), Developments in the Science and Practice of Toxicology, pp. 115–133, Elsevier Science Publishers, B.V., 1983.
3. Felton, J. S., Knize, M. G., Shen, N., Healy, S. K., Andersen, B. D., Bjeldanes, L. F., and Hatch, F. T. Identification of the mutagens in cooked beef. Environ. Health Perspect. (in press) Hagen, 1985.
4. Jägerstad, M., Laser Reuterswärd, A., Öste, R., Dahlqvist, A., Grivas, S., Olsson, K., and Nyhammar, T. Creatinine and Maillard reaction products as precursors of mutagenic compounds formed in fried beef. *In;* G. R. Waller and M. S. Feather (eds.), The Maillard Reaction in Foods and Nutrition, ACS Symposium Series 215, pp. 507–519, American Chemical Society, Washington, D.C., 1983.
5. Jägerstad, M., Grivas, S., Olsson, K., Laser Reuterswärd, A., Negishi, C., and Sato, S. Formation of food mutagens *via* Maillard reactions. *In;* Genetic Toxicology of the

Diet, pp. 155–167, Alan R. Liss, Inc., New York, 1986.
6. Jägerstad, M., Laser Reuterswärd, A., Olsson, R., Grivas, S., Nyhammar, T. Olsson, K., and Dahlqvist, A. Creatin(in)e and Maillard reaction products as precursors of mutagenic compounds: effect of various amino acids. Food Chem., *12*: 255–264, 1983.
7. Wang, P. S. and Odell, G. V. Formation of pyrazines from thermal treatment of some amino-hydroxy compounds. J. Agric. Food Chem., *21*: 868–870, 1973.
8. Jägerstad, M., Olsson, K., Grivas, S., Negishi, C., Wakabayashi, K., Tsuda, M., Sato, S., and Sugimura, T. Formation of 2-amino-3,8-dimethyl-imidazo[4,5-*f*]quinoxaline in a model system by heating creatinine, glycine and glucose. Mutat. Res., *126*: 239–244, 1984.
9. Negishi, C., Wakabayashi, K., Tsuda, M., Sato, T., Saito, H., Maeda, M., and Jägerstad, M. Formation of 2-amino-3,7,8-trimethylimidazo[4,5-*f*]quinoxaline, a new mutagen, by heating a mixture of creatinine, glucose and glycine. Mutat. Res., *140*: 55–59, 1984.
10. Negishi, C., Wakabayashi, K., Yamaizumi, Z., Saito, S., Sugimura, T., and Jägerstad, M. Identification of 4,8-DiMeIQx, a new mutagen. Selected abstracts of papers presented at the 13th Annual Meeting of the Environmental Mutagen Society of Japan, 12–13 October, 1984, Tokyo (Japan). Mutat. Res., *147*: 267–268, 1984.
11. Grivas, S., Nyhammar, T., Olsson, K., and Jägerstad, M. Formation of a new mutagenic DiMeIQx compound in a model system by heating creatinine, alanine and fructose. Mutat. Res., *151*: 177–183, 1985.
12. Grivas, S., Nyhammar, T., Olsson, K., and Jägerstad, M. Isolation and identification of the food mutagens IQ and MeIQx from heated model system of creatinine, glycine and fructose. Food Chem., *20*: 127–136, 1986.
13. Muramatsu, M. and Matsushima, T. Formation of MeIQx and 4,8-DiMeIQx by heating mixtures of creatinine, amino acids and monosaccharides. Selected Abstracts of Papers presented at the 13th Annual Meeting of the Environmental Mutagen Society of Japan, 12–13 October 1984, Tokyo (Japan). Mutat. Res, *147*: 266–267, 1985.
14. Shibamoto, T., Nishimura, O., and Mihara, S. Heterocyclic compounds found in cooked meats. J. Agric. Food Chem., *28*: 237–243, 1980.
15. Shibamoto, T. Heterocyclic compounds in browning and browning/nitrite model systems: occurrence, formation mechanisms, flavor characteristics and mutagenic activity. *In;* G. Charalambous and G. Inglett (eds.), Instrumental Analysis of Foods, vol. I, pp. 229–278, Academic Press, New York, 1983.
16. Nagao, M., Sato, S., and Sugimura, T. Mutagens produced by heating foods. *In;* G. R. Waller and M. S. Feather (eds.), The Maillard Reaction in Foods and Nutrition, ACS Symposium Series 215, pp. 521–536, American Chemical Society, Washington, D.C., 1983.
17. Hargraves, W. A. and Pariza, M. W. Purification and mass spectral characterization of bacterial mutagens from commercial beef extract. Cancer Res., *43*: 1467–1472, 1983.
18. Hayatsu, H., Matsui, Y., Ohara, Y., Oka, T., and Hayatsu, T. Characterization of mutagen fractions in beef extract and cooked ground beef. Use of blue-cotton for efficient extraction. Gann, *74*: 472–482, 1983.
19. Felton, J. S., Knize, M. G., Wood, C., Wuebbles, B. J., Healy, K. S., Stuermer, D. K., Bjeldanes, L. F., Kimble, B. J., and Hatch, F. T. Isolation and characterization of new mutagens from fried ground beef. Carcinogenesis, *5*: 95–102, 1984.
20. Takahashi, M., Wakabayashi, K., Nagao, M., Yamaizumi, Z., Sato, S., Kinae, N., Tomita, I., and Sugimura, T. Quantification of IQ and MeIQx in beef extracts by

liquid chromatography with electrochemical detection (LCEC). Carcinogenesis, *6* (8): 1195–1199, 1985.

21. Takahashi, M., Wakabayashi, K., Nagao, M., Yamaizumi, Z., Sato, S., Kinae, N., Tomita, I., and Sugimura, T. Identification and quantification of 2-amino-3,4,8-trimethylimidazo[4,5-*f*]quinoxaline (4,8-DiMeIQx) in beef extract. Carcinogenesis, 6: 1537–1539, 1985.
22. Yoshida, D., Saito, Y., and Mizusaki, S. Isolation of 2-amino-3-methylimidazo[4,5-*f*]quinoline as mutagen from the heated product of a mixture of creatine and proline. Agric. Biol. Chem., *48*: 241–243, 1984.
23. Taylor, R., Fultz, E., and Knize, M. Mutagen formation in a model beef boiling system III. Purification and identification of three heterocyclic amine mutagens-carcinogens. J. Environ. Sci. Health, *A20*: 135–148, 1985.
24. Fabiansson, S. and Laser Reuterswärd, A. Low voltage electrical stimulation and post-mortem energy metabolism in beef. Meat Sci., *4*: 205–223, 1985.
25. Sulser, H. Die Extraktstoffe das Fleisches. pp. 1–169. Stuttgart Wissenschaftliche Verlagsgesellschaft, MBH, 1978.
26. Miller, A. J. Precursor effects on imidazoquinoline-type mutagen formation in heated model and muscle systems. Abstracts in Genetic Toxicology of the Diet, 4th ICEM (International Conference on Environmental Mutagens), June 19–22, Copenhagen, 1985.
27. Dolara, P., Commoner, B., Vithayathil, A., Cuca, G., Tuley, F., Madyastha, P., Nair, S., and Kriebel, D. The effect of temperature on the formation of mutagens in heated beef stock and cooked ground beef. Mutat. Res., *60*: 231–237, 1979.
28. Spingarn, N. E. and Weisburger, J. H. Formation of mutagens in cooked foods. I. Beef. Cancer Lett., *7*: 259–264, 1979.
29. Pariza, M. W., Ashoor, S. H., Chu, F. S., and Luad, D. B. Effect of temperature and time on mutagen formation in pan-fried hamburger. Cancer Lett., *7*: 63–69, 1979.
30. Bjeldanes, L. F., Morris, M. N., Timourian, H., and Hatch, F. Mutagen formation in fried ground beef. J. Agric. Food Chem., *31*: 18–21, 1983.
31. Holtz, E., Skjöldebrand, C., Jägerstad, M., Laser Reuterswärd, A., and Isberg, P. The effect of recipes on crust formation and mutagenicity in meat products during baking. J. Food Technol., *20*: 57–66, 1985.
32. Grivas, S. and Jägerstad, M. Mutagenicity of some synthetic quinolines and quinoxalines related to IQ, MeIQ or MeIQx in Ames test. Mutat. Res., *137*: 29–32, 1984.
33. Sjödin, P. and Jägerstad, M. A balance study of ^{14}C-labeled 3H-imidazo[4,5-*f*]quinolin-2-amines (IQ and MeIQ) in rats. Food Chem. Toxicol., *22*: 207–210, 1984.
34. Barnes, W. S., Maiello, J., and Weisburger, J. H. *In vitro* binding of the food mutagen 2-amino-3-methylimidazo[4,5-*f*]quinoline to dietary fibers. J. Natl. Cancer. Inst., *70*: 757–760, 1983.
35. Sjödin, P. B., Nyman, M. E., Nilsson, L., Asp, N-G., and Jägerstad, M. Binding of ^{14}C-labeled food mutagens (IQ, MeIQ, MeIQx) by dietary fiber *in vitro*. J. Food Sci., *50*: 1680–1684, 1985.

DIET, NUTRITION AND CANCER, Y. HAYASHI ET AL. (EDS.), JAPAN SCI. SOC. PRESS, TOKYO/VNU SCI. PRESS, UTRECHT, PP. 97-105, 1986

Carcinogenicities in Mice and Rats of IQ, MeIQ, and MeIQx

Hiroko OHGAKI, Hirokazu HASEGAWA, Tamami KATO, Miki SUENAGA, Shigeaki SATO, Shozo TAKAYAMA, and Takashi SUGIMURA

Biochemistry Division, National Cancer Center Research Institute, Tokyo 104, Japan

Abstract: The mutagenic heterocyclic amines, 2-amino-3-methylimidazo[4,5-*f*] quinoline (IQ), 2-amino-3,4-dimethylimidazo[4,5-*f*]quinoline (MeIQ), and 2-amino-3,8-dimethylimidazo[4,5-*f*]quinoxaline (MeIQx) are present in broiled fish, fried beef, and beef extract. Their carcinogenicities in CDF_1 mice and F344 rats were tested by their oral administration in the diet.

In mice given diet containing 0.03% IQ, tumors developed in the liver (hepatocellular carcinomas or hepatocellular adenomas), forestomach (squamous cell carcinomas or papillomas), and lung (adenocarcinomas or adenomas) at high incidences. In mice given diet containing 0.04% or 0.01% MeIQ, squamous cell carcinomas and papillomas of the forestomach developed at high incidences. About 40% of the squamous cell carcinomas induced in the forestomach by 0.04% MeIQ metastasized to the liver. Clear dose-response relations were seen in the incidences of tumors in the groups given 0.04% and 0.01% MeIQ. The squamous cell carcinoma-papilloma ratios were higher in 0.04% groups than in 0.01% groups. Female mice treated with 0.04% and 0.01% MeIQ showed significantly higher incidences of liver tumors than controls. The experiment on the carcinogenicity of MeIQx at a dose of 0.06% in mice is still in progress but by experimental week 74, 4 of 16 males and 7 of 18 females autopsied were found to have liver tumors.

Rats given diet containing 0.03% IQ showed high incidences of hepatocellular carcinomas, adenocarcinomas of the small and large intestines, and squamous cell carcinomas of the Zymbal gand, clitoral gland, and skin. Except for the liver, the target organs of IQ in CDF_1 mice and F344 rats were different.

It is generally accepted that environmental factors, especially dietary factors, are important in the development of human cancers (1–3). Charred parts of broiled fish and beefsteak were found to be mutagenic to *Salmonella typhimurium* TA98 and TA100 in 1977 (*4*, *5*). Heated beef or beef extract was also found to be mutagenic (*6*). Subsequently, several mutagenic heterocyclic amines in pyrolysates of amino acids and protein were isolated and identified using the *S. typhimurium* mutation assay.

These compounds were 3-amino-1,4-dimethyl-5*H*-pyrido[4,3-*b*]indole (Trp-P-1) and 3-amino-1-methyl-5*H*-pyrido[4,3-*b*]indole (Trp-P-2) from a tryptophan pyrolysate (*7*), 2-amino-6-methyldipyrido[1,2-*a*: 3′, 2′-*d*]imidazole (Glu-P-1) and 2-aminodipyrido[1,2-*a*: 3′,2′-*d*]imidazole (Glu-P-2) from a glutamic acid pyrolysate (*8*), and 2-amino-3-methyl-α-carboline (MeAαC) and 2-amino-α-carboline (AαC), which were isolated from a soybean globulin pyrolysate (*9*) and also found in cigarette smoke condensates (*10, 11*). These heterocyclic amines have been demonstrated to be carcinogenic in mice (*12, 13*) and rats (*14, 15*) when given orally in the diet.

Other types of heterocyclic amines, aminoimidazoquinoline and aminoimidazoquinoxaline compounds, have also been isolated from pyrolysates of proteinous foods: 2-amino-3-methylimidazo[4,5-*f*]quinolin (IQ) and 2-amino-3,4-dimethylimidazo [4,5-*f*]quinoline (MeIQ) were isolated from broiled sardines (*16, 17*), and 2-amino-3,8-dimethylimidazo[4,5-*f*]quinoxaline (MeIQx) was isolated from fried beef (*18*). Sardine broiled at the usual cooking temperature contained 4.9 ng/g of IQ and 16.6 ng/g of MeIQ (*19*). IQ and MeIQx were also found to be present in beef extract (*20–22*) and cooked beef (*20, 23*), in which they could account for more than 20 or 30% of the total mutagenic activities (*21, 23*). Later, MeIQx, 2-amino-3,4,8-trimethylimidazo[4,5-*f*]quinoxaline (4,8-DiMeIQx), and 2-amino-3,7,8-trimethylimidazo[4,5-*f*]quinoxaline (7,8-DiMeIQx) were purified from heated solutions of creatinine, sugar, and amino acids (*24–28*). 4,8-DiMeIQx was found in bacteriological-grade beef extract (*29*). IQ was also shown to be produced by heating creatine with proline (*30*). The mutagenic activities of these compounds were more than 100,000 revertants/μg in a test with *S. typhimurium* TA98 and S9 mix (*31*). Because these aminoimidazoquinoline and aminoimidazoquinoxaline compounds are strong mutagens and may be present in a wide variety of broiled foods, their carcinogenicities required investigation.

This article summarizes recent results on the carcinogenicities of IQ, MeIQ, and MeIQx in CDF_1 mice and IQ in F344 rats.

Experimental Protocols for Carcinogenicity of Heterocyclic Amines

Synthetic IQ, MeIQ, and MeIQx were obtained from Nard Institute, Osaka, Japan. These heterocyclic amines were added to a pellet diet (CE-2; CLEA Japan, Tokyo) at concentrations of 0.01% to 0.06%.

CDF_1 mice [(BALB/cAnN×DBA/2N)F_1] and F344 rats of both sexes, purchased from Charles River Japan, Atsugi, Kanagawa, were used. At the start of the experiment, the mice were 6–7 weeks old, and the rats were 8 weeks old. Groups of 40 males and 40 females were given diet containing one of the heterocyclic amines continuously throughout the experiment. Control animals were given basal diet.

Animals that died or became moribund during the experiment were autopsied and all organs were carefully examined histologically.

The significance of differences in the incidences of tumors in experimental and control groups was examined by the χ^2 test.

TABLE 1. Induction of Tumors in the Liver, Forestomach, and Lung by IQ in CDF_1 Mice

	Sex	Effective no.	No. of mice with liver tumors				No. of mice with forestomach tumors			No. of mice with lung tumors		
			Hepato-cellular adenoma	Hepato-cellular carcinoma	Others	Total (%)	Papilloma	Squamous cell carcinoma	Total (%)	Adenoma	Adeno-carcinoma	Total (%)
IQ (0.03%)	M	39	8	8	0	16 (41)	11	5	16 (41)	13	14	27 (69)
	F	36	5	22	0	27 (75)	8	3	11 (31)	7	8	15 (42)
None	M	33	2	0	1	3 (9)	1	0	1 (3)	4	3	7 (21)
	F	38	0	0	3	3 (8)	0	0	0 (0)	3	4	7 (18)

TABLE 2. Induction of Forestomach and Liver Tumors by MeIQ in CDF_1 Mice

	Sex	Effective no.	No. of mice with forestomach tumors				No. of mice with liver tumors			
			Papilloma	Squamous cell carcinoma	Sarcoma	Total (%)	Hepato-cellular adenoma	Hepato-cellular carcinoma	Other	Total (%)
MeIQ (0.04%)	M	38	5	30 [11][a]	0	35 (92)	5	1	1	7 (18)
	F	38	9	24 [9]	1	34 (89)	11	16	0	27 (71)
MeIQ (0.01%)	M	38	4	3	0	6 (18)	8	3	0	11 (29)
	F	36	8	11 [1]	0	19 (53)	4	0	0	4 (11)
Control	M	29	0	0	0	0 (0)	2	1	1	4 (14)
	F	40	0	0	0	0 (0)	0	0	0	0 (0)

[a] No. of mice with metastases of squamous cell carcinomas to the liver.

Carcinogenicity Experiments in CDF$_1$ Mice

1. Carcinogenicity of IQ

CDF$_1$ mice were given diet containing 0.03% IQ. High incidences of liver tumors (hepatocellular carcinomas and hepatocellular adenomas) and forestomach tumors (squamous cell carcinomas and papillomas) were observed in the experimental groups in weeks 72 to 96. Females were more susceptible than males to the induction of liver tumors. Significantly higher incidences of lung tumors (adenocarcinomas and adenomas) than controls were observed in the experimental groups. Data are summarized in Table 1 (*32*).

2. Carcinogenicity of MeIQ

CDF$_1$ mice were given diets containing 0.04% and 0.01% MeIQ for 91 weeks. As summarized in Table 2, the target organs of MeIQ in CDF$_1$ mice were the forestomach and liver (*33*). In groups given 0.04% MeIQ, forestomach tumors were found in 92% of the males and 89% of the females. Most forestomach tumors were squamous cell carcinomas invading the muscle layer or the subserosa. Keratin pearl formation was frequent in squamous cell carcinomas. Metastases of squamous cell carcinomas to the liver were observed in 11 of 30 males and 9 of 24 females given diet containing 0.04% MeIQ. Metastases of squamous cell carcinomas were also observed in the regional lymph nodes and kidney of a few mice. Lymphatic spread of squamous cell carcinomas to the glandular stomach was also frequent. In groups given 0.01% MeIQ, the incidences of forestomach tumors were less than in the groups given 0.04% MeIQ. Clear dose response relations were observed in the incidences of forestomach tumors. Squamous cell carcinoma-papilloma ratios were higher in the groups given 0.04% than in the groups given 0.01% MeIQ. The cumulative incidences of forestomach tumors in these groups are shown in Fig. 1. Forestomach tumors developed slightly sooner in animals of both sexes given 0.04% MeIQ than in those given 0.01% MeIQ.

Liver tumors were observed in 18% of the males and 71% of the females given

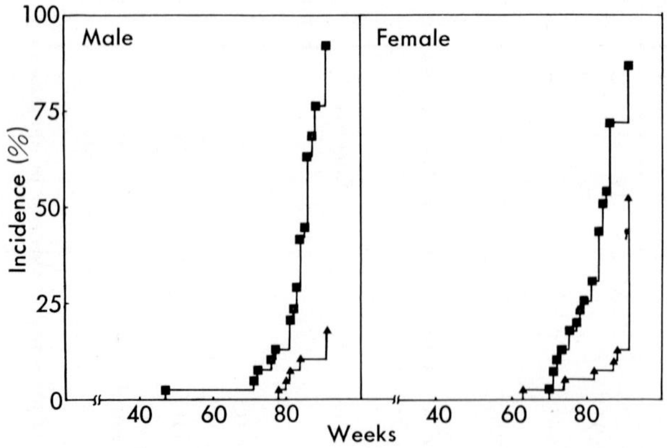

FIG. 1. Cumulative incidences of forestomach tumors induced by MeIQ. ▲, 0.01% MeIQ; ■, 0.04% MeIQ.

TABLE 3. Induction of Liver Tumors by MeIQx in CDF_1 Mice by Week 74 of the Experiment

	Sex	Initial no.	No. of mice examined	No. of mice with liver tumors	
				Hepatocellular adenoma	Hepatocellular carcinoma
MeIQx (0.06%)	M	40	16	0	4
	F	40	18	4	8
None	M	40	19	0	0
	F	40	4	0	0

0.04% MeIQ and 29% of the males and 11% of the females given 0.01% MeIQ. In females given 0.04% and 0.01% MeIQ the incidences were significantly higher than those in controls. The incidence of liver tumors also showed a clear dose-response relation ($p<0.01$). There was no significant difference in the incidence of liver tumors between male experimental and male control groups.

3. Carcinogenicity of MeIQx

The experiment on the carcinogenicity of MeIQx is still in progress. The concentration of MeIQx in the diet is 0.06%, and the data obtained in week 74 are shown in Table 3. A high incidence of liver tumors has been observed, especially in females given MeIQx, from experimental week 57 (our unpublished data).

Carcinogenicity Experiments in F344 Rats

1. Carcinogenicity of IQ

F344 rats were given diet containing 0.03% IQ for 104 weeks. Tumors developed in the liver, small and large intestines, Zymbal gland, skin, and clitoral gland at high incidences (Table 4) (*34*). Most liver tumors were hepatocellular carcinomas, which showed either a mixed trabecular and adenocarcinoma pattern or a pure trabecular pattern. Intestinal tumors were multiple and showed polypoid growth. Most tumors of the large intestine were located in the colonic mucosa 3–8 cm above the anorectal junction. Histologically these intestinal tumors were adenocarcinomas. Tumors in the Zymbal glands and skin were squamous cell carcinomas. Clitoral gland tumors were squamous cell carcinomas or sebaceous adenocarcinomas. Male

TABLE 4. Induction of Tumors by IQ in F344 Rats

	Sex	Effective no.	No. of rats with tumors (%)					
			Liver (Hepato-cellular carcinoma)	Intestine Small (Adenocarcinoma)	Intestine Large (Adenocarcinoma)	Zymbal gland (Squamous cell carcinoma)	Skin (Squamous cell carcinoma)	Clitoral gland (Squamous cell carcinoma)
IQ (0.03%)	M	40	27 (68)	12 (30)	25 (63)	36 (90)	17 (43)	—
	F	40	18 (45)	1 (3)	9 (23)	27 (68)	3 (8)	20 (50)
None	M	50	1 (2)	0	0	0	0	—
	F	50	0	0	0	0	0	0

rats were more susceptible than females to induction of tumors in the liver, intestine, Zymbal gland, and skin.

Except for the liver, the target organs of IQ in CDF_1 mice and F344 rats were different.

DISCUSSION

Data were summarized on the carcinogenicities of IQ, MeIQ, and MeIQx. The results together with previous results on Trp-P-1, Trp-P-2, Glu-P-1, Glu-P-2, MeAαC, and AαC (*12–15*) show that all the heterocyclic amines from cooked foods so far examined were carcinogenic to CDF_1 mice and/or F344 rats. IQ is also reported to be carcinogenic to Sprague-Dawley rats, inducing tumors of the mammary gland, liver, and ear duct (*35*). These carcinogenicity studies show the value of short term mutation tests with *S. typhimurium* TA98 and TA100 as screening tests for carcinogens.

The target organs of these heterocyclic amines varied according to the chemicals and the species of animals tested, except for the liver, which was a common target organ for all of them. IQ is reported to be activated by one type of cytochrome P-450, P448 IIa, to become a 2-N-hydroxy derivative (*36*). It has been suggested that the ultimate forms of IQ, MeIQ, and MeIQx are esters of N-hydroxy derivatives (*37*, *38*). The different organ specificities might therefore be explained in part by differences in the metabolic activations of the heterocyclic amines in the various organs. In addition, recently the H-*ras* oncogene was reported to be activated in hepatocellular carcinomas induced by IQ in F344 rats (*39*). This finding also suggests a possible mechanism for the role of heterocyclic amines in cancer development.

Carcinogenic heterocyclic amines are actually present in various cooked foods (*16*, *18–23*) and cigarette smoke condensates (*10*, *11*, *40*). Therefore, these compounds may be involved in development of human cancer. To assess the actual carcinogenic risk of heterocyclic amines to humans, it will be necessary to investigate the factors modulating their mutagenicity and carcinogenicity and to determine their intake by humans.

ACKNOWLEDGMENTS

This work was supported by Grants-in-Aid for Cancer Research from the Ministry of Health and Welfare and the Ministry of Education, Science and Culture of Japan.

REFERENCES

1. Wynder, E. L. and Gori, G. B. Contribution of the environment to cancer incidence. J. Natl. Cancer Inst., *58*: 825–832, 1977.
2. Doll, R. Strategy for detection of cancer hazards to man. Nature, *265*: 589–596, 1977.
3. Haenszel, W., Kurihara, M., Segi, M., and Lee, R.K.C. Stomach cancer among Japanese in Hawaii. J. Natl. Cancer Inst., *49*: 969–988, 1972.
4. Sugimura, T., Nagao, M., Kawachi, T., Honda, M., Yahagi, T., Seino, Y., Sato, S., and Matsukura, N. Mutagens-carcinogens in foods, with special reference to highly mutagenic pyrolytic products in broiled foods. *In;* H. H. Hiatt, J. D. Watson, and J. A. Winsten (eds.), Origins of Human Cancer, Book C, pp. 1561–1577, Cold Spring

Harbor Laboratory, New York, 1977.
5. Nagao, M., Honda, M., Seino, Y., Yahagi, T., and Sugimura, T. Mutagenicities of smoke condensates and the charred surface of fish and meat. Cancer Lett., *2*: 221–226, 1977.
6. Commoner, B., Vithayathi, A. J., Dolora, P., Nair, S., Madyastha, P., and Cuca, G. C. Formation of mutagens in beef and beef extract during cooking. Science, *201*: 913–916, 1978.
7. Sugimura, T., Kawachi, T., Nagao, M., Yahagi, T., Seino, Y., Okamoto, T., Shudo, K., Kosuge, T., Tsuji, K., Wakabayashi, K., Iitaka, Y., and Itai, A. Mutagenic principle(s) in tryptophan and phenylalanine pyrolysis products. Proc. Jpn. Acad., *54*: 58–61, 1977.
8. Yamamoto, T., Tsuji, K., Kosuge, T., Okamoto, T., Shudo, K., Takeda, K., Iitaka, Y., Yamaguchi, K., Seino, Y., Yahagi, T., Nagao, M., and Sugimura, T. Isolation and structure determination of mutagenic substances in L-glutamic acid pyrolysate. Proc. Jpn. Acad. Ser. B, *54*: 248–250, 1978.
9. Yoshida, D., Matsumoto, T., Yoshimura, R., and Matsuzaki, T. Mutagenicity of amino-α-carbolines in pyrolysis product of soybean globulin. Biochem. Biophys. Res. Commun., *83*: 915–920, 1978.
10. Yoshida, D. and Matsumoto, T. Amino-α-carbolines as mutagenic agents in cigarette smoke condensate. Cancer Lett., *10*: 141–149, 1980.
11. Matsumoto, T., Yoshida, D., and Tomita, H. Determination of mutagens, amino-α-carbolines in grilled foods and cigarette smoke condensate. Cancer Lett., *12*: 105–110, 1981.
12. Matsukura, N., Kawachi, T., Morino, K., Ohgaki, H., Sugimura, T., and Takayama, S. Carcinogenicity in mice of mutagenic compounds from a tryptophan pyrolyzate. Science, *213*: 346–347, 1981.
13. Ohgaki, H., Matsukura, N., Morino, K., Kawachi, T., Sugimura, T., and Takayama, S. Carcinogenicity in mice of mutagenic compounds from glutamic acid and soybean globulin pyrolysates. Carcinogenesis, *5*: 815–819, 1984.
14. Takayama, S., Masuda, M., Mogami, M., Ohgaki, H., Sato, S., and Sugimura, T. Induction of cancers in the intestine, liver and various other organs of rats by feeding mutagens from glutamic acid pyrolysate. Gann, *75*: 207–213, 1984.
15. Takayama, S., Nakatsuru, Y., Ohgaki, H., Sato, S., and Sugimura, T. Carcinogenicity in rats of a mutagenic compound, 3-amino-1,4-dimethyl-5*H*-pyrido[4,3-*b*]indole, from tryptophan pyrolysate. Jpn. J. Cancer Res. (Gann), *76*: 815–817, 1985.
16. Kasai, H., Yamaizumi, Z., Wakabayashi, K., Nagao, M., Sugimura, T., Yokoyama, S., Miyazawa, T., Spingarn, N. E., Weisburger, J. H., and Nishimura, S. Potent novel mutagens produced by broiling fish under normal conditions. Proc. Jap. Acad. Ser. B, *56*: 278–283, 1980.
17. Kasai, H., Nishimura, S., Wakabayashi, K., Nagao, M., and Sugimura, T. Chemical synthesis of 2-amino-3-methylimidazo[4,5-*f*]quinoline (IQ), a potent mutagen isolated from broiled fish. Proc. Jap. Acad. Ser. B, *56*: 382–384, 1980.
18. Kasai, H., Yamaizumi, Z., Shiomi, T., Yokoyama, S., Miyazawa, T., Wakabayashi, K., Nagao, M., Sugimura, T., and Nishimura, S. Structure of a potent mutagen isolated from fried beef. Chem. Lett., 485–488, 1981.
19. Yamaizumi, Z., Kasai, H., Nishimura, S., Edmonds, C. G., and McCloskey, J. A. Stable isotope dilution quantification of mutagens in cooked foods by combined liquid chromatography-thermospray mass spectrometry. Mutat. Res., *173*: 1–7, 1986.

20. Hayatsu, H., Matsui, Y., Ohara, Y., Oka, T., and Hayatsu, T. Characterization of mutagenic fractions in beef extract and in cooked ground beef. Use of blue-cotton for efficient extraction. Gann, *74*: 472–482, 1983.
21. Hargraves, W. A. and Pariza, M. W. Purification and mass spectral characterization of bacterial mutagens from commercial beef extract. Cancer Res., *43*: 1467–1472, 1983.
22. Turesky, R. J., Wishnok, J. S., Tannenbaum, S. R., Pfund, R. A., and Buchi, B. H. Qualitative and quantitative characterization of mutagens in commercial beef extract. Carcinogenesis, *4*: 863–866, 1983.
23. Felton, J. S., Knize, M. G., Wood, C., Wuebbles, B. J., Healy, S. K., Stuermer, D. H., Bjeldanes, L. F., Kimble, B. J., and Hatch, F. T. Isolation and characterization of new mutagens from fried ground beef. Carcinogenesis, *5*: 95–102, 1984.
24. Jägerstad, M., Olsson, K., Grivas, S., Negishi, C., Wakabayashi, K., Tsuda, M., Sato, S., and Sugimura, T. Formation of 2-amino-3,8-dimethylimidazo[4,5-*f*]quinoxaline in a model system by heating creatinine, glycine and glucose. Mutat. Res., *126*: 239–244, 1984.
25. Negishi, C., Wakabayashi, K., Yamaizumi, Z., Saitô, H., Sato, S., Sugimura, T., and Jägerstad, M. Identification of 4,8-DiMeIQx, a new mutagen. Selected Abstracts of Papers presented at the 13th Annual Meeting of the Environmental Mutagen Society of Japan, 12–13 October 1984, Tokyo (Japan). Mutat. Res., *147*: 267–268, 1985.
26. Muramatsu, M. and Matsushima, T. Precursors of heated-food mutagens (IQ, MeIQ and MeIQx). II. Abstracts of Papers presented at the 43th Annual Meeting of the Japanese Cancer Association, October 1984, Fukuoka, p. 29.
27. Grivas, S., Nyhammar, T., Olsson, K., and Jägerstad, M. Formation of a new mutagenic DiMeIQx compound in a model system by heating creatinine, alanine and fructose. Mutat. Res., *151*: 177–183, 1985.
28. Negishi, C., Wakabayashi, K., Tsuda, M., Sato, S., Sugimura, T., Saitô, H., Maeda, M., and Jägerstad, M. Formation of 2-amino-3,7,8-trimethylimidazo[4,5-*f*]quinoxaline, a new mutagen, by heating a mixture of creatinine, glucose and glycine. Mutat. Res., *140*: 55–59, 1984.
29. Takahashi, M., Wakabayashi, K., Nagao, M., Yamaizumi, Z., Sato, S., Kinae, N., Tomita, I., and Sugimura, T. Identification and quantification of 2-amino-3,4,8-trimethylimidazo[4,5-*f*]quinoxaline (4,8-DiMeIQx) in beef extract. Carcinogenesis, *6*: 1537–1539, 1985.
30. Yoshida, D., Saito, Y., and Mizusaki, S. Isolation of 2-amino-3-methylimidazo[4,5-*f*]-quinoline as mutagen from the heated product of a mixture of creatine and proline. Agric. Biol. Chem., *48*: 241–243, 1984.
31. Sugimura, T. Carcinogenicity of mutagenic heterocyclic amines formed during the cooking process. Mutat. Res., *150*: 33–41, 1985.
32. Ohgaki, H., Kusama, K., Matsukura, N., Morino, K., Hasegawa, H., Sato, S., Takayama, S., and Sugimura, T. Carcinogenicity in mice of a mutagenic compound, 2-amino-3-methylimidazo[4,5-*f*]quinoline, from broiled sardine, cooked beef and beef extract. Carcinogenesis, *5*: 921–924, 1984.
33. Ohgaki, H., Hasegawa, H., Kato, T., Suenaga, M., Ubukata, M., Sato, S., Takayama, S., and Sugimura, T. Induction of tumors in the forestomach and liver of mice by feeding 2-amino-3,4-dimethylimidazo[4,5-*f*]quinoline (MeIQ). Proc. Jpn. Acad. *61*: 137–139, 1985.
34. Takayama, S., Nakatsuru, Y., Masuda, M., Ohgaki, H., Sato, S., and Sugimura, T. Demonstration of carcinogenicity in F344 rats of 2-amino-3-methylimidazo[4,5-*f*]-

quinoline from broiled sardine, fried beef and beef extract. Gann, *75*: 467–470, 1984.

35. Tanaka, T., Barnes, W. S., Williams, G. M., and Weisburger, J. H. Multipotential carcinogenicity of the fried food mutagen 2-amino-3-methylimidazo[4,5-*f*]quinoline in rats. Jpn. J. Cancer Res. (Gann), *76*: 570–576, 1985.
36. Yamazoe, Y., Shimada, M., Kamataki, T., and Kato, R. Microsomal activation of 2-amino-3-methylimidazo[4,5-*f*]quinoline, a pyrolysate of sardine and beef extracts, to a mutagenic intermediate. Cancer Res., *43*: 5768–5774, 1983.
37. Nagao, M., Fujita, Y., Wakabayashi, K., and Sugimura, T. Ultimate forms of mutagenic and carcinogenic heterocyclic amines produced by pyrolysis. Biochem. Biophys. Res. Commun., *114*: 626–631, 1983.
38. Saito, K., Yamazoe, Y., Kamataki, T., and Kato R. Mechanism of activation of proximate mutagens in Ames' tester strains: The acetyl-CoA dependent enzyme in *Salmonella typhimurium* TA98 deficient in TA98/1,8-DNP_6 catalyzes DNA-binding as the cause of mutagenicity. Biochem. Biophys. Res. Commun., *116*: 141–147, 1983.
39. Ishikawa, F., Takaku, F., Nagao, M., Ochiai, M., Hayashi, K., Takayama, S., and Sugimura, T. Activated oncogenes in a rat hepatocellular carcinoma induced by 2-amino-3-methylimidazo[4,5-*f*]quinoline. Jpn. J. Cancer Res. (Gann), *76*: 425–428, 1985.
40. Yamashita, M., Wakabayashi, K., Nagao, M., Sato, S., Yamaizumi, Z., Takahashi, M., Kinae, N., Tomita, I., and Sugimura, T. Detection of 2-amino-3-methylimidazo-[4,5-*f*]quinoline (IQ) in cigarette smoke condensate. Jpn. J. Cancer Res. (Gann), *77*: 419–422, 1986.

DIET, NUTRITION AND CANCER, Y. HAYASHI ET AL. (EDS.),
JAPAN SCI. SOC. PRESS, TOKYO/VNU SCI. PRESS, UTRECHT, PP. 107–118, 1986

Mutagenic Nitropyrenes in Foods

Yoshinari OHNISHI, Takemi KINOUCHI, Hideshi TSUTSUI, Motoo UEJIMA, and Keiko NISHIFUJI

Department of Bacteriology, School of Medicine, The University of Tokushima, Tokushima 770, Japan

Abstract: Identification of mutagenic factors in foods is of concern because they may represent carcinogens to man. Cooked foods, especially their basic fractions containing heterocyclic amines, have high mutagenictity; the neutral fractions containing mutagenic nitropyrenes, however, have not been studied in detail. The mutagenicity of various grilled foods—10 vegetables, 4 fish, and 4 kinds of meat with and without sauce—was studied. The concentration of 1-nitropyrene was measured after reduction by a specific nitroreductase purified from *Bacterioides fragilis*. 1-Nitropyrene was detected in grilled corn, horse-mackerel, and mackerel, and accounted for less than 10% of the total mutagenicity of the crude extracts in the *Salmonella* mutation test using strain TA98 in the absence of S9 mix. The mutagenicity of the diethyl ether-soluble basic fractions of meat grilled without a marinating sauce was very high. However, the sauce decreased the mutagenicity of the basic fractions and increased the mutagenicity of the neutral fractions. Moreover, considerable amounts of 1-nitropyrene were detected in pork and yakitori (grilled chicken) grilled with the sauce. The neutral fractions of yakitori grilled for 3, 5, and 7 min contained 3.8, 19, and 43 ng, respectively, of 1-nitropyrene per gram of yakitori, accounting for 3.0, 2.7, and 1.3%, respectively, of the total extract mutagenicity. We conclude that formation of 1-nitropyrene in the yakitori is due to pyrene produced by incomplete combustion of fat in the chicken, its nitration at acidic pH by nitrogen dioxide emitted by burning of cooking gas, and some components of the marinating sauce. Antimutagenic activity of edible mushrooms against 3-amino-1,4-dimethyl-5*H*-pyrido[4,3-*b*]indol (Trp-P-1) was also studied.

The majority of cancer deaths are attributed to diet and cigarette smoking (*1*) and certain mutagenic chemicals play a predominant role in the etiology of cancer (*2–4*). Mutagenic substances are produced during cooking of different types of foods and food components (*5–9*). The greatest amount of mutagenic activity is formed during pyrolysis of amino acids and proteins (*10*, *11*). These compounds, heterocyclic amines shown in Table 1 as 3-amino-1,4-dimethyl-5*H*-pyrido[4,3-*b*]-

TABLE 1. Mutagenicity of Various Mutagens (*17–21*)

Mutagen	Test strain	S9	Mutagenicity ($His^+/\mu g$)
1,8-Dinitropyrene	TA98	−	870,000
MeIQ	TA98	+	661,000
1,6-Dinitropyrene	TA98	−	629,000
1,3-Dinitropyrene	TA98	−	496,000
IQ	TA98	+	433,000
MeIQx	TA98	+	145,000
1,3,6-Trinitropyrene	TA98	−	121,000
Trp-P-2	TA98	+	104,000
Orn-P-1[a]	TA98	+	56,800
1-Nitro-3-acetoxypyrene	TA98	−	54,800
Glu-P-1	TA98	+	49,000
AF-2[b]	TA100	−	42,000
1,3,6,8-Tetranitropyrene	TA98	−	40,800
Trp-P-1	TA98	+	39,000
Aflatoxin B_1	TA100	+	28,000
4-Nitroquinoline 1-oxide	TA100	−	9,900
MNNG[c]	TA100	−	9,350
1-Nitro-3-hydroxypyrene	TA98	−	3,770
Glu-P-2	TA98	+	1,900
1-Nitropyrene	TA98	−	1,830
Benzo[a]pyrene	TA100	+	660
AαC	TA98	+	300
MeAαC	TA98	+	200
N,N-Dimethylnitrosamine	TA100	+	0.23
N,N-Diethylnitrosamine	TA100	+	0.15
Ethyl methanesulfonate	TA100	−	0.081

[a] 4-amino-6-methyl-1*H*-2,5,10,10*b*-tetra azafluoranthene, [b] 2-(2-furyl)-3-(5-nitro-2-furyl)acrylamide, [c] *N*-methyl-*N'*-nitro-*N*-nitrosoguanidine.

indole (Trp-P-1), 3-amino-1-methyl-5*H*-pyrido[4,3-*b*]indole (Trp-P-2), 2-amino-6-methyldipyrido[1,2-*a*: 3′, 2′-*d*]imidazole (Glu-P-1), 2-aminodipyrido[1,2-*a*: 3′, 2′-*d*]imidazole(Glu-P-2), 2-amino-3-methylimidazo[4,5-*f*]quinoline (IQ), 2-amino-3,4-dimethylimidazo[4,5-*f*]quinoline (MeIQ), 2-amino-3,8-dimethylimidazo[4,5-*f*]quinoxaline (MeIQx), 2-amino-9*H*-pyrido[2,3-*b*]indole or 2-amino-α-carboline (AαC), and 2-amino-3-methyl-9*H*-pyrido[2,3-*b*]indole or 2-amino-3-methyl-α-carboline (MeAαC), were isolated from the organic solvent-soluble basic fractions of extracts of cooked foods (*12–16*). They showed strong frame-shift mutagenicity for *Salmonella typhimurium* strain TA98 in a microsome metabolic activation system (Table 1) and their carcinogenicity has also been demonstrated (*22*). Nevertheless, not all the mutagenic activity of cooked food could be recovered in the basic fraction containing heterocyclic amines, and the other fractions have not been studied in detail.

Dinitropyrenes (diNPs) are the most potent mutagens (Table 1), and three diNPs, MeIQ, and IQ are now called supermutagens. These five highest mutagens are carcinogenic (*22–27*). Although the mutagenicity of 1-nitropyrene (1-NP) is lower than that of diNPs, it is still 22,600-fold higher than that of a standard mutagen, ethyl methanesulfonate (Table 1). These NPs have been detected in environmental pollutants such as airborne particulates (*28, 29*), car exhaust emissions (*20, 28, 30–34*), used crankcase oil (*35*), emissions of kerosene heaters (*34, 36*), photocopies *37, 38*), coal fly ash (*39*), and wastewater from gasoline stations (*35, 40*), and even in tea leaves (*41*). They are direct-acting mutagens in the Ames *Salmonella* mutation test as shown in Table 1. Their widespread occurrence is not surprising because NPs are readily formed by exposure of pyrene to nitrogen dioxide (*42*). If pyrene formed during incomplete combustion of organic material, especially fat, in foods is exposed to nitrogen dioxide in burning urban gas, mutagenic nitro derivatives are readily induced. Therefore, the mutagenicity of the neutral fractions of foods grilled directly over a city gas flame and their NP content were measured.

Mutagenicity and NP Content of Grilled Foods

1. Grilled vegetables

We grilled many kinds of vegetables directly over a city gas flame for various lengths of time to obtain materials good for eating. The edible parts were homogenized in ethanol by a kitchen mixer for 10 min and extracted with benzene-ethanol (4: 1) by ultrasonication. The organic material in the crude extracts was fractionated into diethyl ether-soluble neutral, acidic, and basic fractions. The mutagenicity of these fractions was measured with *S. typhimurium* strains TA98, TA100, TA98NR, and TA98/1,8-DNP$_6$ in the presence and absence of a 9,000×g supernatant from livers of phenobarbital- and 5,6-benzoflavone-treated Sprague-Dawley rats (S9 mix). Some of this mutagenicity is shown in Table 2. Generally, the crude extracts of these vegetables showed high mutagenicity for strain TA98 in the presence of S9 mix and the mutagenicity of the basic fractions was highest among the neutral, acidic, and basic fractions. Since 1-NP and 1,6-diNP appear in the neutral fraction, the muta-

TABLE 2. Mutagenicity and NP Concentration of Grilled Vegetables

Vegetable	Fat in edible part[b] (%)	Grilling time (min)	Revertants/plate/g of grilled food[a] TA98				1-NP (ng/g of grilled food)	% Mutagenicity of 1-NP (TA98, −S9)
			(−)S9		(+)S9			
			Crude extract	Neutral fraction	Crude extract	Basic fraction		
Cabbage	0.1	4	0.8	0.8	26	27	<0.01	<2.11
Green pepper	0.1	1.5	5.3	0.5	128	25	<0.01	<0.32
Eggplant	0.1	2	26.1	0	613	58	n.d.[c]	n.d.
Onion	0.1	5	4.0	0.5	88	46	<0.01	<0.42
Pumpkin	0.1	3	5.8	0.8	49	16	<0.02	<0.58
Carrot	0.2	2.5	21.2	0	66	8	n.d.	n.d.
Potato	0.2	3	4.7	0	28	11	n.d.	n.d.
Sweet potato	0.2	5	4.0	0.7	30	11	<0.03	<1.27
Mushroom (Shiitake)	0.3	2	67.5	3.1	1,000	598	<0.01	<0.03
Corn	1.4	3	10.5	1.6	172	46	0.53	8.37

[a] Mutagenicity was determined in duplicate and the number of spontaneous revertants was subtracted.
[b] Data from ref. *43*. [c] Not determined.

genicity of neutral fractions for strain TA98 without S9 mix is also shown in Table 2, though it was low.

For quantitative analysis of 1-NP and 1,6-diNP, fractions corresponding to 1-NP and 1,6-diNP in the neutral fractions were collected by high-performance liquid chromatography (HPLC) and incubated with a mixture containing a specific nitroreductase (nitroreductase I or III, respectively) purified from *Bacteroides fragilis* (*44, 45*). The reduced material was applied on HPLC and the fluorescence of 1-aminopyrene or 1,6-diaminopyrene was measured quantitatively. The detection limits for 1-NP and 1,6-diNP in the system used were 1.1×10^{-2} and 1.3×10^{-2} pmol, respectively (*20, 35*). We found no NPs in grilled vegetables except corn. The concentration of 1-NP in the grilled corn was 0.53 ng per gram of corn, accounting for 8.4% of the total mutagenicity of the crude extract for strain TA98 in the absence of S9 mix. The fat content of corn is very high compared with other vegetables (Table 2). This may induce ease of burning and formation of pyrene on the surface of the corn when subjected to high temperature (550–750°C) over an open gas flame (*46*).

TABLE 3. Mutagenicity and NP Concentration of Grilled Fish

Fish[b]	Fat in edible part[c] (%)	Revertants/plate/g of grilled food[a] TA98 (−)S9 Crude extract	(−)S9 Neutral fraction	(+)S9 Crude extract	(+)S9 Basic fraction	1-NP (ng/g of grilled food)	% Mutagenicity of 1-NP (TA98, −S9)
Tongue sole	1.2	159	12	1,570	573	<0.04	<0.04
Yellowtail	16.1	4	19	716	437	<0.03	<1.18
Horse-mackerel	6.9	83	25	1,930	1,410	0.35	0.7
Mackerel	16.5	8	6	1,280	1,290	0.45	10.0

[a] Mutagenicity was determined in duplicate and the number of spontaneous revertants was subtracted.
[b] Grilling time of all samples was 3 min. [c] Data from ref. *43*.

TABLE 4. Mutagenicity and NP Concentration of Grilled Meat

Meat[b]	Sauce	Fat in edible part[c] (%)	Revertants/plate/g of grilled food[a] TA98 (−)S9 Crude extract	(−)S9 Neutral fraction	(+)S9 Crude extract	(+)S9 Basic fraction	ng/g of grilled food 1-NP	ng/g of grilled food 1,6-diNP	% Mutagenicity (TA98, −S9) 1-NP	% Mutagenicity (TA98, −S9) 1,6-diNP
Beef	−	19.6	264	218	4,010	2,440	<0.01	<0.01	<0.006	<1.06
	+		175	209	1,280	1,600	0.50	<0.01	0.48	<1.60
Pork	−	7.4	61	36	3,400	3,390	0.66	<0.01	1.84	<4.62
	+		158	187	2,990	576	3.13	<0.01	3.37	<1.77
Bacon	−	11.9	2,840	7,160	8,330	6,620	11.90	0.10	0.71	0.99
White meat of chicken	−	0.7	107	57	753	478	<0.01	n.d.[d]	<0.016	n.d.
	+		114	46	953	284	<0.01	n.d.	<0.015	n.d.
Chicken (Yakitori)	−	16.5	286	87	2,700	1,200	0.09	<0.01	0.05	<0.98
	+		846	687	1,430	878	1.51	0.043	0.30	1.42

[a] Mutagenicity was determined in duplicate and the number of spontaneous revertants was subtracted.
[b] Grilling time of all samples was 5 min. [c] Data from ref. *43*. [d] Not determined.

2. *Grilled fish*

Among the four tester strains described above, strain TA98 showed the highest mutagenic response to grilled fish in the presence of S9 mix. Moreover, the mutagenicity of the basic fractions of the grilled fish in the presence of S9 mix was higher than that of the other fractions. 1-NP was detected in grilled horse-mackerel and mackerel although the mutagenicity of their neutral fractions was low for strain TA98 in the absence of S9 mix (Table 3).

3. *Grilled meat*

The mutagenicity of the crude extracts of beef, pork, and chicken grilled without sauce was very high for strain TA98 in the presence of S9 mix (Table 4). Although the basic fractions of these meats also showed high mutagenicity in the same assay system, probably because of the presence of amino acid and protein pyrolysates, the basic fractions of meat grilled with the sauce, (a commercial marinating sauce [No. 1] for yakitori [grilled chicken]), had decreased mutagenicity for strain TA98 in

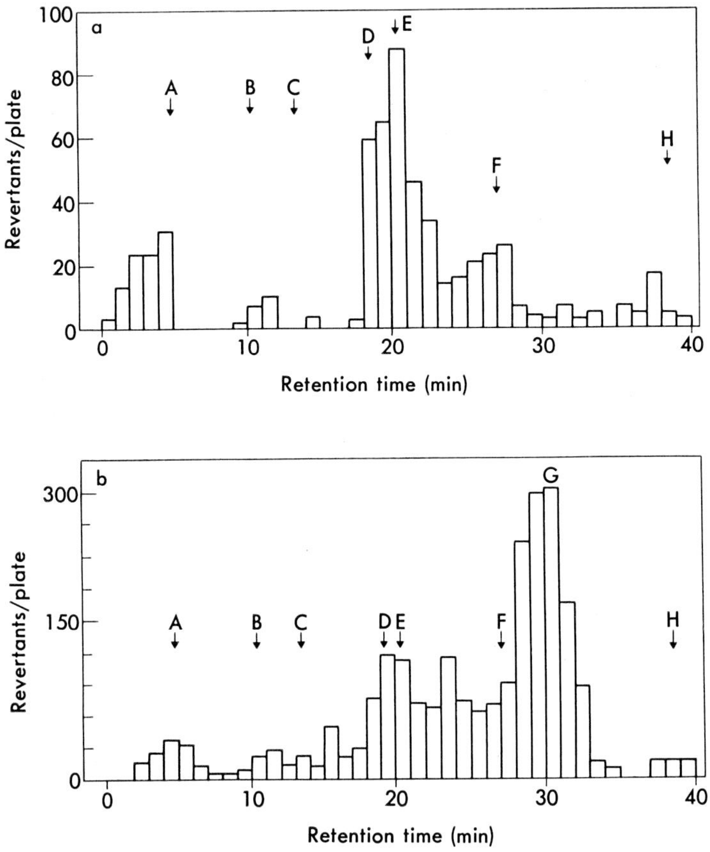

FIG. 1. Mutagenicity of the diethyl ether-soluble neutral fractions of (a) yakitori and (b) bacon grilled for 7 and 5 min, respectively, on HPLC. Arrows indicate the retention time of standard chemicals; A, nitro-Trp-P-2; B, 1-nitro-6/8-acetoxypyrene; C, 1-nitro-3-acetoxypyrene; D, 1,6-dinitropyrene; E, 1,8-dinitropyrene; F, 1-nitropyrene; G, unknown, H, 3-nitrofluoranthene.

the presence of S9 mix. Moreover, the neutral fractions of the meat, especially chicken, grilled with the sauce, showed higher mutagenicity for strain TA98 in the absence of S9 mix than that of the meat grilled without the sauce. The low mutagenicity of the neutral fraction of the yakitori for strains TA98NR and TA98/1,8-DNP_6 in the absence of S9 mix (data no shown) suggested that the fraction contains 1-NP and diNPs. In fact, the neutral fraction of the yakitori contained 1.51 ng of 1-NP and 0.043 ng of 1,6-diNP per gram of yakitori. The presence of 1-NP in the yakitori was also proved by the fluorescent spectrum and the mass spectrum of 1-aminopyrene produced after enzymatic reduction of the corresponding 1-NP fraction. Since the amount of 1,6-diNP was too little to be detected by the same procedures, the mutagenicity of each fraction after HPLC of the neutral fraction was measured (Fig. 1a). The fractions corresponding to 1,6-diNP and 1,8-diNP had the highest peaks of mutagenicity.

The mutagenicity of grilled bacon was extremely high; however, this is not the normal cooking procedure (Table 4). The mutagenicity pattern of the neutral fraction on HPLC shows the highest, and unknown, peak eluted after the fraction corresponding to 1-NP (Fig. 1b). The crude extract of pan-broiled bacon, the normal procedure, showed 391 and 1,950 His^+ revertants from strain TA98 in the absence and presence, respectively, of S9 mix.

The mutagenicity of grilled white chicken meat was low (Table 4) probably due to its low fat content.

Formation of 1-NP in Yakitori

Yakitori is a popular food in Japan and is generally cooked over an open city gas flame. We determined the mutagenicity of yakitori cooked with and without the marinating sauce (No. 1) for various time periods and the concentration of 1-NP in the neutral fraction (*47*). Five min of grilling was adequate for making good yakitori. The mutagenicity of yakitori without the sauce for strain TA98 in the presence of S9 mix increased more with time than that in the absence of S9 mix. In contrast, the mutagenicity with the sauce was higher in strain TA98 in the absence of S9 mix than in its presence. The concentration of 1-NP in yakitori cooked without the sauce for 7 min was 1.4 ng per gram of yakitori. However, the marinated yakitori grilled for 3, 5, and 7 min contained 3.8, 19.1, and 43 ng, respectively, of 1-NP per g, accounting for 3.0, 2.7, and 1.3%, respectively, of the total mutagenicity of the extract in strain TA98 in the absence of S9 mix. 1,6-DiNP was not detected in these samples. We do not know why the mutagenicity and 1-NP content of both types of yakitori are higher than those shown in Table 4. The reason may be the difference in moisture and fat content of the commercial chicken, temperature of the meat surface or strength of the gas flame. Actually, the values for the charred part of the yakitori grilled for 5 min were greater than those in Table 4.

The main components of the sauce are mirin (a kind of sake), honey, and soy sauce. In order to determine the components of the sauce responsible for 1-NP formation, we marinated bite-sized pieces of raw chicken in various components of the sauce and solutions of different pH and grilled them for 7 min on the same gas range

(*47*). These pieces of yakitori showed mutagenic activity, but not a proportional concentration of 1-NP. Mirin is the most effective component of the sauce in regard to 1-NP formation (3.1 ng per gram of yakitori), but the amount of 1-NP is less than that in the yakitori cooked with the complete sauce (10.7 ng per gram of yakitori). When the pH of sauce No. 1 (pH 4.7) was changed to 7.6 with sodium hydroxide, the concentration of 1-NP formed decreased from 10.7 to 5.8 ng per gram of yakitori. However, the pH of the marinade is not acidic enough to produce 1-NP; yakitori grilled with 0.15 N H_2SO_4 (pH 1.7) contained only 0.9 ng per gram of yakitori. Since other commercial sauces, sauces No. 2 and No. 3, were not effective in 1-NP formation (2.7 and 1.2 ng per g), some other minor, sauce components and/or composition of components in addition to acidity of the sauce are important for 1-NP formation.

Although grilling the yakitori with the sauce increased the direct-acting mutagenicity, grilling the sauce alone with glass wool on the same gas range did not produce highly active direct-acting mutagens nor 1-NP, indicating that pyrene formation during incomplete combustion of the chicken fat is another important factor (*47*).

Since 1-NP was not formed when chicken was grilled with the sauce on an electric range, exhaust gas, especially nitrogen dioxide, emitted by burning of cooking gas must be important for nitration of pyrene. When water in a kettle was heated on a gas range, the concentration of nitrogen dioxide increased from 11 to 263 ppb in a room (28 m^3) used for grilling chicken (*47*). A low concentration of nitrogen dioxide may be produced by grilling the chicken. However, grilling chicken after boiling water did not increase the concentration of nitrogen dioxide in the gas kitchen although it greatly increased the mutagenicity of indoor air particles, from 6.7 and 5.0 to 244 and 429 His^+ revertants per m^3 of room air for strain TA98 in the absence and presence, respectively, of S9 mix.

In summary, pyrene produced by incomplete combustion of fat in the chicken, its nitration by nitrogen dioxide emitted by the burning of cooking gas, acidity of the food, and some components of the sauce are important factors in the 1-NP formation in yakitori.

Estimation of Human Intake of NPs

If one consumes 100 g of yakitori cooked with marinating sauce No. 1 for 5 min every 4 weeks, he will in 20 years have consumed at most 494 μg of 1-NP and 4.9 μg of diNPs (if the concentration of diNPs is 1/100 that of 1-NP). In the case of sauces No. 2 and No. 3 the human intake is 31 and 70 μg of 1-NP and 0.31 and 0.71 μg of diNPs, respectively, in 20 years. We have already estimated that in 20 years a human inhales at most 120 μg of 1-NP and 4.6 μg of diNPs from outdoor air polluted with diesel engine exhausts, and 892 ng of 1-NP and 456 ng of diNPs from indoor air polluted with a kerosene heater (if a kerosene heater is used for 8 hr a day 3 months a year in a 28-m^3 room) (*48*).

Evaluation of whether or not such intake is hazardous for human health is difficult even though we know that 2 mg and 40 mg of 1-NP in mice and rats, respectively, is noncarcinogenic (*23*, *25*) and that 40 μg of 1,8-diNP in rat subcutaneous tissue (*25*) and 150 μg of 1,6-diNP in rat lung (*27*) is carcinogenic. However, in the case of yakitori the mutagenicity of 1-NP and 1,6-diNP in the neutral fraction was at most a small percentage of the total mutagenicity of the crude extract. Therefore, the other nitro compounds such as other nitro-polycyclic aromatic hydrocarbons and nitro derivatives of heterocyclic amines may be important. Since the basic fractions of grilled meat, especially without the sauce, had very high mutagenicity, we must avoid producing and eating highly mutagenic heterocyclic amines in grilled meat.

Antimutagenic Activity of Mushrooms

We eat not only mutagenic but also antimutagenic substances in various foods. Since the strongest mutagens in cooked foods are heterocyclic amines, we measured the antimutagenic acivity of various edible mushrooms against Trp-P-1 (Fig. 2). The acidic amino acid fraction of *Flammulina velutipes* (enokitake) containing glutathione was the most effective. The diethyl ether-soluble acidic fraction of the mushroom also decreased the mutagenicity of IQ, MeIQ, MeIQx, and Trp-P-1. This fraction contained linoleic acid and unknown antimutagenic substances. We

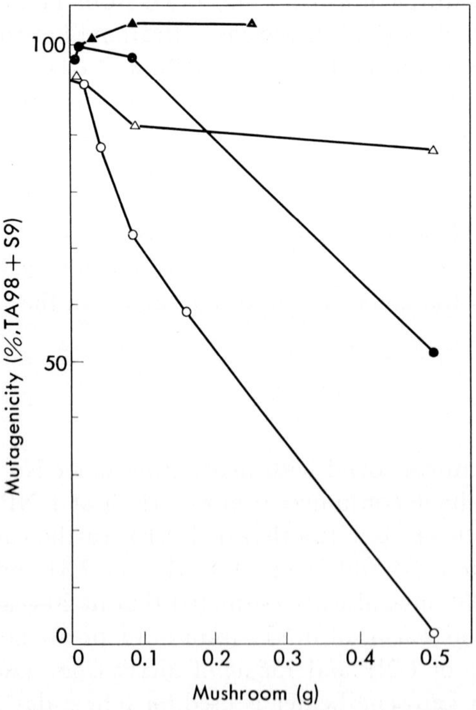

FIG. 2. Inhibitory effects of acidic amino acid fractions extracted from edible mushrooms on the mutagenicity of Trp-P-1 (25 ng/plate). 100% mutagenicity means 789 His+ revertants per plate from *S. typhimurium* strain TA98 in the presence of S9 mix. ▲, *Agaricus bisporus* (Tsukuritake); △, *Lyophyllum ulmarium* (Shirotamogitake); ●, *Pholiota nameko* (Nameko); ○, *Flammulina velutipes* (Enokitake).

must identify the effective components and the mechanism of their antimutagenesis in the future.

ACKNOWLEDGMENTS

We are grateful to Dr. Tadashi Ogawa for his valuable advice on the fractionation experiments with mushrooms and to Dr. Shigeaki Sato for supplying IQ, MeIQ, and MeIQx. This work was supported by grants-in-aid for cancer research from the Ministry of Education, Science and Culture and the Ministry of Health and Welfare of Japan and by funds from the Nissan Science Foundation.

REFERENCES

1. Doll, R. and Peto, R. The causes of cancer: quantitative estimates of avoidable risks of cancer in the United States today. J. Natl. Cancer Inst., *66*: 1191–1308, 1981.
2. McCann, J., Choi, E., Yamasaki, E., and Ames, B. N. Detection of carcinogens as mutagens in the *Salmonella*/microsome test: assay of 300 chemicals. Proc. Natl. Acad. Sci. U.S.A., *72*: 5135–5139, 1975.
3. Purchase, I.F.M., Longstaff, E., Ashby, J., Styles, J. A., Anderson, D., Lefevre, P. A., and Westwood, F. R. Evaluation of six short term tests for detecting organic chemical carcinogens and recommendations for their use. Nature, *264*: 624–627, 1976.
4. Nagao, M., Sugimura, T., and Matsushima, T. Environmental mutagens and carcinogens. Annu. Rev. Genet., *12*: 117–159, 1978.
5. Nagao, M., Honda, M., Seino, Y., Yahagi, T., and Sugimura, T. Mutagenicities of smoke condensates and the charred surface of fish and meat. Cancer Lett., *2*: 221–226, 1977.
6. Commoner, B., Vithayathil, A. J., Dolara, P., Nair, S., Madyastha, P., and Cuca, G. C. Formation of mutagens in beef and beef extract during cooking. Science, *201*: 913–916, 1978.
7. Weisburger, J. H. Mechanism of action of diet as a carcinogen. Cancer, *43*: 1987–1995, 1979.
8. Spingarn, N. E., Slocum, L. A., and Weisburger, J. H. Formation of mutagens in cooked foods. II. Foods with high starch content. Cancer Lett., *9*: 7–12, 1980.
9. Bjeldanes, L. F., Morris, M. M., Timourian, H., and Hatch, F. T. Effects of meat composition and cooking conditions on mutagen formation in fried ground beef. J. Agric. Food Chem., *31*: 18–21, 1983.
10. Matsumoto, T., Yoshida, D., Mizusaki, S., and Okamoto, H. Mutagenic activity of amino acid pyrolysates in *Salmonella typhimurium* TA98. Mutat. Res., *48*: 279–286, 1977.
11. Kosuge, T., Tsuji, K., Wakabayashi, K., Okamoto, T., Shudo, K., Iitaka, Y., Itai, A., Sugimura, T., Kawachi, T., Nagao, M., Yahagi, T., and Seino, Y. Isolation and structure studies of mutagenic principles in amino acid pyrolysates. Chem. Pharm. Bull., *26*: 611–619, 1978.
12. Sugimura, T., Kawachi, T., Nagao, M., Yahagi, T., Seino, Y., Okamoto, T., Shudo, K., Kosuge, T., Tsuji, K., Wakabayashi, K., Iitaka, Y., and Itai, A. Mutagenic principle(s) in tryptophan and phenylalanine pyrolysis products. Proc. Jpn. Acad., *53*: 58–61, 1977.
13. Yamamoto, T., Tsuji, K., Kosuge, T., Okamoto, T., Shudo, K., Takeda, K., Iitaka, Y., Yamaguchi, K., Seino, Y., Yahagi, T., Nagao, M., and Sugimura, T. Isolation

and structure determination of mutagenic substances in L-glutamic acid pyrolysate. Proc. Jpn. Acad., *54B*: 248–250, 1978.

14. Yoshida, D., Matsumoto, T., Yoshimura, R., and Matsuzaki, T. Mutagenicity of amino-α-carbolines in pyrolysis products of soybean globulin. Biochem. Biophys. Res. Commun., *83*: 915–920, 1978.
15. Kasai, H., Yamaizumi, Z., Wakabayashi, K., Nagao, M., Sugimura, T., Yokoyama, S., Miyazawa, T., Spingarn, N. E., Weisburger, J. H., and Nishimura, S. Potent novel mutagens produced by broiling fish under normal conditions. Proc. Jpn. Acad., *56B*: 278–283, 1980.
16. Kasai, H., Yamaizumi, Z., Shiomi, T., Yokoyama, S., Miyazawa, T., Wakabayashi, K., Nagao, M., Sugimura, T., and Nishimura, S. Structure of a potent mutagen isolated from fried beef. Chem. Lett., 485–488, 1981.
17. McCann, J., Spingarn, N. E., Kobori, J., and Ames, B. N. Detection of carcinogens as mutagens: bacterial tester strains with R factor plasmids. Proc. Natl. Acad. Sci. U.S.A., *72*: 979–983, 1975.
18. Sugimura, T. and Sato, S. Mutagens-carcinogens in foods. Cancer Res., *43*: 2415s–2421s, 1983.
19. Rosenkranz, H. S. and Mermelstein, R. Mutagenicity and genotoxicity of nitroarenes. All nitro-containing chemicals were not created equal. Mutat. Res., *114*: 217–267, 1983.
20. Manabe, Y., Kinouchi, T., and Ohnishi, Y. Identification and quantification of highly mutagenic nitroacetoxypyrenes and nitrohydroxypyrenes in diesel-exhaust particles. Mutat. Res., *158*: 3–18, 1985.
21. Tokiwa, H. and Ohnishi, Y. Mutagenicity and carcinogenicity of nitroarenes and their sources in the environment. CRC Crit. Rev. Toxicol., *17*: 23–60, 1986.
22. Sugimura, T. Carcinogenicity of mutagenic heterocyclic amines formed during the cooking process. Mutat. Res., *150*: 33–41, 1985.
23. Tokiwa, H., Otofuji, T., Horikawa, K., Kitamori, S., Otsuka, H., Manabe, Y., Kinouchi, T., and Ohnishi, Y. 1,6-Dinitropyrene: mutagenicity in *Salmonella* and carcinogenicity in BALB/c mice. J. Natl. Cancer Inst., *73*: 1359–1363, 1984.
24. Ohgaki, H., Negishi, C., Wakabayashi, K., Kusama, K., Sato, S., and Sugimura, T. Induction of sarcomas in rats by subcutaneous injection of dinitropyrenes. Carcinogenesis, *5*: 583–585, 1984.
25. Ohgaki, H., Hasegawa, H., Kato, T., Negishi, C., Sato, S., and Sugimura, T. Absence of carcinogenicity of 1-nitropyrene, correction of previous results, and new demonstration of carcinogenicity of 1,6-dinitropyrene in rats. Cancer Lett., *25*: 239–245, 1985.
26. Takayama, S., Ishikawa, T., Nakajima, H., and Sato, S. Lung carcinoma induction in Syrian golden hamsters by intratracheal instillation of 1,6-dinitropyrene. Jpn. J. Cancer Res. (Gann), *76*: 457–461, 1985.
27. Maeda, T., Izumi, K., Otsuka, H., Manabe, Y., Kinouchi, T., and Ohnishi, Y. Induction of squamous cell carcinoma in the rat lung by 1,6-dinitropyrene. J. Natl. Cancer Inst., *76*: 693–701, 1986.
28. Gibson, T. L. Sources of direct-acting nitroarene mutagens in airborne particulate matter. Mutat. Res., *122*: 115–121, 1983.
29. Tokiwa, H., Kitamori, S., Nakagawa, R., and Matamala, L. Demonstration of powerful mutagenic dinitropyrene in airborne particulate matter. Mutat. Res., *121*: 107–116, 1983.
30. Pederson, T. C. and Siak, J.-S. The role of nitroaromatic compounds in the direct-

acting mutagenicity of diesel particle extracts. J. Appl. Toxicol., *1*: 54–60, 1981.
31. Rosenkranz, H. S. Direct-acting mutagens in diesel exhaust: magnitude of the problem. Mutat. Res., *101*: 1–10, 1982.
32. Salmeen, I., Durisin, A. M., Prater, T. J., Riley, T., and Shuetzle, D. Contribution of 1-nitropyrene to direct-acting Ames assay mutagenicities of diesel particulate extracts. Mutat. Res., *104*: 17–23, 1982.
33. Xu, X. B., Nachtman, J. P., Jin, Z. L., Wei, E. T., and Rappaport, S. M. Isolation and identification of mutagenic nitro-PAH in diesel-exhaust particulates. Anal. Chem. Acta, *136*: 163–174, 1982.
34. Ohnishi, Y., Kinouchi, T., Manabe, Y., Tsutsui, H., Otsuka, H., Tokiwa, H., and Otofuji, T. Nitro compounds in environmental mixtures and foods. *In;* M. D. Waters, S. S. Sandhu, J. Lewtas, L. Claxton, G. Strauss and S. Nesnow (eds.), Short-Term Bioassays in the Analysis of Complex Environmental Mixtures IV, pp. 195–204, Plenum Publishing Corp., New York, 1985.
35. Manabe, Y., Kinouchi, T., Wakisaka, K., Tahara, I., and Ohnishi, Y. Mutagenic 1-nitropyrene in wastewater from oil-water separating tanks of gasoline stations and in used crankcase oil. Environ. Mutagen., *6*: 669–681, 1984.
36. Tokiwa, H., Nakagawa, R., and Horikawa, K. Mutagenic/carcinogenic agents in indoor pollutions; the dinitropyrenes generated by kerosene heaters and fuel gas and liquefied petroleum gas burners. Mutat. Res., *157*: 39–47, 1985.
37. Löfroth, G., Hefner, E., Alfheim, I., and Møller, M. Mutagenic activity in photocopies. Science, *209*: 1037–1039, 1980.
38. Rosenkranz, H. S., McCoy, E. C., Sanders, D. R., Butler, M., Kiriazides, D. K., and Mermelstein, R. Nitropyrenes: isolation and identification, and reduction of mutagenic impurities in carbon black toners. Science, *209*: 1039–1043, 1980.
39. Harris, W. R., Ches, E. K., Okamoto, D., Remsen, J. F., and Later, D. W. Contribution of nitropyrene to the mutagenic activity of coal fly ash. Environ. Mutagen., *6*: 131–144, 1984.
40. Ohnishi, Y., Kinouchi, T., Manabe, Y., and Wakisaka, K. Environmental aromatic nitro compounds and their bacterial detoxification. *In;* M. D. Waters, S. S. Sandhu, J. Lewtas, L. Claxton, N. Chernoff, and S. Nesnow (eds.), Short-Term Bioassays in the Analysis of Complex Environmental Mixtures III, pp. 527–539, Plenum Publishing Corp., New York, 1983.
41. Dennis, M. J., Massey, R. C., McWeeny, D. J., and Knowles, M. E. Estimation of nitropolycyclic aromatic hydrocarbons in food. Food Addit. Contam., *1*: 29–37, 1984.
42. Tokiwa, H., Nakagawa, R., Morita, K., and Ohnishi, Y. Mutagenicity of nitro derivatives induced by exposure of aromatic compounds to nitrogen dioxide. Mutat. Res., *85*: 195–205, 1981.
43. Standard tables of food composition in Japan. Resources Council, Science and Technology Agency, Japan, 1982 (in Japanese).
44. Kinouchi, T., Manabe, Y., Wakisaka, K., and Ohnishi, Y. Biotransformation of 1-nitropyrene in intestinal anaerobic bacteria. Microbiol. Immunol., *26*: 993–1005, 1982.
45. Kinouchi, T. and Ohnishi, Y. Purification and characterization of 1-nitropyrene nitroreductases from *Bacteroides fragilis*. Appl. Environ. Microbiol., *46*: 596–604, 1983.
46. Larsson, B. K., Sahlberg, G. P., Eriksson, A. T., and Busk, L. Å. Polycyclic aromatic hydrocarbons in grilled food. J. Agric. Food Chem., *31*: 867–873, 1983.
47. Kinouchi, T., Tsutsui, H., and Ohnishi, Y. Detection of 1-nitropyrene in yakitori (grilled chicken). Mutat. Res., 1986 (in press).

48. Ohnishi, Y., Kinouchi, T., Manabe, Y., and Tsutsui, H. Importance of nitroarene research in human health. Environ. Mutagen Res. Commun., *6*: 29–37, 1984 (in Japanese).

DIET, NUTRITION AND CANCER, Y. HAYASHI ET AL. (EDS.),
JAPAN SCI. SOC. PRESS, TOKYO/VNU SCI. PRESS, UTRECHT, PP. 119–137, 1986

Occurrence and Detection of Natural Mutagens and Modifying Factors in Food Products

Jan C. M. van der HOEVEN

Notox Toxicological Research and Consultancy v.o.f., 5231 DD's-Hertogenbosch, The Netherlands

Abstract: Various food products of plant origin were investigated for the occurrence of natural mutagens using the *Salmonella*/microsome assay. In general, food plants were freeze-dried and subsequently extracted with a number of solvents. Solvents were evaporated and the residues obtained were tested for mutagenicity. In addition to S9-mix, gut flora extracts were applied for metabolic activation.

From bracken fern (*Pteridium aquilinum*) a novel mutagen, designated Aquilide A, was isolated and its chemical structure was identified. Aquilide A requires activation to become mutagenic. This activation occurs spontaneously at pH levels above 6–7. Activated Aquilide A was found to be genotoxic in cultured mammalian cells.

Natural mutagens were detected in 4 out of 6 vegetables investigated. In addition, broad beans (*Vicia faba*) were found to be mutagenic after treatment with nitrite. All mutagenic vegetables showed marked intercultivar variations. From lettuce and string beans quercetin was isolated (after chemical hydrolysis) and in ruhbarb emodin, an anthraquinon, was detected. The mutagnic activity of these two compounds was further investigated using cultured mammalian cells. Quercetin and emodin responded negative or weakly positive in the systems applied.

The genotoxic properties of a number of pyrrolizidin alkaloids, which are reported to occur in various flowering plants and as a result occur in honey and some herbal preparations, were studied using a cocultivation system of V79 Chinese hamster cells and primary cultures of chick embryo hepatocytes (PCCEH/V79). All four pyrrolizidine alkaloids investigated were found to be potent inducers of SCEs in this test system.

Anti-mutagenic effects of butylated hydroxyanisole (BHA) and indole-3-carbinol (I3C) were detected using the PCCEH/V79 cocultivation system. This indicates that the cocultivation system described can be a valuable tool for the screening of various products for potential anti-carcinogenic properties. Extracts of lettuce and string beans, and a number of natural chemicals were found to reduce the mutagenic activity of cigarette smoke condensate and benzo (a) pyrene (BaP) as detected in the *Salmonella*/microsome assay. Intercultivar variation with respect to the antimutagenic activity observed was less pronounced than the variation noted for the mutagenic activity of these vegetables. Measures which may result in a reduction of the exposure to a number of natural mutagens are discussed.

The development of short term mutagenicity tests like the *Salmonella*/microsome assay (*1*) has made it possible to screen large numbers of chemicals and complex mixtures for mutagenic and consequently potential carcinogenic activity. During the first years after the introduction of the *Salmonella*/microsome assay this test was almost exclusively used to test the mutagenicity of synthetic chemicals. However, since the discovery of the mutagenicity of heated food (*2*) and the discovery of the mutagenicity of ubiquitary occurring flavonoids (*3*) these short term tests have also been used to screen food products for mutagenicity and to isolate and identify the mutagenic compounds involved. Although some attention has been paid to the occurrence of natural mutagens in food products of plant origin (*e.g.*, 4–13) not many systematic and in depth studies have been carried out on this subject, this in strong contrast with the mutagens produced during the heating of food. Despite the fact that some natural chemicals contained in food plants are clearly carcinogenic (*14*), the potential risk posed by consumption of certain food products of plant origin is not taken seriously. This seems mainly due to the general belief that natural products are healthy *per se* and the perception that it is very difficult if not impossible to change food habits and to lower actual levels of natural mutagens/carcinogens in food.

This paper summarizes studies performed during the last several years at the Department of Toxicology of the Agricultural University of Wageningen on the detection and occurrence of natural mutagens and modifying factors in food products of plant origin. Possible measures to prevent people from exposure to a number of natural mutagens are discussed.

Mutagenicity of Bracken Fern (*Pteridium aquilinum*)

Bracken fern is a carcinogenic plant which occurs throughout the world, in some countries covering vast areas. For instance, it is estimated that in England and Wales (U.K.) 161,875 ha of land are covered by or infested with bracken (*15*). In some areas like Japan, bracken is consumed as a food plant and large amounts, several thousands of tons, are imported yearly for this purpose.

The possible exposure of people to the carcinogen(s) occurring in bracken is not restricted to direct consumption. Other possible exposure routes are the consumption of milk or dairy products from cattle fed with or grazing on bracken and the consumption of drinking water contaminated by compounds from bracken leached into the soil. A recent discovery is that the spores of bracken fern, of which many of thousands per frond are spread into the air, are also carcinogenic (*16*). As a result, exposure to carcinogens produced by bracken, may also occur by inhalation.

Some years ago we described the detection of mutagenic extracts prepared from freeze-dried bracken fern (*17*). Mutagenic activity was detected in methanol extracts. A prerequisite for the detection is an alkaline treatment of the substances present in the methanol extract. This was accomplished by evaporation of the methanol, dissolution of the residue in water and subsequent addition of concentrated NaOH. After several min (pH>10) the solution was filtered over a RP18 Seppak cartridge on which the mutagenic substance was concentrated. The mutagen was removed from the cartridge by elution with dimethyl sulphoxide (DMSO). The DMSO solutions were tested for mutagenicity using tester strains TA98 and TA100 without metabolic activation. It was found that by direct contact (several min) of the *Salmonella typhimurium* bacteria and the alkaline activated bracken mutagen, the sensitivity of the assay could be increased. Using this direct contact method we were able to detect mutagenic activity of very small amounts of freeze-dried bracken: after a contact time of 4 min more than one thousand revertants per plate were induced by alkaline treated extracts originating from 1 mg of freeze-dried bracken (*18*). The methanol extract was purified further by column chromatography using polyamide as adsorbent and water as a solvent. The first UV absorbing fraction contained more than 50% of the original mutagenic activity detected after alkaline treatment. This first fraction (P1) was further purified using HPLC with an RP18 column and methanol/water (26/74; v/v) as a solvent. One single UV absorbing peak designated H11 contained all the mutagenic acitivity which was recovered from the polyamide purification step. Further analysis using mass spectrometry revealed the presence of a single compound with a molecular weight of 398. Analysis of ^1H and ^{13}C NMR data resulted in the discovery of a new compound which we designated Aquilide A (Fig. 1) (*19*). Under alkaline conditions Aquilide A is converted into a reactive compound designated 3A (Activated Aquilide A). 3A is rapidly, at least under acidic conditions, converted into pterosin B, a nonmutagenic compound (*20*). Conversion into pterosin B also takes place by acidic treatment of the parent compound Aquilide A.

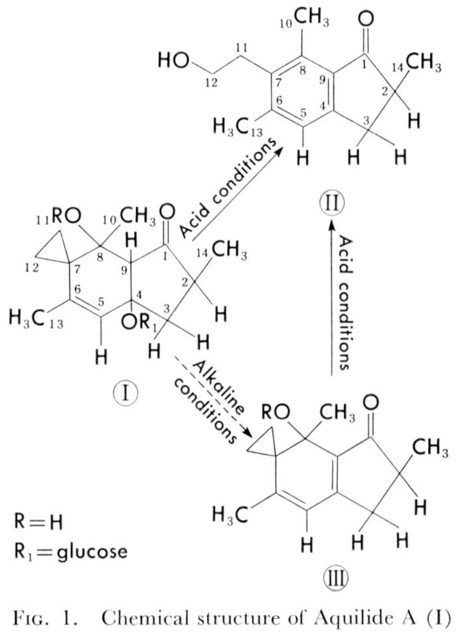

FIG. 1. Chemical structure of Aquilide A (I) and conversion into pterosin B (II), or 3A (III).

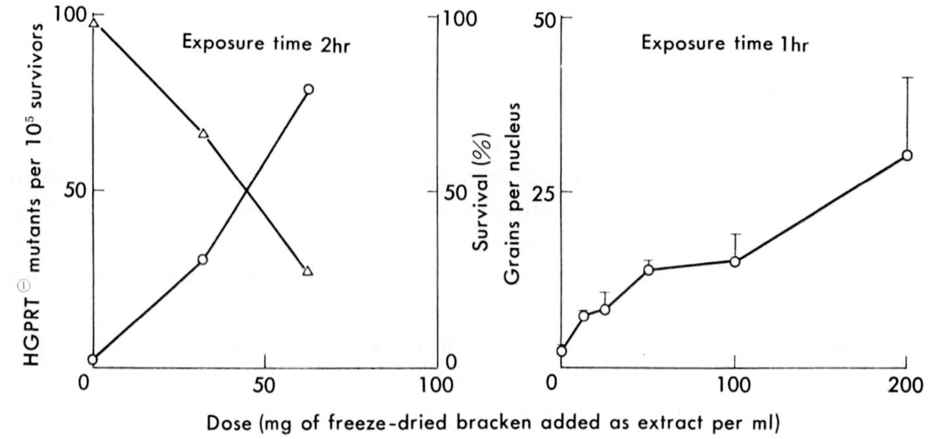

FIG. 2. Induction of HGPRT-deficient mutants in V79 Chinese hamster cells and induction of unscheduled DNA synthesis in human fibroblasts by exposure to activated Aquilide A. Dose is expressed as mg of freeze-dried bracken added as extract per ml since the amount of substance contained in the purified fractions was not determined. Purification was performed as described (*49*), such purified fractions contain Aquilide A only. Directly before testing these fractions were activated at alkaline conditions. Presuming that 1 mg of freeze-dried bracken contains 25 μg Aquilide A, the genotoxic effects shown are induced by exposure to concentrations of 3A in the range of 0–275 μg/ml. ○, mutants; △, survival.

To further evaluate the genotoxic properties of Aquilide A after alkaline treatment the compound was tested in various cultured mammalian cell systems. As we reported in 1980 (*17*), alkaline pretreated methanol extract from bracken is a potent inducer of SCEs in V79 Chinese hamster cells. The same property was observed for the purified compound. As can be seen from Fig. 2, alkaline pretreated Aquilide A is also a potent inducer of HGPRT-deficient mutants in the same cell line. Exposure of cultures of human fibroblasts (HAN-cells) resulted in an increase of UDS as detected by autoradiography (Fig. 2). The outcome of these studies made it very probable that Aquilide A indeed is a carcinogen. Hirono and co-workers (*21*) found a compound with the same structure as Aquilide A in a carcinogenic extract from bracken. Further studies with the purified compound, which they designated ptaquiloside, revealed that this substance indeed induced typical tumors in rat such as those observed after exposure to whole bracken (*22*). They also reported that this compound induces the acute cattle poison syndrome which is known to occur after ingestion of large quantities of bracken by cows (*23*).

To determine whether 3A is also formed under more physiological conditions than pH levels$>$10, Aquilide A was dissolved in water at a pH of 7.4. Conversion of Aquilide A was subsequently detected by determination of the UV/VIS spectrum at different intervals after start of the incubation. Within several min substantial changes in the spectrum between 240–350 nm were observed. Other evidence for the formation of 3A from Aquilide A was derived from NMR experiments. In these experiments, a dilute solution of NaOH in D_2O was added to a solution of Aquilide A in methanol-D_4 to obtain a pH of 7–8. The reaction occurring under these condi-

TABLE 1. ^{1}H and ^{13}C NMR Data on 3A

Position	^{1}H NMR[a,b]	^{13}C NMR[a]
1	2.58	212.6
2	2.28 and 2.96	42.5, $^{1}J_{C-H}=129$
3		36.7, $^{1}J_{C-H}=131$
4	6.23	[d]
5		120.8, $^{1}J_{C-H}=161$
6		[d]
7		32.9
8		70.8
9		[d]
10	1.35	23.4, $^{1}J_{C-H}=129$
11	0.76[c] and 1.06[c]	9.9,[e] $^{1}J_{C-H}=161$[f]
12	1.15[c] and 1.48[c]	10.3,[e] $^{1}J_{C-H}=161$[f]
13	1.84	20.4, $^{1}J_{C-H}=127$
14	1.25	17.1, $^{1}J_{C-H}=129$

[a] The ^{1}H and ^{13}C NMR spectra were recorded on a Bruker CXP-300 spectrometer operating at 300.066 and 75.460 MHz, respectively. For the assignment of the ^{13}C resonances selective decoupling experiments were used. Methanol-D_4 was used as a solvent. Chemical shifts are in ppm from Me_4Si. The ^{1}H and ^{13}C NMR data were obtained at pH 7–8 by addition of a diluted NaOH solution to the methanol-D_4 solution under nitrogen atmosphere at a temperature of $-10°C$. [b] In agreement with data reported by Niwa *et al.* (*24*). [c] May be interchanged (*25*). [d] Values for C4, C6, and C9 are 157.8, 169.5, and 133.5; these values cannot be assigned to the specific atoms. [e] May be interchanged (*25*). [f] From ref. *26*.

tions was studied using ^{1}H and ^{13}C NMR techniques. These data show high field signals which are characteristic of the cyclopropyl ring. It was also observed that H-9 ($\delta=2.73$) disappeared and that the glucoside bond was broken resulting in specific signals for free glucose ($\delta=4.58$ and $\delta=5.21$). The data obtained for 3A formed in this way are in accordance with those reported by Niwa *et al.* (*24*). Further evidence for the structure of 3A was obtained from the ^{13}C NMR spectrum (Table 1). Acidification of the alkaline solution of 3A with deuterated sulfuric acid immediately resulted in a complete reaction of 3A into pterosin B. That this activation into a mutagenic compound indeed takes place at a pH of 7.4 was shown by an experiment in which Aquilide A was incubated at this pH level for 4 hr and subsequently tested for mutagenic activity. The mutagenicity in this case was comparable with that observed after the standard alkaline treatment (pH>12, 10 min.).

The above described subtle chemical activation of Aquilide A under alkaline conditions represents a very interesting phenomenon not previously encountered in the field of mutagenesis. This property of Aquilide A provides an explanation for the organ specificity observed with respect to the carcinogenic effects of bracken fern. Since Aquilide A is converted into a direct acting mutagen at elevated pH levels, it probably will predominantly produce tumors at the sites where such alkaline conditions exist. Data on cattle grazing on bracken infested land show that they have increased frequencies of esophageal and bladder tumors (*27, 28*). At these sites pH levels are rather high. The pH in the esophagus is determined by the pH of the saliva which is 8.1–8.2 for cattle. The pH of urine of cattle is 7.5–8.5. Under both conditions the pH is high enough to chemically activate Aquilide A. In case of the

TABLE 2. Detection of Mutagenic Activity of Urine of Rats Fed with Bracken

Amount of urine added per plate (ml)	Treatment	Rat A		Rat B	
		TA98 (rev./plate[a])	TA100 (rev./plate[a])	TA98 (rev./plate[a])	TA100 (rev./plate[a])
0.09	—	15	40	10	433
0.90	—	66	260	65	0
0.09	Alkaline treatment	335	496	216	806
0.90	Alkaline treatment	[b]	[b]	[b]	[b]

Two 3 month-old male Wistar rats were fed a diet containing 33% dried bracken for 72 hr. The urine produced during the last 48 hr was collected. The urine of each rat was split into two volumes of about 9 ml. One part was put over a Seppak C18 cartridge. The compounds remaining on the cartridge were removed by elution with 1 ml of methanol. This was evaporated and the residue was dissolved in DMSO and subsequently tested for mutagenicity using the direct contact method. The other part was treated with NaOH for 10 min at pH 13 and then treated in the same way as described for the first part. The urine of rat A was collected at room temperature, whereas the urine of rat B was collected at 0°C. Rat A consumed 26 g feed, rat B 16 g feed. The pH of rat A amounted to 6.9, that of rat B was 6.8. [a] Revertants per plate; spontaneous numbers are subtracted. [b] This dose was very toxic towards the bacteria, no survival.

bladder, the original Aquilide A is of course not present since such a glucoside cannot be absorbed from the intestine. One possibility is that the glucose moiety of Aquilide A, not activated in the esophagus, is split off by enzymes from the bacterial flora in the rumen of cattle. The aglycon produced, designated Aquiline A, can be absorbed, transported by the blood and conjugated in the liver resulting in a compound with similar properties as Aquilide A. Within the bladder, under alkaline conditions, activation into 3A may finally occur causing bladder tumors. Rats fed with bracken fern are reported to develop high tumor incidences in the small intestine and bladder. Within the intestine the tumors are predominantly found at the terminal 20 cm section of the ileum. This is the site at which highest pH levels are observed in the rat intestine. The Aquilide A that is not converted into 3A reaches the lower parts of the intestine where it is hydrolyzed by the glycosidases produced by the gut flora. In a manner similar to that described for cattle a conjugate of Aquiline A is excreted into the urine and may be activated into 3A within the bladder. If conjugated Aquiline A is excreted in rat urine, the urine should be mutagenic without alkaline treatment since the pH is about 7. Additional alkaline treatment should increase the mutagenicity since it is not expected that all conjugated Aquiline A will be activated at this pH level. The results summarized in Table 2 indeed show this pattern. For the human situation it is most likely that bracken fern will produce tumors in the oral cavity or the esophagus. Tumors within the human bladder are less likely since the pH of human urine is usually below 7.

Detection of Mutagenic Properties of Extracts of Certain Vegetables

To determine the occurrence of natural mutagens in food plants commonly consumed in The Netherlands six different vegetables were studied. Several cultivars of each vegetable grown under known and completely identical conditions were obtained from horticultural experimental stations. After harvesting, the vegetables

TABLE 3. Mutagenic Activity of Vegetables Commonly Consumed in The Netherlands

Vegetable	Number of cultivars tested	Mutagenic response	Inter-cultivar variation factor[a]
Lettuce	5	+	7
Rhubarb	4	+	12
String beans	4	+	3
Paprika	3	+	2
Spinach	6	—	—
Brussels sprouts	4	—	—

[a] For details see ref. *30*.

were treated as usual for consumption in The Netherlands. All vegetables were rinsed with running tap water and non-edible parts were removed. String beans, rhubarb, Brussels sprouts, and spinach were cooked. Lettuce and paprika received no further culinary treatment. Then the vegetables were frozen followed by freeze-drying. The dried material was extracted using different solvents, starting with apolar solvents and ending up with water. Solvents were evaporated at reduced air pressure. The residues were dissolved in DMSO before testing in the Ames assay; strains TA98 and TA100 and in some cases TA1537 were used. In addition to liver homogenate, gut flora extracts from rats was used for metabolic activation, especially to hydrolyze naturally occurring glycosides. Four out of six vegetables tested showed clear mutagenic activity (Table 3). A very interesting finding is that all mutagenic vegetables showed a marked intercultivar variation. The highest variations were observed for rhubarb: about 10-fold. The variation in mutagenic property may be the result of plant breeding activities by which susceptibility towards various infectious plant diseases is decreased by increasing concentrations of natural anti-microbial substances and other natural pesticides. By plant breeding also completely new natural substances can be introduced into edible cultivars. An example of such plant breeding activity is the creation of a cultivar with a resistancy to leaf aphids (*Nasonovia ribis nigri*). A cultivated lettuce (*Lactuca sativa*) was crossed with a lettuce type occurring in the wild (*Lactuca virosa*). The common Dutch name for this latter type is "gifsla" which means "toxic lettuce." Extracts of *Lactuca virosa* have been used in the past as a sedative. The resistant cultivar did not contain more mutagens, as detected with the methods used; on the contrary, the mutagenic activity was somewhat decreased. The compounds thought to be responsible for the sedative action of *Lactuce virosa* are sesquiterpene lactones. A compound belonging to this class of compounds, hymenovin (*29*), has been reported to be mutagenic. Purification of the mutagenic compound(s) from the methanol fractions of lettuce and string beans revealed the presence of quercetin glycosides. The amount of quercetin detected is responsible for over 75% of the mutagenic activity observed. From the combined light petroleum and chloroform fractions of rhubarb, emodin was isolated as the single contributor to the mutagenic activity found.

The mutagenic properties of the polyphenolic compounds quercetin and emodin were further evaluated using *in vitro* mammalian systems. The results of these studies are summarized in Table 4. Although both compounds are clearly positive in the

TABLE 4. Genotoxicity in Cultured Mammalian Cells of Quercetin and Emodin, Two Natural Polyphenolic Compounds Which Are Mutagenic in the *Salmonella*/Microsome Assay[a]

Cell line	Endpoint	Metabolic activation system	Result	
			Quercetin	Emodin
V79	SCEs	—	—	—
V79	SCEs	S9-mix	—	—
V79	SCEs	PCCEH[b]	—	—
V79	HGPRT	—	—	—
V79	HGPRT	S9-mix	—	—
V79	HGPRT	PCCEH[b]	—	—
L5178Y	TK[c]	—	±	—[d]
L5178Y	TK	S9-mix	—	—[d]
L5178Y	HGPRT	—	—	—[d]
L5178Y	HGPRT	S9-mix	—	—[d]

[a] See refs. *31, 32.* [b] Primary cultures of chick embryo hepatocytes. [c] Thymidine kinase. [d] These results have not been published in detail.

Salmonella/microsome assay they were found unable to induce mutations or SCEs in cultured mammalian cells. Both compounds inhibit cell division at concentrations in the range of 1–10 μg/ml in cultured mammalian cells, thus they are considered to be able to penetrate into these mammalian cells. Their metabolism and/or reaction with mammalian cells and mammalian DNA apparently differs from that in bacteria and with bacterial DNA. Quercetin has been found to be non-carcinogenic in the majority of the studies performed. Emodin has not previously been investigated for carcinogenic activity.

Mutagenicity of Fava Beans

Fava beans have been reported to become mutagenic after treatment with nitrite at acidic conditions (*33*) and Correa and co-workers (*34*) reported a positive correlation between the consumption of large amounts of fava beans and gastric cancer in areas of Colombia where drinking water contains relatively high concentrations of nitrate.

We have investigated the mutagenicity of a number of cultivars of nitrite treated fava beans (*35*) grown under identical conditions in The Netherlands. As was observed for other vegetables, fava beans (nitrite treated) showed a marked intercultivar variation in mutagenic properties (see Fig. 3). Since their mutagenic activity after nitrosation seems to be influenced to a great extent by plant breeding, it may be possible to introduce varieties which do not contain the pre-mutagenic precursor. Fava beans are consumed in considerable amounts in many parts of the world, therefore, introduction of such cultivars could lead to an important reduction of the exposure to a gastric carcinogen. This operation may be very successful since it can be carried out in such a way that people do not have to change their food pattern.

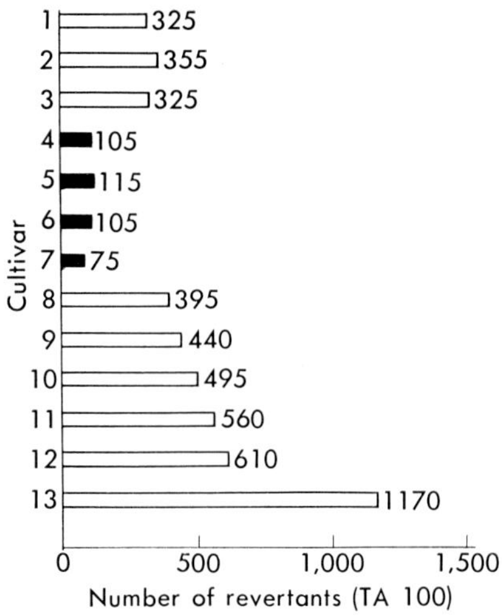

FIG. 3. Intercultivar variation in mutagenic activity of various cultivars of fava beans after treatment with nitrite at pH 2.0.

Primary Cultures of Chick Embryo Hepatocytes as a Metabolic System to Detect Genotoxic and Anti-genotoxic Properties of Chemicals Present in Food Products

Most chemical carcinogens found in food products require metabolic activation to reactive metabolites before they can be detected as genotoxins in *in vitro* assays. The metabolic activation systems often applied for this purpose are rat liver homogenate and primary cultures of rat hepatocytes. Important drawbacks of the first system are generation of non-relevant metabolites, inability to perform conjugation reaction, fast decrease of metabolic capacity and cytotoxicity towards mammalian cells. Disadvantages of the latter systems are the rapid (although less rapid than observed for liver homogenate) decrease in cytochrome P-450 content (*36*), problems in obtaining completely sterile cultures and skill required to obtain high viability.

From the literature and experience in our laboratory it was known that primary cultures of chick embryo hepatocytes (PCCEH) are a useful tool in *in vitro* biotransformation studies (*37–39*). In contrast with rat hepatocytes the enzyme δ-aminolevuline synthase, which is the enzyme rate limiting in the hepatic heme biosynthesis, is inducible in chick embryo hepatocytes (*40–42*). As a result, the cytochrome P-450 content remains at a high level in PCCEH whereas that of rat hepatocytes is diminished to less than 20% of the original amount within a day after the onset of these cultures (*36*). In addition, various cytochrome P-450 isoenzymes are inducible in PCCEH. Compounds like phenobarbital, β-naphtoflavone, 3-methylcholanthrene, and Aroclor 1254 are reported to induce cytochrome P-450 and mixed function oxygenases in PCCEH up to levels 2–10 times the background levels (*37*). To investigate whether PCCEH are suitable as a biotransformation system in genotoxicity studies varioius experiments were performed with known carcinogens requiring metabolic activation. As target cells, V79 Chinese hamster lung cells were used. Therefore, the cocultivation system was designated the PCCEH/V79 system. Two end-

TABLE 5. Induction of SCEs in V79 Chinese Hamster Cells after Treatment with DMN in the Presence of Liver Homogenate (LH, Prepared from Aroclor 1254 Pretreated Male Wistar Rats) or PCCEH

DMN concentration (μg/ml)	SCEs per chromosome±S.E.M.			
	LH		PCCEH	
	Exp. 1	Exp. 2	Exp. 1	Exp. 2
0	0.30±0.02	0.28±0.02	0.22±0.01	0.22±0.01
7	0.35±0.02	n.t.[a]	0.32±0.03	0.25±0.02
37	n.t.	0.30±0.02	n.t.	0.56±0.05
74	0.53±0.04	0.43±0.02	1.37±0.09	1.22±0.09
370	0.64±0.03	0.60±0.04	1.98±0.11	1.92±0.12

In general, 6 to 10 15-day old chick embryos (Hubbard strain) were used for each experiment. The liver was perfused by injection of 10 ml of HEPES buffered (20 mM, pH 7.4) Hanks' balanced salt solution (HBSS) containing 0.04% EDTA. After perfusion the discolored liver was removed carefully without disruption of the gallbladder. The liver tissue was then placed in a petri dish containing the perfusion fluid and cut into small pieces. These pieces were transferred into an Erlenmeyer flask with HEPES buffered HBSS containing 1 U dispase II per ml and incubated at 37°C for 10 min under gentle agitation. Dissociated cells were removed by separation of the turbid solution from the remaining tissue pieces. The remaining tissue pieces were treated two more times in the same way. The cells obtained were sedimented by centrifugation at 200 *g* for 4 min. The erythrocytes present in the cell pellet obtained were lysed by resuspending the pellet in buffered ammonium chloride (*41*) at room temperature followed by centrifugation at 200 *g* for 4 min. Finally, the hepatocytes were resuspended in Williams E medium supplemented with 10% newborn calf serum and 2 mM glutamine. The total yield was usually $5\text{–}15 \cdot 10^6$ viable hepatocytes per liver. About $1\text{–}3 \cdot 10^5$ hepatocytes were inoculated per 6 cm petri dish. Non viable cells were removed after 2–4 hr by refreshing the tissue culture medium. A day later V79 cells were inoculated on top of the hepatocytes. Other protocols where the hepatocytes were inoculated on top of cultured V79 cells have also been used with similar results. At the time both cell types were attached and detached and dead cells were removed, the test compounds were added. After exposure for 18–40 hr the cells were trypsinated and treated as described previously (*31*) for detection of SCEs and/or HGPRT-deficient mutants. [a] Not tested.

TABLE 6. Response of Various Premutagens in the PCCEH/V79 Test System: Induction of SCEs and HGPRT-deficient Mutants

Test compound	Induction of SCEs		Induction of HGPRT-deficient mutants	
	Results	Effective range (μg/ml)	Results	Effective range (μg/ml)
BaP	+	0.2–5.0	+	0.5–10.0
3MC	+	0.1–1.0	+	0.1–1.0
CSC[a]	±	16–64	+	16–64
HLT[b]	+	0.5–5.0	–	0 (–10)[c]
DMN	+	7–370	+	125–500

[a] For details see ref. *43*. [b] Results on SCEs were reported previously (*44*). [c] Up to this level no increase in the number of mutants was observed.

prints were used: induction of SCEs and induction of HGPRT-deficient mutants.

Details of the methods used are given in the description under Table 5 in which an example is offered of the results obtained with dimethylnitrosamine (DMN). A summary of the results obtained with benzo(a)pyrene (BaP), 3-methylcholanthrene (3 MC), cigarette-smoke-condensate (CSC), heliotrine (HLT), and DMN

TABLE 7. Important Features of Primary Cultures of Chick Embryo Hepatocytes as a Metabolic System in Genotoxicity Studies

I. Easy to prepare:
Total time required to obtain $50–100 \cdot 10^6$ hepatocytes about 2 hr.

II. Stability:
The cytochrome P450 content remains at a high level, even after prolonged (>72 hr) culture times without the addition of stimulating factors.

III. Reproducibility:
Results in genotoxicity studies are less vulnerable for inter-individual variability since a number of embryos (5–10) are used to obtain a suspension of $(50–100 \cdot 10^6)$ hepatocytes for one experiment.

IV. Viability:
In contrast with, *e.g.*, rat hepatocytes, the viability of primary chick embryo hepatocytes in general is high (>95% as determined with the trypan blue dye exclusion test).

V. Induction of cytochrome P450:
Induction of various cytochrome P450 iso-enzymes is possible in culture. This creates the possibility of using these hepatocytes for studies of interaction of various chemicals, *e.g.*, antigenotoxic properties.

is given in Table 6. From these results it can be concluded that the PCCEH/V79 system is valuable to detect the genotoxicity of carcinogens. The main advantages of PCCEH as a metabolic activation system are summarized in Table 7.

1. Detection of the genotoxic properties of pyrrolizidine with alkaloids using the PCCEH/V79 system

Many flowering plants have been found to contain carcinogenic pyrrolizidine alkaloids (PA) (*45, 46*). Some plants contain a number of different alkaloids (*47*). These alkaloids generally induce liver tumors although lung, kidney, brain, and pancreas tumors are also reported to be induced in experimental animals by PA exposure (*48, 49*). Despite the high number of PA present in plants and the potent carcinogenic and toxic properties of a number of them, no fast and/or general analytical or biological methods are available to systematically screen products for the presence of known or unknown carcinogenic PA.

Since PA are poorly detected as genotoxins using the *Salmonella*/microsome assay (*50*) we applied the PCCEH/V79 system to investigate their genotoxic properties.

TABLE 8. Induction of SCEs in V79 Chinese Hamster Lung Cells by 4 PA after Co-cultivation with Primary Chick Embryo Hepatocytes[a]

Concentration	SCEs per chromosome (±S.E.M.)			
	Heliotrine	Monocrotaline	Seneciphylline	Senkirkine
0	0.30±0.02	0.29±0.02	0.22±0.02	0.27±0.02
0.6	0.48±0.02	n.t.	1.46±0.06	0.95±0.08
1.2	0.83±0.03	0.54±0.04	1.76±0.14	1.28±0.09
2.5	1.70±0.08	0.74±0.05	2.05±0.10	1.45±0.10
3.5	n.t.[b]	1.04±0.10	2.36±0.15	n.t.
5.0	1.70±0.10	1.26±0.10	n.t.	1.86±0.16

[a] For details see ref. *44*. [b] Not tested.

As shown in Table 6, HLT was found a potent inducer of SCEs whereas no HGPRT-deficient mutants were induced. Three other mutagenic PA were therefore tested in the PCCEH/V79 test using SCEs as an endpoint. The results obtained with these four PA are summarized in Table 8. All PA investigated induced significant increases in SCEs at the 1 μg-level. Seneciphylline induced 1.46 SCEs per chromosome after exposure to 0.6 μg/ml! Since many plants have been reported to contain PA with genotoxic properties, food products derived directly or indirectly from PA-containing plants may also contain these PA. For example, honey produced in Australia in areas where *Echium plantigineum* is found may contain up to 1 mg/kg of echimidine (*51*). Milk from cattle fed with hay contaminated with *Senecio alpinus* contained seneciphylline and senecionine (*52*). Screening of suspected food products like honey and milk can easily be carried out with the PCCEH/V79 test system.

2. Detection of antigenotoxic properties of chemicals using the PCCEH/V79 cocultivation system

Known anticarcinogens like the synthetic antioxidant butylated hydroxyanisole (BHA) and the naturally occurring chemical indole-3-carbinol (I3C) most probably inhibit carcinogenesis by inducing the drug metabolizing enzymes involved in detoxification processes. Such properties are difficult to investigate *in vitro*. The liver homogenate used in the standard Ames assay cannot be used since detoxifying enzymes in particular no longer function under these conditions. The cytochrome P-450 system of rat hepatocytes is known to be rather unstable under *in vitro* conditions. So, even when induction of the detoxifying enzymes took place *in vivo* these two metabolizing systems are not very well suited for *in vitro* studies. To determine the feasibility of the PCCEH/V79 system for studying antigenotoxic activity of chemicals we investigated the activity of BHA and I3C using this test system.

Preliminary results obtained on I3C with this test system are summarized in Table 9. With both compounds, I3C and BHA (data not shown), a clear reduction

TABLE 9. Effect of I3C on the Induction of SCEs by BaP in V79 Chinese Hamster Cells Co-cultivated with Primary Chick Embryo Hepatocytes

BaP concentration (μg/ml)	I3C concentration (μg/ml)	SCEs per chromosome (averages[a) ±S.E.M.)
10	0	0.75±0.03
10	2.5	0.52±0.03
10	5.0	0.46±0.03
10	10.0	0.49±0.03
0	10.0	0.25±0.02
0	0	0.21±0.02

The hepatocytes were prepared as described in the paragraph on the genotoxic effects of PA. About 10^5 hepatocytes were inoculated per 6 cm petri dish. Directly after removal of the dead and unattached hepatocytes BHA or I3C was added to the cultures. Eighteen hr later the medium was refreshed and about 10^6 V79 Chinese hamster cells were added to the cultures of hepatocytes. After 20 hr of incubation the cultures were rinsed with Hanks' balanced salt solution and the cells were trypsinated. Further procedures to obtain sister chromatid differentiation were as described before. [a)] Averages from two independent experiments; each experiment was performed in duplicate, 100 metaphases were scored.

in the number of SCEs induced by exposure to BaP was observed. This indicates that the PCCEH/V79 system can be a valuable tool in testing compounds for potential anticarcinogenic properties.

Detection of Antimutagenic Properties of Chemicals and Extracts of Vegetables Using the Salmonella/Microsome Assay

To screen various chemicals occurring in food on antimutagenic properties they were tested by the *Salmonella*/microsome assay in the presence and absence of BaP or CSC. Results obtained with 10 chemicals are summarized in Fig. 4. Clear antimutagenic effects were observed with ellagic acid, riboflavin, and chlorophyllin. The effects observed were similar for BaP and CSC indicating a general mechanism of inhibition, probably an interaction between antimutagen and enzyme(s) in the liver homogenate. Chlorophyll *a* and chlorophyll *b* showed a less clear inhibition of CSC- and BaP-induced mutagenicity. Ascorbic acid, β-carotene, tocopherol acetate, chlorogenic acid, and BHA responded negative.

To investigate whether vegetables contain antimutagenic substances and, if so, whether intercultivar variations are present, extracts of cultivars of lettuce and string beans were tested. Extracts were tested in the presence and absence of CSC. All tests were carried out with TA98 in the presence of S9-mix. All extracts tested showed antimutagenic properties: petroleum ether, chloroform, methanol, and water extracts. The strongest antimutagenic effects were found for both lettuce and string beans in the petroleum ether and chloroform extracts. A typical example of results obtained with string beans is shown in Table 10. In contrast with the mutagenic properties observed, the antimutagenic activity of extracts of string beans showed no marked intercultivar variation. Similar results were obtained with lettuce.

TABLE 10. Antimutagenic Effects of Extracts of String Beans toward CSC Tested in TA98 in the Presence of S9-mix

Cultivar	Percentage of inhibition[a]	
	Petroleum ether extract	Chloroform extract
Hazet	52	56
Helda	58	57
Infra	52	52
Romore	51	45

[a] Percentage of inhibition: $1-\frac{\text{actual number}}{\text{expected number}}\times 100\%$. Expected number: number of revertants in the presence of CSC only increased with the number of those in the presence of the vegetable extract alone. Actual number: number of revertants observed when the vegetable extract together with CSC was added to the plates. Spontaneous numbers are subtracted.

Possible Preventive Measures

Using the data described in this paper it should be possible in specific cases to reduce human exposure to natural mutagens/carcinogens. Considering the energy

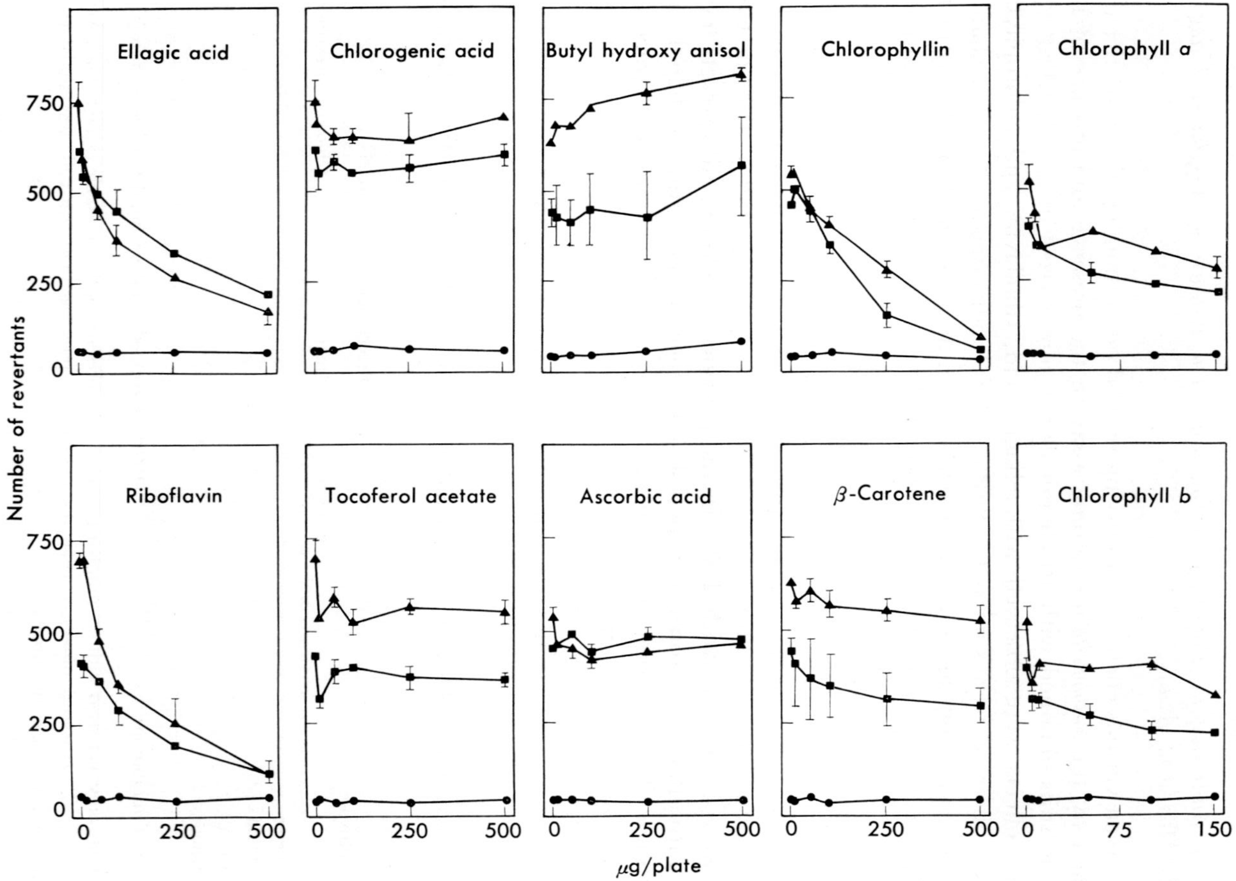

FIG. 4. Modifying effects of various chemicals on the mutagenic activity of BaP (▲) or CSC (■) in TA98 in the presence of S9-mix (*53*).

and money that are spent to prevent and reduce human exposure to synthetic carcinogens and mycotoxins like aflatoxin B_1, attempts in this direction seem justified. In the case of bracken fern the authorities of countries like Japan should encourage their people to prepare bracken in such a way that all carcinogenic activity is destroyed; this is possible by alkaline and subsequent acidic treatment. Bracken fern processed commercially should be screened by National Food Agencies on the occurrence of Aquilide A. In areas where dairy products and drinking water can be contaminated by the bracken carcinogen chemical analyses should be regularly performed. In places where large numbers of spores can be introduced into the air people like farmers and foresters should be educated to wear masks during times of the year sporulation occurs. In areas where bracken fern forms a real health risk it should be eradicated. It should be noted that the large occurrence of bracken in certain parts of the world is a result of human activities. In countries like Brazil large deforested areas are now infested with bracken fern (*13*).

With respect to natural mutagens occurring in food plants, all vegetables on the market should be investigated for their presence. Newly developed cultivars should, like newly developed chemicals, be screened for the presence of mutagens. In contrast with many synthetic chemicals these newly developed cultivars are meant to be consumed in important quantities. Therefore, it is difficult to understand why activities in the field of plant breeding are not yet regulated. There is no reason to expect that natural pesticides are less dangerous than synthetic ones.

Since consumption of fava beans is high in a number of countries around the world (*e.g.*, Colombia and countries around the Mediteranean) and some epidemiological evidence exists for a positive correlation between consumption of high amounts of fava beans and gastric cancer, carcinogenicity studies with this food product should be carried out. If the results with respect to intercultivar variation of mutagenic properties are confirmed in these studies, plant breeding programs should be carried out to develop low mutagenic cultivars or cultivars which completely lack this activity. Introduction of such newly developed cultivars may subsequently solve the problem. Regular attention should be given to the occurrence of PA in various food products. Herbal products and honey should be screened for this purpose. Since a number of different mutagenic/carcinogenic PA may occur in a single food item PCCEH/V79-test may be a valuable tool.

Identification and isolation of new antimutagens/anticarcinogens can also contribute to prevention of induction of cancer in the human population. If exposure to large amounts of carcinogens is unavoidable (*e.g.*, by addiction to cigarette smoking), identification of volatile anticarcinogens and subsequent use of the compounds in cigarettes may dramatically reduce the risk of lung cancer among smokers.

ACKNOWLEDGMENTS

I would like to thank Prof. Dr. J. H. Koeman and other members and students of the Department of Toxicology of the Agricultural University, Wageningen, The Netherlands who made these studies possible and contributed to the investigations mentioned in this paper. Mr. A. van Veldhuizen, Dr. H.A.J. Holterman, Dr. M. A. Posthumus, and Prof. Dr. A. de Groot of the Laboratory of Organic Chemistry of the Agricultural University are acknowledged for their help in resolving the structure of Aquilide A and 3A.

REFERENCES

1. Ames, B. N., McCann, J., and Yamasaki, E. Methods for detecting carcinogens and mutagens with the *Salmonella*/mammalian microsome mutagenicity test. Mutat. Res., *31*: 347–364, 1975.
2. Nagao, M., Honda, M., Seino, Y., Yahagi, T., and Sugimura, T. Mutagenicities of smoke condensates and the charred surface of fish and meat. Cancer Lett., *2*: 221–226, 1977.
3. Bjeldanes, L. F. and Chang, G. W. Mutagenic activity of quercetin and related compounds. Science (Wash. D.C.), *197*: 577–578, 1977.
4. Fukuoka, M., Yoshihira, K., Natori, S., Sakamoto, K., Iwahara, S., Hosaka, S., and Hirono, I. Characterization of mutagenic principles and carcinogenicity of dill weed and seed. J. Pharmacobio-dyn., *3*: 236–244, 1980.
5. Hashimoto, Y., Shudo, K., Okamoto, T., Nagao, M., Takahashi, Y., and Sugimura, T. Mutagenicities of 4-hydroxy-1,4-benzoxazinones naturally occurring in maize plants and of related compounds. Mutat. Res., *66*: 191–194, 1979.
6. Knuutinen, J. and Von Wright, A. The mutagenicity of Lactarius mushrooms. Mutat. Res., *103*: 115–118, 1982.
7. Sterner, O., Bergman, R., Kesler, E., Magnusson, G., Nilsson, L., Wickberg, B., and Zimerson, E. Mutagens in larger fungi. I. Forty-eight species screened for mutagenic activity in the *Salmonella*/microsome assay. Mutat. Res., *101*: 269–281, 1982.
8. Sterner, O., Bergman, R., Franzen, C., Kesler, E., and Nilsson, L. Mutagens in larger fungi. II. The mutagenicity of commercially pickled Lactarius necator in the *Salmonella* assay. Mutat. Res., *104*: 233–237, 1982.
9. Stolz, D. R., Stavric, B., Krewski, D., Klassen, R., Bendall, R., and Junkins, B. Mutagenicity screening of foods. I. Results with beverages. Environ. Mutagen., *4*: 477–492, 1982.
10. Stolz, D. R., Stavric, B., Stapley, R., Klassen, R., Bendall, R., and Krewski, D. Mutagenicity screening of foods. II. Results with fruits and vegetables. Environ. Mutagen., *6*: 343–354, 1984.
11. Tamura, G., Gold, C., Ferro-Luzzi, A., and Ames, B. N. Fecalase: a model for the activation dietary glycosides to mutagens by intestinal flora. Proc. Natl. Acad. Sci. U.S.A., *77*: 4961–4965, 1980.
12. Subden, R. E., Krizus, A., and Rancourt, D. Mutagen content of table wines made from various grape species and hybrid cultivars. Food Chem. Toxicol., *22*: 309–313, 1984.
13. Ivie, G. W., Holt, D. L., and Ivey, M. Natural toxicants in human foods: psoralens in raw and cooked parsnip root. Science, *213*: 909–920, 1981.
14. International Agency for Research on Cancer. IARC Monographs on the Evaluation of Carcinogenic Risk of Chemicals to Humans. Some Naturally Occurring Substances, IARC, Lyon, 1976.
15. Taylor, J. A. Bracken: an increasing problem and a threat to health. Outlook Agric., *10*: 298–304, 1980.
16. Evans, H. C. Personal communication.
17. Van der Hoeven, J.C.M. and Van Leeuwen, F. E. Isolation of a mutagenic fraction from bracken (*Pteridium aquilinum*). Mutat. Res., *79*: 377–380, 1980.
18. Van der Hoeven, J.C.M., Lagerwey, W. J., Meeuwissen, C.A.J.M., Hauwert, P.C.M., Voragen, A.G.J., and Koeman, J. H. Mutagens in food products of plant origin. *In;* M. Sorsa and H. Vainio (eds.), Mutagens in Our Environment, pp. 327–338, A. R.

Liss Inc., New York, 1982.

19. Van der Hoeven, J.C.M., Lagerwey, W. J., Posthumus, M. A., Van Veldhuizen, A., and Holterman, H.A.J. Aquilide A, a new mutagenic compound isolated from bracken fern (*Pteridium aquilinum (L.)* Kuhn). Carcinogenesis, *4*: 1587–1590, 1983.
20. Fukuoka, M., Kuroyanagi, M., Yoshira, K., Natori, S., Nagao, M., Takahashi, Y., and Sugimura, T. Chemical and toxicological studies on bracken fern, *Pteridium aquilinum* var. *latiusculum*. IV. Surveys on bracken constituents by mutagen test. J. Pharm. Dyn., *1*: 324–331, 1978.
21. Hirono, I., Yamada, K., Niwa, H., Shizuri, Y., Ojika, M., Hosaka, S., Yamaji, T., Wakamatsu, K., Kigoshi, H., Niiyama, K., and Yosaki, Y. Separation of carcinogenic fraction of bracken fern. Cancer Lett., *21*: 239–246, 1984.
22. Hirono, I., Aiso, S., Yamaji, T., Mori, H., Yamada, K., Niwa, H., Ojika, M., Wakamatsu, K., Kigoshi, H., Niiyama, K., and Uosaki, Y. Carcinogenicity of ptaquiloside isolated from bracken. Gann, *75*: 833–836, 1984.
23. Hirono, I., Kono, Y., Takahashi, K., Yamada, K., Niwa, H., Ojika, M., Kigoshi, H., Niiyama, K., and Uosaki, Y. Reproduction of acute bracken poisoning in a calf with ptaquiloside, a bracken constituent. Vet. Rec., *115*: 375–378, 1984.
24. Niwa, H., Ojika, M., Wakamatsu, K., Yamada, K., Hirono, I., and Matsushita, K. Ptaquiloside, a novel norsesquiterpene glucoside from bracken, *Pteridium aquilinum* var. *latiusculum*. Tetrahedron Lett., *24*: 4117–4120, 1983.
25. Bradshaw, A.P.W., Hanson, A. R., and Sadler, I. H. Studies in terpenoid biosynthesis. Part 26. Application of ^{1}H and ^{11}C NMR spectroscopy to the biosynthesis of the illudin sesquiterpenoids. J. Chem. Soc. Perkin. Trans I., 2445–2448, 1982.
26. Stothers, J. B. Carbon-13 NMR Spectroscopy, p. 333, Academic Press, New York, 1972.
27. Jarret, W.F.H., McNeil, P. E., Grimshaw, W.T.R., Selman, I. E., and McIntyre, W.I.M. High incidence area of cattle cancer with a possible interaction between an environmental carcinogen and a papilloma virus. Nature, *274*: 215–217, 1978.
28. Bryan, G. T. and Pamukcu, A. M. Sources of carcinogens and mutagens in edible plants: production of urinary bladder and intestinal tumors by bracken fern (*Pteridium aquilinum*). *In;* H. F. Stich (ed.), Carcinogens and Mutagens in the Environment, vol. I, Food Products, pp. 75–82, CRC Press, Inc., Boca Raton, Fl, 1982.
29. MacGregor, J. T. Mutagenic activity of hymenovin, a sesquiterpene lacton from Western bitterweed. Food Cosmet. Toxicol., *15*: 225–227, 1977.
30. Van der Hoeven, J.C.M., Lagerwey, W. J., Bruggeman, I. M., Voragen, F. G., and Koeman, J. H. Mutagenicity of extracts of some vegetables commonly consumed in The Netherlands, J. Agric. Food Chem., *31*: 1020–1026, 1983.
31. Van der Hoeven, J.C.M., Bruggeman, I. M., and Debets, F.M.H. Genotoxicity of quercetin in cultured mammalian cells. Mutat. Res., *136*: 9–21, 1984.
32. Bruggeman, I. M. and Van der Hoeven, J.C.M. Lack of activity of the bacterial mutagen emodin in HGPRT and SCE assay with V79 Chinese hamster cells. Mutat. Res., *138*: 219–224, 1984.
33. Piacek-Llanes, B. G. and Tannenbaum, S. R. Formation of an activated N-nitroso compound in nitrite treated fava beans (*Vicia faba*). Carcinogenesis, *3*: 1379–1384, 1983.
34. Correa, P., Cuello, C, Fajardo, L. F., Haenszel, W., Bolanos, O., and de Ramirez, B. Diet and gastric cancer: nutrition survey in a high-risk area. J. Natl. Cancer Inst., *70*: 673–678, 1983.
35. Van der Hoeven, J.C.M., Lagerwey, W. J., Van Gastel. A., Huitink, J., De Dreu, R.,

and Van Broekhoven, L. W. Intercultivar difference with respect to mutagenicity of fava beans (*Vicia faba L.*) after incubation with nitrite. Mutat. Res., *130*: 391–394, 1984.

36. Paine, A. J. The maintenance of cytochrome P-450 in liver cell culture: a prerequisite to the study of drug metabolism and toxicity. ATLA Abstr., *8*: 86–91, 1980.
37. Althaus, F. R., Sinclair, J. F., Sinclair, P. R., and Meyer, U. A. Drug-mediated induction of cytochrome(s) P-450 and drug metabolism in cultured hepatocytes maintained in chemically defined medium. J. Biol. Chem., *25*: 2148–2153, 1979.
38. Debets, F.M.H., Hamers, W.J.H.M.B., and Strik, J.J.T.W.A. Metabolism as a prerequisite for the prophyrinogenic action of polyhalogenated aromatics, with special reference to hexachlorobenzene and polybrominated biphenyls. Int. J. Biochem., *12*: 1019–1024, 1980.
39. Sinclair, J. F., Sinclair, P. R., and Bonkowsky, H. L. Hormonal requirements for the induction of cytochrome P-450 in hepatocytes cultured in a serum-free medium. Biochem. Biophys. Res. Commun., *86*: 710–717, 1979.
40. Granick, S. The induction *in vitro* of the synthesis of δ-aminolevulinic acid synthetase in chemical porphyria: a response to certain drugs, sex hormones and foreign chemicals. J. Biol. Chem., *241*: 1359–1375, 1966.
41. Sassa, S. and Kappas, A. Induction of δ-aminolevulinate synthase and porphyrins in cultured liver cells maintained in chemically defined medium. J. Biol. Chem., *252*: 2428–2436, 1977.
42. Giger, U. and Meyer, U. A. Induction of δ-aminolevulinate synthase and cytochrome P-450 hemoproteins in hepatocyte culture. J. Biol. Chem., *256*: 11182–11190, 1981.
43. Jongen, W.M.F., Hakkert, B. C., and Van der Hoeven, J.C.M. Genotoxicity testing of cigarette-smoke condensate in the SCE and HGPRT assays with V79 Chinese hamster cells. Food Chem. Toxicol., *23*: 603–607, 1985.
44. Bruggeman, I. M. and Van der Hoeven, J.C.M. Induction of SCEs by some pyrrolizidine alkaloids in V79 Chinese hamster cells co-cultivated with chick embryo hepatocytes. Mutat. Res., *142*: 209–212, 1985.
45. Bull, L. B., Culvenor, C.C.J., and Dick, A. T. The pyrrolizidine alkaloids, North Holland, Amsterdam, 1968.
46. International Agency for Research on Cancer. Monographs on the Evaluation of Carcinogenic Risk of Chemicals to Humans. Some Food Additives, Feed Additives and Naturally Occurring Substances, IARC, Lyon, 1976.
47. Niwa, H., Ishiwata, H., and Yamada, K. Separation and determination of macrocyclic pyrrolizidine alkaloids of the otonecine type present in the edible plant *Petasites japonicus* by reversed-phase high-performance liquid chromatography. J. Chromatogr., *257*: 146–150, 1983.
48. Schoental, R. Carcinogens in plants and micro-organisms. *In;* C. E. Searle (ed.), Chemical Carcinogens, ACS Monograph 173, pp. 626–689, Washington, D.C., 1976.
49. Schoental, R. Health hazards of pyrrolizidine alkaloids, a short review. Toxicol. Lett., *10*: 323–326, 1982.
50. Yamanaka, H., Nagao, M., Sugimura, T., Furuya, T., Shirai, A., and Matsushima, T. Mutagenicity of pyrrolizidine alkaloids in the *Salmonella*/mammalian microsome test. Mutat. Res., *68*: 211–216, 1979.
51. Culvenor, C.C.J., Edgar, J. A., and Smith, L. W. Pyrrolizidine alkaloids in honey from *Echium plantagineum* (*L.*). J. Agric. Food Chem., *29*: 958–960, 1981.
52. Schmid, P. P., Zweifel, U., Lüthy, J., and Schlatter, C. Determination and identification of senecio alkaloids in food of animal origin. Interdisciplinary conference on food

toxicology, Zürich, Switzerland, 13–15 October, 1982.

53. Terwel, L. and Van der Hoeven, J.C.M. Antimutagenic activity of some naturally occurring compounds towards cigarette-smoke condensate and benzo(a)pyrene in the *Salmonella*/microsome assay. Mutat. Res., *152*: 1–4, 1985.

DIET, NUTRITION AND CANCER, Y. HAYASHI ET AL. (EDS.),
JAPAN SCI. SOC. PRESS, TOKYO/VNU SCI. PRESS, UTRECHT, PP. 139-145, 1986

Human Carcinogenic Risk in the Use of Bracken Fern

Iwao HIRONO

Department of Pathology, Fujita-Gakuen Health University School of Medicine, Toyoake 470-11, Japan

Abstract: Young bracken fern is used as a human food in Japan and other countries. However, it has been demonstrated that bracken is carcinogenic to cattle and laboratory animals such as rats, mice, and guinea pigs. Rats fed a diet containing bracken fern developed tumors of the ileum and urinary bladder. Mammary cancer was also induced in Sprague-Dawley rats fed a bracken diet. Study of the human cancer risk of this plant is thus of great importance. The fern is usually used as a foodstuff in Japan after it is treated with plain boiling water or boiling water containing wood ash or sodium bicarbonate; sometimes, bracken is pickled in salt. Although the carcinogenic activity of processed bracken thus prepared was reduced markedly, weak activity was still retained. This paper deals with the carcinogenicity of unprocessed and processed bracken fern to laboratory animals and its human cancer risk.

The research on naturally occurring carcinogens of plant origin was first started on carcinogenic pyrrolizidine alkaloid in plants of the tribe Senecioneae about thirty years ago. Subsequently, carcinogenicity was found in cycad, bracken fern, comfrey, and several other plants; their carcinogenic principles were also clearly demonstrated. All these plants have long been used as human food or herbal remedies in Japan and other countries; their carcinogenic activity has therefore attracted great attention as environmental carcinogens present in daily life. The most important problem is to elucidate the possible role of these plant materials as causative principles of human cancer. Young bracken fern is known to be the most popular edible wild plant in Japan. Its carcinogenicity, the influence of cooking on the carcinogenic activity, the possible human hazard of bracken carcinogen and assessment of the plant's human cancer risk are discussed in this paper.

Carcinogenicity of Bracken Fern

Bracken fern, *Pteridium aquilinum*, is widely distributed in many parts of the world. It is well known that livestock are frequently affected after ingestion of a sufficient quantity of the plant, and the problem of its toxicity has been recognized in the fields of stock-breeding and dairy farming since the end of the last century (*1–4*). The predominant feature of cattle bracken-poisoning is severe damage to bone marrow activity, which gives rise to leucopenia, thrombocytopenia, the hemorrhagic syndrome, and hematuria. Based on these findings, Evans and Mason (*5*) speculated that the causative principle contained in bracken fern is probably radiomimetic and started a feeding experiment in rats. They reported in 1965 that rats fed a diet containing bracken fern developed multiple ileal adenocarcinomas. Subsequently, the simultaneous induction of urinary bladder tumors and the occurrence of mammary carcinoma in Sprague-Dawley rats fed a bracken diet have been reported by Pamukcu and Price (*6*) and Hirono *et al.* (*7*, *8*), respectively. It was also confirmed that rats fed a bracken diet developed not only ileal epithelial tumors, such as adenoma and adenocarcinoma, but also ileal sarcoma (*7*).

The nature of the carcinogen contained in bracken fern has not been elucidated. Recently, a carcinogenic substance, ptaquiloside (Fig. 1), a novel norsesquiterpene glucoside of the illudane type was successfully isolated (*9–12*) and was demonstrated to be the substance causing bracken poisoning in cattle (*13*). Young braken fern is used as a human food and sometimes as an herbal remedy in Japan, China, Korea, and other countries. The top of the frond of the young fern is still curled and in the crosier or fiddlehead stage of growth when eaten; these young bracken ferns were also confirmed to be carcinogenic (*7*). The carcinogenic activity of bracken fern is influenced by geographic conditions and the incidence of intestinal tumors varied from 72% to 100% depending on collection site (Table 1). It is conceivable that weather conditions also influence this carcinogenic activity. Furthermore, the relationship between the administration of a bracken diet and incidence of tumors was studied

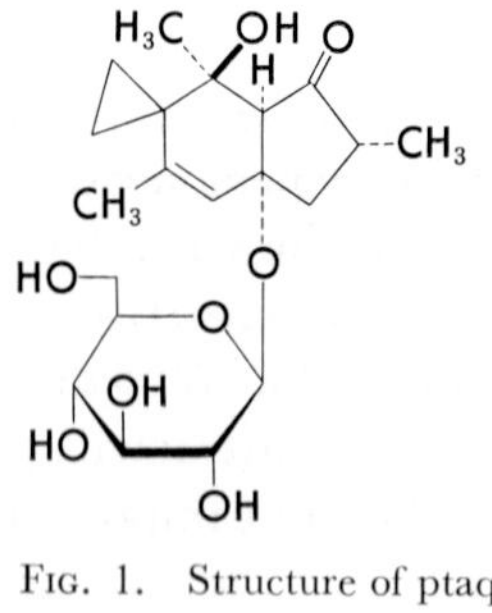

FIG. 1. Structure of ptaquiloside.

TABLE 1. Carcinogenicity in ACI Rats from Bracken Collected at Different Sites

Bracken[a] diet	Collection of bracken		Administration period (days)	Incidence of intestinal tumors
	Site	Season		
Proportion of bracken to basal diet 1: 2 by weight	Hokkaido, Hitaka	May–June	120	24/24 (100.0%)
	Gifu, Okumyogata	April–June	,,	10/12 (83.3%)
	,, Takayama	April–June	,,	13/18 (72.2%)

[a] Young bracken fern

TABLE 2. Carcinogenic Activity of Bracken Fern in ACI Rats

Bracken[a] diet	Administration period (days)	Total amount of dry bracken ingested per rat	Estimate as fresh bracken	Incidence of intestinal tumors
Proportion of bracken to basal diet 1:2 by weight	120	400 g	2.8–4.0 kg	24/24 (100%)
	60	200 g	1.4–2.0 kg	18/23 (78%)
	20	67 g	470–670 g	10/20 (50%)

[a] Young bracken fern.

TABLE 3. Difference in Susceptibility to Young Bracken Fern Carcinogenicity among Different Species of Animals

Bracken diet	Animal	Age	Administration period (days)	Incidence of intestinal tumors
Proportion of bracken to basal diet 1:2 by weight	Rat (ACI)	4 weeks	120	72–100%
	,, (Wistar)	5 ,,	480	94%
	Mouse (C57BL/6)	4 ,,	120	32%
	,, (dd)	4 ,,	120	0 (lung adenoma 70%)
	Hamster	6 ,,	120	16%
	Guinea pig[a]	200 g body weight	300	83%

[a] From ref. *18*.

in ACI rats fed a diet containing bracken powder in a weight proportion of 1 part bracken to 2 parts basal diet. The tumor incidence in these animals fed for 120 days, 60 days, and 20 days was 100%, 78%, and 50%, respectively. Thus, as short a period as 20 days caused intestinal tumors in 50% of the animals (Table 2), suggesting that the carcinogenicity of bracken fern is considerable. Carcinogenic not only to rats (*5–8*) and cattle (*14*), but also to mice (*15, 16*), hamsters (*15, 17*), guinea pigs (*15, 18*), and Japanese quail (*15*), the highest susceptibility to the fern was observed in rats, and the lowest in hamsters. Mice of dd strain did not develop intestinal tumors but had a 70% incidence of lung adenoma. Guinea pigs developed intestinal and urinary bladder tumors in nearly the same high incidence as rats (Table 3).

Young bracken fern is generally used as a food in Japan after being processed by one of the following treatments: 1. Fresh bracken is immersed in plain boiling water or in boiling water containing wood ash or sodium bicarbonate; 2. Fresh bracken is pickled in salt, and then immersed in boiling water and washed with running water before use. Although reduced markedly, weak carcinogenic activity of processed bracken thus prepared was retained (*19*). The carcinogenic activity of processed bracken was studied in ACI rats. Whereas tumor incidence in rats fed a diet containing unprocessed bracken was 78.5%, it was 25%, 10%, and 4.7% in rats fed diets containing processed bracken treated with wood ash, sodium bicarbonate, and sodium chloride, respectively. The tumor incidence in those fed a diet containing processed bracken treated with plain boiling water was 66.6%, while the same unprocessed bracken diet induced tumors in 100%. It is thus evident that treatment with wood ash or sodium bicarbonate remarkably reduces the carcinogenic activity of bracken fern compared to treatment with plain boiling water. This may be due to the decomposition of ptaquiloside, a bracken carcinogen, by the alkalinity of the wood ash or sodium bicarbonate used in the treatment (*9*). Not only bracken carcinogen but also other naturally occurring carcinogens of green plant origin can be eliminated by treatment with boiling water. These include cycasin, a carcinogen contained in cycads, and carcinogenic pyrrolizidine alkaloids such as petasitenine, symphytine, and senkirkine, which are contained in petasites, comfrey, and coltsfoot, respectively (*20*, *21*). Symphytine and senkirkine as pure alkaloids are almost insoluble in boiling water; however, these pyrrolizidine alkaloids can be eliminated from plant materials in which they exist by treatment with boiling water. The mechanism involved in such elimination is yet unknown (*22*).

Epidemiology of the Carcinogenicity of Bracken Fern

Kamon and Hirayama (*23*) made an epidemiologic survey of cancer in a mountainous area of central Japan where residents eat large amounts of bracken fern. They reported a significantly higher risk of esophageal cancer in people who ate either hot tea gruel or bracken fern every day; the risk was particularly high when both types of food were consumed. It was also reported by Hirono et al. (*24*) that rats fed a diet containing 30% bracken powder for 33 weeks developed tumors of the upper alimentary tract.

The possible human hazard of bracken fern carcinogen has been indicated by several researchers (*25–27*), especially its transfer into milk. A calf given milk from a cow receiving a sublethal dietary bracken supplement showed a hematological response which would be typical of a calf directly consuming bracken at a low rate (*26*). Pamukcu *et al.* (*27*) studied the carcinogenicity of the milk of bracken fern-fed cows. The milk obtained was fed to rats as either fresh or freeze-dried powdered milk mixed with a grain diet. Groups receiving both fresh and powdered diets developed small intestinal, renal, or urinary bladder carcinomas, while none of the rats fed either fresh or powdered milk from cows receiving a normal (non-bracken fern) diet displayed neoplasia of those organs. Milk from cows that have eaten bracken fern may thus be hazardous to humans. Pamukcu and Bryan (*28*) reported that indirect human exposure to bracken fern through milk and dairy products is most likely to occur in areas of the world such as Turkey, Bulgaria, Yugoslavia, and Colombia, and other regions where grazing cattle ingest the plant, especially under free range conditions, and where the dried fern is used for bedding in winter. It has also been speculated that rain may leach carcinogens from the plant and eventually transport it into human water supplies. Recently, Evans (*29*) has been studying the carcinogenicity of bracken spores. Bracken is a sporophyte, and spores are shot out by a catapult mechanism from the sporangium. It is obvious that they can become airborne and inhaled in the same way as pollen grains and mold spores. Evans has therefore been testing bracken spores by giving them orally to mice. The experiment has not yet been completed.

The Japanese probably consume the largest amount of bracken fern of any peoples in the world. The development of transportation and the technical improvement of storage in recent years have made it possible to obtain natural products such as bracken fern and other wild plants in any season and anywhere in Japan. The time-honored ways of cooking bracken fern, *e.g.*, treatment with wood ash or sodium bicarbonate, are being forgotten among the young generation. Moreover, one of the factors making it easy to obtain the fern is its import in large amounts from foreign countries, the Soviet Union, the People's Republic of China, and North Korea. An average of 13,000 tons of bracken fern is imported each year (from "Imports of commodities by country" published by the Customs Bureau of the Ministry of Finance, Japan). All are imported as salt bracken and enter the market in small packages.

Assessment of Human Cancer Risk of Bracken Fern

It is generally difficult to estimate the human cancer risk of a carcinogen based on the results of carcinogenicity examinations of laboratory animals for the following reasons: the differences in life span and susceptibility to carcinogens between animals and human beings, multiple causal factors of human cancer, and the presence of factors inhibiting human carcinogenesis. Moreover, the dose and duration of exposure of human beings to a certain carcinogen cannot be accurately assessed. One approach to quantitative risk assessment of carcinogens is to estimate the virtually safe dose (VSD), *i.e.*, carcinogenic doses at extremely low risk levels, by downward extrapolations of animal dose-response data. Various mathematical models such as the one-hit model, probit model, logit model, multi-hit model, or Weibull model have been proposed for this purpose. For estimation of the VSD of bracken fern, carcinogenicity examination should be performed in rats fed a diet containing bracken at various low concentrations; however, such an experiment has not yet been attempted. Even if the VSD were estimated, there would still be puzzling problems such as the probable multi-causal nature of human cancer and the high susceptibility to certain carcinogens in aged or diseased people, in addition to the difficulty in selecting the most appropriate mathematical model. It is now clear from epidemiological data that most human cancers are induced by environmental causes. It is probable that even a small amount of carcinogen can induce tumors when effective enhancing agents are present in the environment. The possibility of cancer development by a certain carcinogen is increased when modifiers, such as the habitual use of a specific food or medicine or occupational conditions, provide a favorable conditions for that development, even though the amount of carcinogen exposure is small.

Using the scoring system proposed by Squire (*30*) for ranking animal carcinogens, bracken and ptaquiloside were tested as 86 and 90 respectively, thereby being ranked in the first class and subject to restriction or ban. It may be more important, however, to estimate the amount and duration of the ingestion of bracken fern by humans to evaluate human risk. It is considered best to avoid the use of these plants as food or herbal remedies. Reduction of the total amount of carcinogens in the diet including carcinogenic natural products may eventually contribute to the prevention of cancer.

REFERENCES

1. Evans, W. C., Evans, E.T.R., and Hughes, L. E. Studies on bracken poisoning in cattle. Part 1. Br. Vet. J., *110*: 295–306, 1954.
2. Evans, W. C., Evans, I. A., Chamberlain, A. G., and Thomas, A. J. Studies on bracken poisoning in cattle. Part VI. Br. Vet. J., *115*: 83–85, 1959.
3. Rosenberger, G. Nature, manifestations, cause and control of chronic enzootic haematuria in cattle. Vet. Med. Rev., No. 2/3, 189–206, 1971.
4. Evans, W. C., Patel, M. C., and Koohy, Y. Acute bracken poisoning in homogastric and ruminant animals. Proc. R. Soc. Edinb., *81B*: 29–64, 1982.
5. Evans, I. A. and Mason, J. Carcinogenic activity of bracken. Nature (Lond.), *208*: 913–914, 1965.
6. Pamukcu, A. M. and Price, J. M. Induction of intestinal and urinary bladder cancer in rats by feeding bracken fern (*Pteris aquilina*). J. Natl. Cancer Inst., *43*: 275–281, 1969.
7. Hirono, I., Shibuya, C., Fushimi, K., and Haga, M. Studies on carcinogenic properties of bracken, *Pteridium aquilinum*. J. Natl. Cancer Inst., *45*: 179–188, 1970.
8. Hirono, I., Aiso, S., Hosaka, S., Yamaji, T., and Haga, M. Induction of mammary cancer in CD rats fed bracken diet. Carcinogenesis, *4*: 885–887, 1983.
9. Niwa, H., Ojika, M., Wakamatsu, K., Yamada, K., Hirono, I., and Matsushita, K. Ptaquiloside, a novel norsesquiterpene glucoside from bracken, *Pteridium aquilinum* var. *latiusculum*. Tetrahedron Lett., *24*: 4117–4120, 1983.
10. Niwa, H., Ojika, M., Wakamatsu, K., Yamada, K., Ohba, S., Saito, Y., Hirono, I., and Matsushita, K. Stereochemistry of ptaquiloside, a novel norsesquiterpene glucoside from bracken, *Pteridium aquilinum* var. *latiusculum*. Tetrahedron Lett., *24*: 5371–5372, 1983.
11. Hirono, I., Yamada, K., Niwa, H., Shizuri, Y., Ojika, M., Hosaka, S., Yamaji, T., Wakamatsu, K., Kigoshi, H., Niiyama, K., and Uosaki, Y. Separation of carcinogenic fraction of bracken fern. Cancer Lett., *21*: 239–246, 1984.
12. Hirono, I., Aiso, S., Yamaji, T., Mori, H., Yamada, K., Niwa, H., Ojika, M., Wakamatsu, K., Kigoshi, H., Niiyama, K., and Uosaki, Y. Carcinogenicity in rats of ptaquiloside isolated from bracken. Gann, *75*: 833–836, 1984.
13. Hirono, I., Kono, Y., Takahashi, K., Yamada, K., Niwa, H., Ojika, M., Kigoshi, H., Niiyama, K., and Uosaki, Y. Reproduction of acute bracken poisoning in a calf with ptaquiloside, a bracken constituent. Vet. Rec., *115*: 375–378, 1984.
14. Pamukcu, A. M., Göksoy, S. K., and Price, J. M. Urinary bladder neoplasms induced by feeding bracken fern (*Pteris aquilina*) to cows. Cancer Res., *27*: 917–924, 1967.
15. Evans, I. A. The radiomimetic nature of bracken toxin. Cancer Res., *28*: 2252–2261, 1968.
16. Hirono, I., Sasaoka, I., Shibuya, C., Shimizu, M., Fushimi, K., Mori, H., Kato, K., and Haga, M. Natural carcinogenic products of plant origin. Gann Monogr. Cancer Res., *17*: 205–217, 1975.
17. Hirono, I. Natural carcinogenic products of plant origin. CRC Crit. Rev. Toxicol., *8*: 235–277, 1981.
18. Ushijima, J., Matsukawa, K., Yuasa, A., and Okada, M. Toxicities of bracken fern in guinea pigs. Jpn. J. Vet. Sci., *45*: 593–602, 1983.
19. Hirono, I., Shibuya, C., Shimizu, M., and Fushimi, K. Carcinogenic activity of processed bracken used as human food. J. Natl. Cancer Inst., *48*: 1245–1250, 1972.
20. Nishida, K., Kobayashi, A., and Nagahama, T. Cycasin, a new toxic glycoside of

Cycas revoluta Thunb. I. Isolation and structure of cycasin. Bull. Agric. Chem. Soc. Jpn., *19*: 77–84, 1955.

21. Niwa, H., Ishiwata, H., and Yamada, K. Separation and determination of macrocyclic pyrrolizidine alkaloid of the otonecine type present in the edible plant *Petasites japonicus* by reversed-phase high-performance liquid chromatography. J. Chromatogr., *257*: 146–150, 1983.
22. Yamada, K. Personal communication.
23. Kamon, S. and Hirayama, T. Epidemiology of cancer of the oesophagus in Miye, Nara and Wakayama prefecture with special reference to the role of bracken fern. Proc. Jpn. Cancer Assoc., *34*: 211, 1975.
24. Hirono, I., Hosaka, S., and Kuhara, K. Enhancement by bracken of induction of tumors of the upper alimentary tract by N-propyl-N-nitrosourethan. Br. J. Cancer, *46*: 423–427, 1982.
25. Evans, I. A., Widdop, B., Jones, R. S., Barber, G. D., Leach, H., Jones, D. L., and Mainwaring-Burton, R. The possible human hazard of the naturally occurring bracken carcinogen. Biochem. J., *124*: 28 p–29 p, 1971.
26. Evans, I. A., Jones, R. S., and Mainwaring-Burton, R. Passage of bracken fern toxicity into milk. Nature (Lond.), *237*: 107–108, 1972.
27. Pamukcu, A. M., Ertürk, E., Yalciner, S., Milli, U., and Bryan, G. T. Carcinogenic and mutagenic activities of milk from cows fed bracken fern (*Pteridium aquilinum*). Cancer Res., *38*: 1556–1560, 1978.
28. Pamukcu, A. M. and Bryan, G. T. Bracken fern, a natural urinary bladder and intestinal carcinogen. *In;* E. C. Miller et al. (eds.), Naturally Occurring Carcinogens-Mutagens and Modulators of Carcinogenesis, pp. 89–99 Jpn. Sci. Soc. Press, Tokyo/ Univ. Park Press, Baltimore, 1979.
29. Evans, I. A. Bracken carcinogenicity. *In;* C. E. Searle (ed.), Chemical Carcinogens, vol. 2, ACS Monogr., 182, pp. 1171–1204, American Chemical Society, Washington, D.C., 1984.
30. Squire, R. A. Ranking animal carcinogens: a proposed regulatory approach. Science, *214*: 877–880, 1981.

MODULATION OF CARCINOGENESIS BY DIETARY COMPONENTS IN EXPERIMENTAL ANIMALS

DIET, NUTRITION AND CANCER, Y. HAYASHI ET AL. (EDS.),
JAPAN SCI. SOC. PRESS, TOKYO/VNU SCI. PRESS, UTRECHT, PP. 149–158, 1986

Suppression of Carcinogenesis by Retinoids: Interactions with Peptide Growth Factors and Their Receptors as a Key Mechanism

Michael B. Sporn and Anita B. Roberts

Laboratory of Chemoprevention, Division of Cancer Etiology, National Cancer Institute, Bethesda, Maryland 20892, U.S.A.

Abstract: To understand the molecular mechanism of action of retinoids in control of differentiation and carcinogenesis, it is now necessary to consider the interactions of retinoids with oncogenes, as well as with peptide growth factors and their receptors. Experimental studies in these areas are described. In particular, it is shown that retinoic acid inhibits the proliferative effects of transforming growth factor-beta (TGF-beta) and platelet-derived growth factor (PDGF) on fibroblasts that have been transfected with the c-*myc* oncogene. The ability of retinoic acid to induce differentiation in several types of human and murine tumor cells also is associated with the ability of retinoic acid to suppress the expression of either the c-*myc* oncogene, or the related N-*myc* gene.

The usefulness of retinoids for prevention of cancer in experimental animals is now well established (*1*), and it is only a matter of time before the practical application of these agents for the same purpose in men and women will also be widespread. Considering the very large number of both natural and synthetic retinoids that are now available, and the world-wide level of effort that is being devoted to this problem, it would appear inevitable that retinoids will find increasing application for the prevention of human cancer. There are two general possibilities as to how this might happen. The first application would be merely to use very low (and toxicologically essentially safe, without any significant risk) doses of retinoids to provide some baseline level of protection to very large numbers of people who are at relatively low risk for development of cancer; this is essentially a "physiological," epidemiologic approach to cancer prevention. This approach is very costly and difficult to evaluate, since the results of clinical trials with retinoids given at such low doses may take years to obtain. The second application involves the use of retinoids, mostly synthetic, at much higher doses in populations at significantly higher risk for development of cancer; this is essentially a "pharmacological" approach to prevention of cancer. The higher doses of retinoids that are used in this approach of course entail a greater toxicological risk to the human subjects, but this can be ethically justified if the risk of development of cancer is high enough, as in women at ultra-high risk for development of pre-menopausal breast cancer, or men and women at ultra-high risk for development of bladder cancer. Furthermore, advances in the synthetic chemistry of retinoids have led to the development of much more effective and less toxic agents; thus, the opportunities to use synthetic retinoids for human cancer prevention will increase. We have published numerous reviews of these issues and will not discuss them further here (*2–5*).

Molecular Mechanisms of Action of Retinoids

The remainder of this article will deal with the question of the molecular mechanisms of action of the retinoids in cancer prevention. Although it has sometimes been suggested that the "physiological" and "pharmacological" approaches to cancer prevention with retinoids involve separate and different molecular mechanisms, there is no real evidence that this is the case (*6*), and it is scientifically more reasonable to assume that there is a common molecular mechanism (*7*). Further understanding of this mechanism will obviously lead to the more intelligent use of retinoids for cancer prevention.

Since carcinogenesis may be considered essentially as a disorder of cell differentiation, the overall scientific problem of the role of retinoids in prevention of cancer is essentially identical to their role in control of cell differentiation. In two reviews that we have written (*6, 7*), we have suggested that three areas of laboratory investigation, previously separate, are beginning to merge to provide useful new approaches to understanding the mechanism of action of the retinoids in control of cell differentiation. These three areas of laboratory investigation are shown in the Venn diagram in Fig. 1 and are as follows: 1) oncogene studies, 2) studies of peptide growth factors and their receptors, and 3) studies of the mechanism of action of small effector mole-

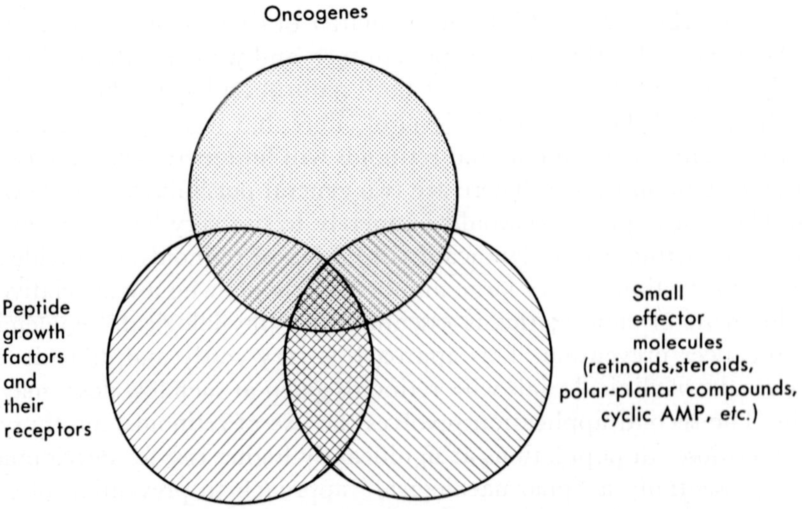

FIG. 1. Venn diagram, showing relationships of oncogenes, peptide growth factors and their receptors, and small effector molecules such as retinoids.

cules, such as retinoids, steroids, and dimethyl sulfoxide, which are capable of altering cellular differentiation. Ultimately, the problem of the molecular mechanisms of action of retinoids in controlling differentiation and carcinogenesis is converging on one of the central problems in all of biology, namely the control of gene expression, and the final elucidation of the role of retinoids in the cancer problem will depend on further advances in molecular and cellular genetics.

There is relatively little information about the role of retinoids in controlling oncogene expression, although several important observations have already been made. One of the best ways to investigate this problem is to study the ability of retinoids to induce differentiation in fully neoplastic tumor cells. The first significant finding in this area was the ability of all-*trans*-retinoic acid to suppress the expression of the c-*myc* oncogene in HL-60 leukemia cells (*8*) or F9 teratocarcinoma cells (*9*), both of which are very sensitive to induction of differentiation by retinoic acid (*10*). Although the specific molecular function for the nuclear protein encoded by the c-*myc* gene has not yet been elucidated, it is presumed that excessive expression of this gene is correlated with the excessive proliferation and arrested differentiation of the HL-60 cell. A second related, but different, *myc* gene, N-*myc*, has also been identified in several tumors (*11*) and may also represent a target for the action of retinoic acid. Indeed, in the case of differentiation of neuroblastoma cells induced by retinoic acid (*12, 13*), it has recently been shown that retinoic acid causes marked inhibition of the expression of the N-*myc* gene, rather than the c-*myc* gene (*14*); this is a very early event that precedes the cell cycle changes and the marked ourgrowth of neurites that are also caused by retinoic acid.

Since one of the principal functions of oncogenes is to modulate the cellular pathways controlled by growth factors and their receptors (*15*), it is not surprising that it has also been demonstrated that retinoic acid can modulate the actions of specific peptide growth factors and their specific membrane-bound receptors in susceptible cells. There have been a particularly large number of studies involving control by retinoids of the cellular effects of epidermal growth factor (EGF) and its homolog, transforming growth factor-alpha (TGF-alpha), and of the expression of the common receptor for both these peptides, generally known as the EGF receptor (this receptor may also rightfully be termed the "TGF-alpha receptor," since it would presently appear that there is no significant difference in the biological activity of the two peptides and that TGF-alpha, rather than EGF, is expressed earlier in embryological development). Since the EGF (TGF-alpha) receptor is partially encoded by the *erb*-B oncogene, studies in this area are of particular relevance to the phenomena of carcinogenesis and differentiation. The earliest studies in this area were done by Jetten and colleagues (*16, 17*, reviewed in ref. *18*), who showed that retinoic acid treatment caused a nearly 10-fold increase in the number of available EGF receptor sites in a variety of cells. The simplest hypothesis for this phenomenon is that retinoic acid enhances the synthesis of the specific mRNA for the EGF receptor, although there are undoubtedly other mechanisms which need to be considered, since this particular receptor is under a number of different physiological controls. With the availability of the specific gene probe for the receptor, it should be possible to obtain a definitive answer to this question in the near future.

Interactions of Retinoids with Peptide Growth Factors

In our own laboratory, we have worked with several different cell lines (*19*, *20*) to study effects of retinoids on peptide growth factors. Our first studies were performed

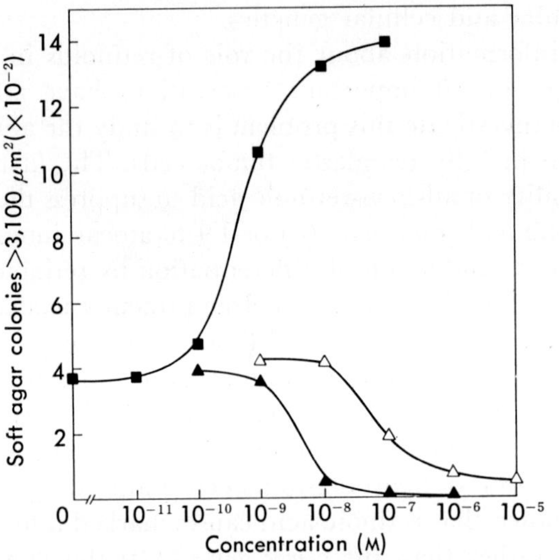

FIG. 2. Effects of retinoic acid and glucocorticoids on anchorage-independent growth of NRK cells in the presence of EGF and TGF-beta. Soft-agar assays were carried out in the presence of EGF (5 ng/ml) and TGF-beta (50 pg/ml). ■, retinoic acid; ▲, dexamethasone; △, deoxycorticosterone. See ref. *19* for details.

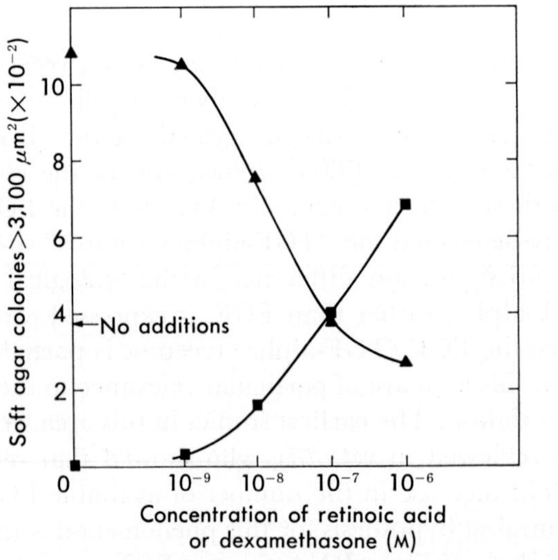

FIG. 3. Antagonistic interaction of retinoic acid and dexamethasone on colony formation of NRK cells in soft agar. The assay was carried out in the presence of EGF (5 ng/ml) and TGF-beta (60 pg/ml). ▲, retinoic acid was held constant at 10^{-7} M while the concentration of dexamethasone was varied; ■, dexamethasone was held constant at 10^{-7} M while the concentration of retinoic acid was varied. Arrow, number of colonies formed in the presence of EGF and TGF-beta alone. See ref. *19* for details.

with the non-neoplastic rat kidney fibroblast cell line, NRK 49F (*19*). These cells require two peptide growth factors in order to grow in an anchorage-independent manner (in soft agar), which is considered an excellent marker of the transformed (neoplastic) phenotype. These two peptides have been termed transforming growth factor-alpha (TGF-alpha) and transforming growth factor-beta (TGF-beta); EGF can substitute equally well for TGF-alpha in this assay. Although TGF-alpha and TGF-beta share a common nomenclature, they are entirely distinct chemical entities, with little in the way of common properties (*21*, *22*). Figure 2 shows that retinoic acid causes a dose-dependent increase in the formation of colonies of NRK cells in soft agar, while the potent glucocorticoid, dexamethasone, suppresses the growth of these cells in soft agar; the effects of retinoic acid and dexamethasone are directly antagonistic, as shown in Fig. 3. As had been shown by Jetten (*18*) for many other cells, retinoic acid markedly enhances the binding of labelled EGF to NRK cells, while dexamethasone decreases this binding, as shown in Fig. 4. Indeed, it was possible to correlate the biological activities (with respect to cell proliferation) of retinoic acid and dexamethasone in this system with their ability to modulate the binding of EGF to its receptor. It is important to emphasize that while the EGF (TGF-alpha)

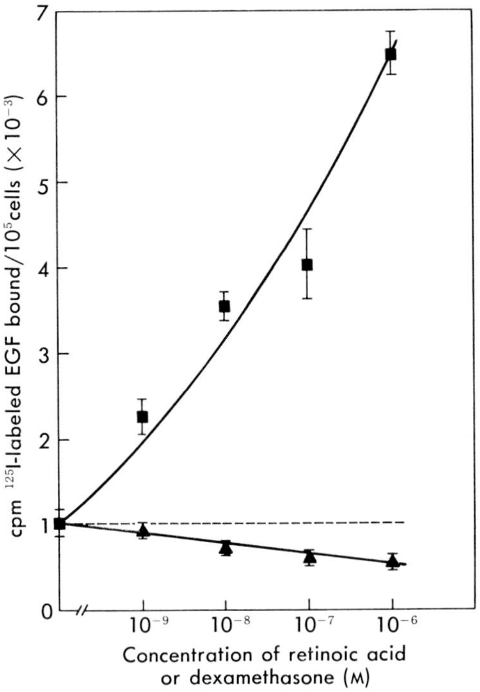

FIG. 4. Concentration dependence of the effect of retinoic acid or dexamethasone on the binding of ^{125}I-labeled EGF to NRK cells. Cells were seeded at a density of 3×10^4 cells/ml/16-mm multiwell dish in Dulbecco's modified Eagle's medium with 10% calf serum. Retinoic acid (■) or dexamethasone (▲) was added 4 hr after seeding, and the incubation was continued for 60 hr prior to the binding assay. Results are reported as the mean±S.D.(bars) of 6 individual determinations. See ref. *19* for details.

receptor has a predominantly growth-stimulating activity in this NRK system, there are other cell lines, such as the human A 431 cell line, in which stimulation of the EGF receptor leads to a predominantly growth-inhibitory response (*23*). Thus one can easily visualize at least one mechanism whereby retinoic acid can be a bifunctional agent with respect to cell proliferation: it can stimulate proliferation when stimulation of the EGF receptor causes enhanced proliferation, and it can inhibit cell proliferation when stimulation of the EGF receptor leads to an inhibition of proliferation. The bifunctionality of the EGF receptor may depend on many variables, including the concentration of ligand (EGF or TGF-alpha) present, or the duration of action of the ligand, so that one cannot arbitrarily say *a priori* that the receptor itself is stimulatory or inhibitory; the activity of the receptor depends on the set of chemical substances, both effectors and receptors, that are operant in the cell at a particular time. We have already shown that a very similar concept applies to the bifunctional action of the effector peptide, TGF-beta (*24*).

We have also intensively studied a second cell system with respect to its response to retinoic acid (*20*), namely the cell line obtained by transfecting Fischer rat 3T3 fibroblasts with the cellular *myc* oncogene (FR 3T3). In other studies, we have shown that these cells are most useful for investigating the relationships between oncogenes and growth factors, and in particular the manner in which growth factors can be bifunctional (*24*) and in which an oncogene such as *myc* can markedly enhance the response of a cell to exogenous growth factors (*25*). We have shown that FR 3T3 cells, transfected with the c-*myc* oncogene, can be induced to grow and form colonies in soft agar by treatment with either EGF alone or the combination of platelet-derived growth factor (PDGF) and TGF-beta; the parent cells, before *myc* transfection, will not respond in this way. This induction of anchorage-independent growth by each of these sets of growth factors involves different cellular pathways that can be distinguished by their sensitivity to retinoic acid. In contrast to its stimulatory effects on the proliferation of NRK cells in soft agar, retinoic acid is markedly inhibitory to the proliferation of *myc*-transfected FR 3T3 cells in an anchorage-independent growth assay. Moreover, the sensitivity of the cells to inhibition by retinoic acid depends on the growth factors that are stimulating proliferation. The formation of colonies of FR 3T3 cells, induced by the combined action of PDGF and TGF-beta is 100-fold more sensitive to inhibition by retinoic acid than is the same colony formation induced by EGF. The experimental data are shown in Figs. 5 and 6. Figure 5a shows that retinoic acid increases the number of cell-surface receptors for EGF in a dose-dependent manner, much the same way that it does in NRK cells; Fig. 5b shows a Scatchard plot that indicates that retinoic acid has no apparent effect on receptor affinity. Figure 6 shows the marked difference between the effects of retinoic acid on the proliferation of NRK cells (see Fig. 1), as compared to FR 3T3 cells, and that its inhibitory effects on the FR 3T3 cells depend on the set of growth factors acting on this *myc*-transfected cell. As seen in Fig. 6a, one pathway of proliferation, induced by EGF alone, is not particularly sensitive to retinoic acid; in fact 10^{-8} M retinoic acid causes no inhibition of this pathway. Another pathway of proliferation, induced by the combination of PDGF and TGF-beta, is exquisitely sensitive to inhibition by retinoic acid, as shown in Fig. 6b; here 10^{-8} M retinoic acid causes almost

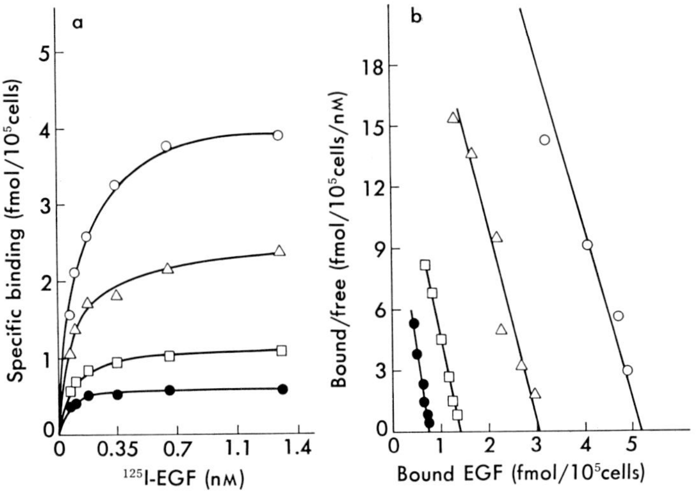

FIG. 5. Retinoic acid increases the number of cell-surface EGF receptors on FR 3T3 *myc*-transfected fibroblast cells. a: binding of ^{125}I-labeled EGF (150 μCi μg^{-1}) to cells grown either in the absence of added retinoic acid (●) or in the presence of 10^{-8} M (□), 10^{-7} M (△), or 10^{-6} M (○) retinoic acid. b: Scatchard plot of the data in a. See ref. *20* for details.

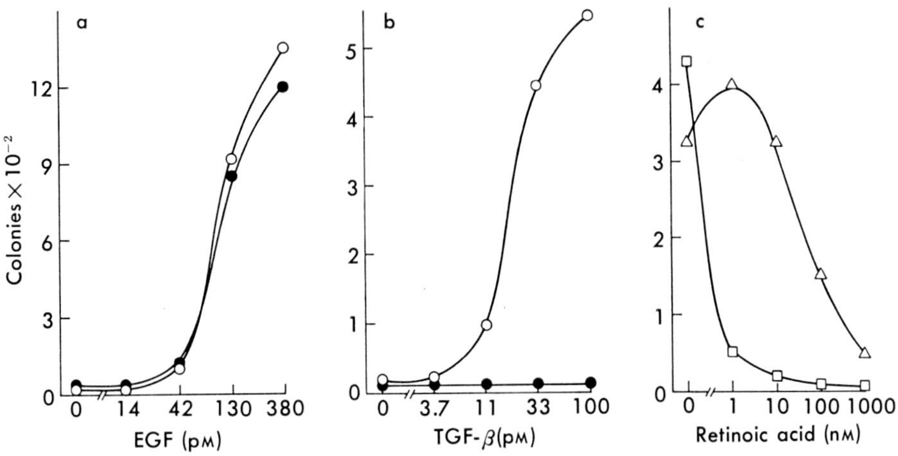

FIG. 6. Selective inhibition by retinoic acid of the anchorage-independent growth of FR 3T3 *myc*-transfected fibroblasts in the presence of PDGF and TGF-beta. a: colony growth assayed in the presence of EGF alone. b: colony growth assayed in the presence of 150 pM PDGF and the concentration of TGF-beta indicated. c: colony growth assayed in the presence of 42 pM EGF alone (△) or the combination of 150 pM PDGF and 100 pM TGF-beta (□) and varying concentrations of retinoic acid. The cells were assayed for the formation of colonies of >62 μm diameter after 7 days incubation in soft agar in Dulbecco's modified medium, 10% calf serum. Purified murine EGF, human PDGF, or human platelet TGF-beta was added at the time of the assay in the presence (●) or absence (○) of 10^{-8} M retinoic acid. See ref. *20* for details.

total inhibition of proliferation. The differences in the dose-responsiveness of the two pathways are shown in Fig. 6c. It is interesting that the inhibitory action of retinoic acid is seen only when the FR 3T3 cells are grown under conditions of anchorage-independence; when these same cells are grown in monolayer culture, retinoic acid either stimulates growth or has no effect. Thus, it would appear that the selective action of retinoic acid cannot be related mechanistically to the proliferation of cells *per se*, but, rather specifically to its effects on processes related to the transformed phenotype and the anchorage-independent growth of cells.

In summary, we have shown that studies of the interactions of retinoids with peptide growth factors and their receptors have already yielded valuable new information about the process of transformation and its suppression. To return to Fig. 1, it is clear that any future hypothesis dealing with the mechanism of action of retinoids in prevention of cancer will need to integrate the role of retinoids, peptide growth factors and their receptors, as well as specific genes that control differentiation and proliferation (this is just another name for oncogenes). As we have noted previously (*6*), retinoid research has suffered for many years from obsolete, shopworn hypotheses that do not fit the data relating to the structure-activity relationships of these substances. In particular, any modern hypothesis of retinoid action must be able to explain the activity of the vast range of substances, including the new retinoidal benzoic acid derivatives (*10*), that can support growth in vitamin A-deficient animals, as well as suppress carcinogenesis. It would appear that new investigative approaches, that employ modern methods of molecular genetics, combined with studies of peptide growth factors and their receptors, will be most useful to explain the mechanism of action of retinoids, as well as to assure the most rational use of these promising agents for the prevention or treatment of human cancer.

ACKNOWLEDGMENT

We thank Sue Perdue for her assistance with the manuscript.

REFERENCES

1. Moon, R. C. and Itri, L. M. Retinoids and Cancer. *In;* M. B. Sporn, A. B. Roberts, and D. S. Goodman (eds.), The Retinoids, vol. 2, pp. 327–371, Academic Press, Orlando, 1984.
2. Sporn, M. B., Dunlop, N. M., Newton, D. L., and Smith, J. M. Prevention of chemical carcinogenesis by vitamin A and its synthetic analogs (retinoids). Fed. Proc., *35*: 1332–1338, 1976.
3. Sporn, M. B. and Newton, D. L. Chemoprevention of cancer with retinoids. Fed. Proc., *38*: 2528–2534, 1979.
4. Sporn, M. B. and Newton, D. L. Retinoids and chemoprevention of cancer. *In;* M. S. Zedeck and M. Lipkin (eds.), Inhibition of Tumor Induction and Development, pp. 71–100, Plenum, New York, 1981.
5. Sporn, M. B. and Roberts, A. B. Biological methods for analysis and assay of retinoids —relationship between structure and activity. *In;* M. B. Sporn, A. B. Roberts, and D. S. Goodman (eds.), The Retinoids, vol. 1, pp. 235–279, Academic Press, Orlando, 1984.
6. Sporn, M. B. and Roberts, A. B. Role of retinoids in differentiation and carcinogenesis. Cancer Res., *43*: 3034–3040, 1983.
7. Roberts, A. B. and Sporn, M. B. Cellular biology and biochemistry of the retinoids.

In; M. B. Sporn, A. B. Roberts, and D. S. Goodman (eds.), The Retinoids, vol. 2, pp. 209–286, Academic Press, Orlando, 1984.

8. Westin, E. H., Wong-Staal, F., Gelmann, E. P., Dalla Favera, R., Papas, T. S., Lautenberger, J. A., Eva, A., Reddy, E. P., Tronick, S. R., Aaronson, S. A., and Gallo, R. C. Expression of cellular homologues of retroviral *onc* genes in human hematopoietic cells. Proc. Natl. Acad. Sci. U.S.A., *79*: 2490–2494, 1982.
9. Campisi, J., Gray, H. E., Pardee, A. B., Dean, M., and Sonenshein, G. E. Cell-cycle control of c-*myc* but not c-*ras* expression is lost following chemical transformation. Cell, *36*: 241–247, 1984.
10. Strickland, S., Breitman, T. R., Frickel, F., Nürrenbach, A., Hädicke, E., and Sporn, M. B. Structure-activity relationships of a new series of retinoidal benzoic acid derivatives as measured by induction of differentiation of murine F9 teratocarcinoma cells and human HL-60 promyelocytic leukemia cells. Cancer Res., *43*: 5268–5272, 1983.
11. Yancopoulos, G. D., Nisen, P. D., Tesfaye, A., Kohl, N. E., Goldfarb M. P., and Alt, F. W. N-*myc* can cooperate with *ras* to transform normal cells in culture. Proc. Natl. Acad. Sci., U.S.A., *82*: 5455–5459, 1985.
12. Sidell, N. Retinoic acid-induced growth inhibition and morphologic differentiation of human neuroblastoma cells *in vitro*. J. Natl. Cancer Inst., *68*: 589–596, 1982.
13. Sidell, N., Altman, A., Haussler, M. R., and Seeger, R. C. Effects of retinoic acid (RA) on the growth and phenotypic expression of several human neuroblastoma cell lines. Exp. Cell Res., *148*: 21–30, 1983.
14. Thiele, C. J., Reynolds, C. P., and Israel, M. A. Decreased expression of N-*myc* precedes retinoic acid-induced morphological differentiation of human neuroblastoma. Nature, *313*: 404–406, 1985.
15. Sporn, M. B. and Roberts, A. B. Autocrine growth factors and cancer. Nature, *313*: 747–751, 1985.
16. Jetten, A. M. Action of retinoids and phorbol esters on cell growth and the binding of epidermal growth factor. Ann. N.Y. Acad. Sci., *359*: 200–217, 1981.
17. Jetten, A. M. Effects of retinoic acid on the binding and mitogenic activity of epidermal growth factor. J. Cell. Physiol., *100*: 235–240, 1982.
18. Jetten, A. M. Retinoids and their modulation of cell growth. *In;* G. Guroff (ed.), Growth and Maturation Factors, vol. 3, pp. 252–293, John Wiley & Sons, New York, 1985.
19. Roberts, A. B., Anzano, M. A., Lamb, L. C., Smith, J. M., and Sporn, M. B. Antagonistic actions of retinoic acid and dexamethasone on anchorage-independent growth and epidermal growth factor binding of normal rat kidney cells. Cancer Res., *44*: 1635–1641, 1984.
20. Roberts, A. B., Roche, N. S., and Sporn, M. B. Selective inhibition of the anchorage-independent growth of *myc*-transfected fibroblasts by retinoic acid. Nature, *315*: 237–239, 1985.
21. Roberts, A. B., Frolik, C. A., Anzano, M. A., and Sporn, M. B. Transforming growth factors from neoplastic and non-neoplastic tissues. Fed. Proc., *42*: 2621–2626, 1983.
22. Roberts, A. B. and Sporn, M. B. Transforming growth factors. Cancer Surv., *4*: 683–705, 1985.
23. Kawamoto, T., Sato, J. D., Le, A., Polikoff, J., Sato, G. H., and Mendelsohn J. Growth stimulation of A431 cells by epidermal growth factor: identification of high-affinity receptors for epidermal growth factor by an anti-receptor monoclonal antibody. Proc. Natl. Acad. Sci. U.S.A., *80*: 1336–1341, 1983.
24. Roberts, A. B., Anzano, M. A., Wakefield, L. M., Roche, N. S., Stern, D. F., and

Sporn, M. B. Type beta transforming growth factor: a bifunctional regulator of cellular growth. Proc. Natl. Acad. Sci. U.S.A., *82*: 119–123, 1985.

25. Stern, D. F., Roberts, A. B., Roche, N. S., Sporn, M. B., and Weinberg, R. A. Differential responsivness of *myc*- and *ras*-transfected cells to growth factors: selective stimulation of *myc*-transfected cells by EGF. Mol. Cell. Biol. *6*: 870–877, 1986.

DIET, NUTRITION AND CANCER, Y. HAYASHI ET AL. (EDS.),
JAPAN SCI. SOC. PRESS, TOKYO/VNU SCI. PRESS, UTRECHT, PP. 159–168, 1986

Significance of L-Ascorbic Acid and Urinary Electrolytes in Promotion of Rat Bladder Carcinogenesis

Shoji FUKUSHIMA, Tomoyuki SHIRAI, Masao HIROSE, and Nobuyuki ITO
First Department of Pathology, Nagoya City University Medical School, Nagoya 467, Japan

Abstract: The present studies report on the significance of L-ascorbic acid (AA) and urinary electrolytes for promotion of rat urinary bladder carcinogenesis. Male F344 rats were given an oral administration of 0.05% N-butyl-N-(4-hydroxybutyl)nitrosamine (BBN) as an initiator for 4 weeks, and were then subjected to treatment with dietary supplements of test chemicals for 32 weeks. Administration of 5% sodium L-ascorbate (SA), the sodium ion form of AA significantly promoted urinary bladder carcinogenesis, whereas administration of 5% AA did not. The urine of rats given SA but not AA was characterized by an apparent elevation of pH, an increase of sodium ion concentration, and increases in the urinary content of total AA and its metabolite, dehydroascorbic acid. Administration of 3% $NaHCO_3$, which induced elevation of pH and increase of sodium ion concentration in the urine, promoted BBN bladder carcinogenesis. When rats were given 5% AA plus 3% $NaHCO_3$, AA enhanced the promoting activity of $NaHCO_3$. Lowering of pH by 1% NH_4Cl clearly reduced the promoting activity of 5% SA when these two compounds were given concurrently. Treatment with 5% AA plus 3% K_2CO_3 promoted BBN bladder carcinogenesis in rats, whereas addition of 5% $CaCO_3$ or 5% $MgCO_3$ to AA did not. These results strongly indicate the important role of urinary sodium or potassium ion concentration and pH in modulating urinary bladder carcinogenesis by AA.

There have been many reports concerning the relationship between vitamins and neoplastic development (*1*, *2*), and ascorbic acid (vitamin C) has attracted particular attention with regard to prevention of cancer (*3*). For example, ascorbic acid is known to block nitrosamine formation by reaction of nitrites with amines (*4*), and has also been shown to inhibit bacterial mutagenesis by N-nitroso compounds (*5*). Therefore, it seems possible that its consumption as a component of foods may help prevent cancer in humans.

Division of the process of chemical carcinogenesis into two stages, initiation and promotion, is now known to apply to many organs, including the urinary bladder (*6–11*). Previously, we studied in a screening experiment the promoting activities of various chemicals in urinary bladder carcinogenesis in male rats (*12*). In these preliminary experiments, sodium L-ascorbate (SA), the ionic form of ascorbic acid, appeared to exert promoting potential for urinary bladder carcinogenesis.

Modification by Sodium L-Ascorbate, L-Ascorbic Acid, Sodium Erythorbate, and Erythorbic Acid on Urinary Bladder Carcinogenesis

Modifying effects of SA (*13*), L-ascorbic acid (AA) (*14*), sodium erythorbate (SE) (*14*), and erythorbic acid (EA) on the development of preneoplastic and neoplastic lesions were examined in the urinary bladder after pretreatment with N-butyl-N-(4-hydroxybutyl)nitrosamine (BBN).

Six week-old male F344 rats (Charles River Japan, Inc.) were used. SA, AA, SE, and EA were all of food additive grade. In experiment 1, groups 1 and 2 were given drinking water containing 0.05% BBN for 4 weeks, and then powdered diet (Oriental M) containing 5% (group 1) or 1% (group 2) SA for 32 weeks. Group 3 was given only BBN for the first 4 weeks. Group 4 was given drinking water without BBN for the first 4 weeks, and then powdered diet containing 5% SA. The total observation period in experiment 1 was 36 weeks. The urinary bladders were examined histologically after inflation with formalin fixative. For quantitative analysis, urinary bladder lesions were counted by light microscopy, the total length of the basement membrane was measured with a color video image processor (VIP-21CH; Olympus-Ikegami Tsushin Co.), and numbers of lesions per 10 cm of basement membrane were calculated.

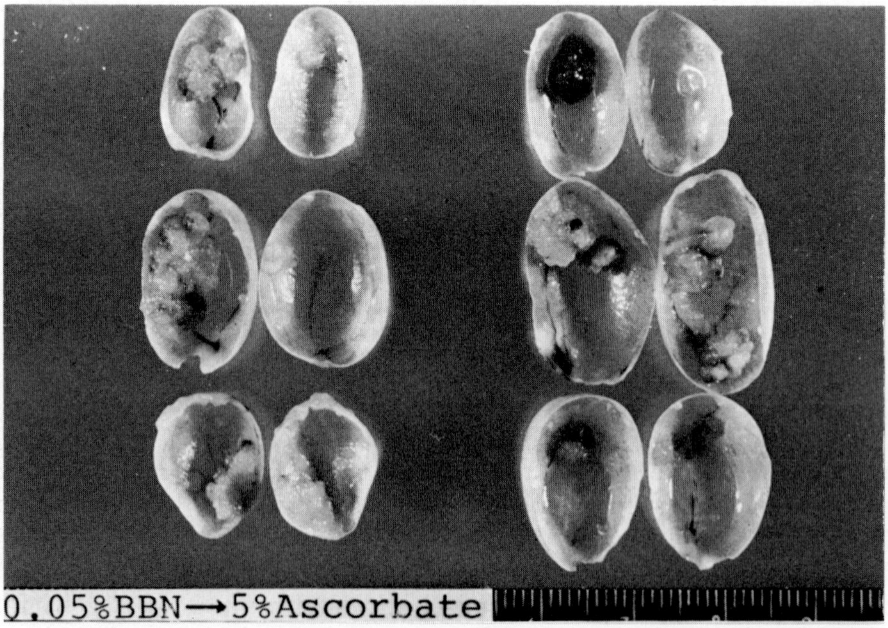

FIG. 1. Macroscopic appearance of the urinary bladder of rats treated with BBN followed by 5% SA. Multiple, large tumors are seen.

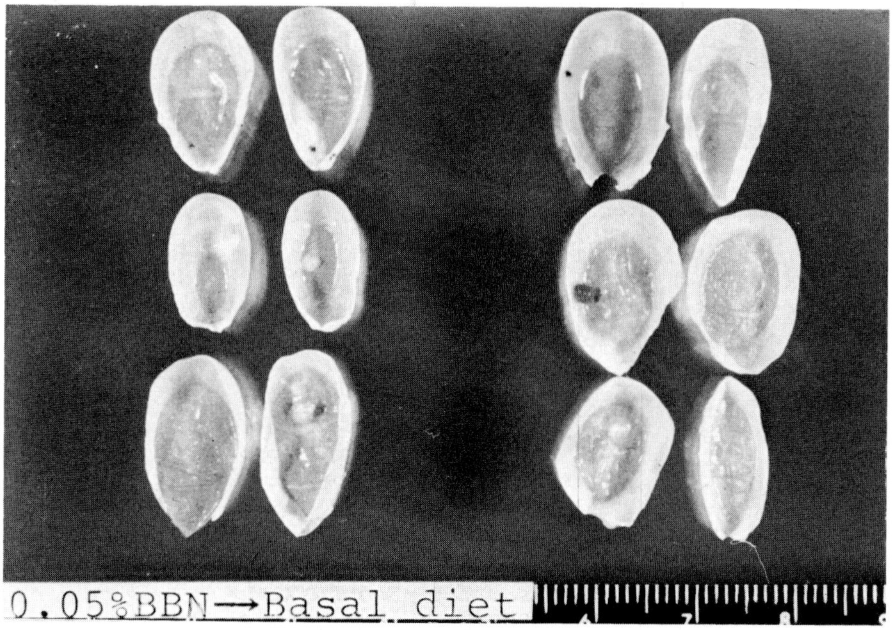

FIG. 2. Macroscopic appearance of the urinary bladder of rats treated with BBN alone as controls. A few tiny tumors are seen.

Macroscopically, rats given 0.05% BBN followed by 5% SA had a greater incidence of tumors of the urinary bladder than the controls (Figs. 1 and 2).

The quantitative data for histopathological lesions of the urinary bladder observed in rats are summarized in Table 1. As described previously (*15, 16*), the lesions found in the urinary bladder epithelium were classified into 4 types: simple hyperplasia, papillary or nodular hyperplasia (PN hyperplasia, preneoplastic hyperplasia) (*15, 16*), papilloma, and carcinoma. Pretreatment with 0.05% BBN followed by 5% SA in group 1 resulted in significantly increased incidences and numbers per 10 cm basement membrane of carcinomas. The incidence and the number of PN hyperplasias and papillomas were also significantly increased by 5% SA. However, 1% SA did not significantly increase the occurrence of any bladder lesions in rats pretreated with BBN, although it demonstrated a tendency to increase the incidence of PN hyperplasia. There were no remarkable changes in the urinary bladder of rats treated with 5% SA alone without prior BBN. These results showed that SA promotes urinary bladder carcinogenesis in rats initiated with BBN.

Treatment with SA resulted in increases in the content of AA and its metabolite, dehydroascorbic acid in the urine (*13*). While these changes might contribute to the induction of carcinoma in the urinary bladder, it was uncertain from these findings whether AA itself exerts promoting activity. We therefore examined the promoting potential of three chemicals related to SA, namely, AA, SE, and EA for two-stage urinary bladder carcinogenesis in rats initiated with BBN. EA is the epimer of AA.

Six week-old male F344 rats were randomly divided into 5 groups of 25 each in experiment 2. They were given drinking water with (groups 1–3) or without (groups

TABLE 1. Histological Findings in the Urinary Bladders of Rats Treated with BBN Followed by Test Chemical

Group	Treatment		No. of rats	PN hyperplasia		Papilloma		Carcinoma	
	BBN	Test chemical		Incidence (%)	No./10 cm of BM[a]	Incidence (%)	No./10 cm of BM	Incidence (%)	No./10 cm of BM
Experiment 1									
1	+	5% SA	25	25 (100)[c]	7.0±3.0[b,c]	23 (92)[c]	3.1±1.6[b,c]	24 (96)[c]	1.4±0.7[b,c]
2	+	1% SA	29	11 (38)	0.6±1.0	12 (41)	0.4±0.6	0 —	0
3	+	—	30	8 (27)	0.3±0.5	12 (40)	0.5±0.5	1 (3)	0.0±0.2
4	–	5% SA	20	0 —	0	0 —	0	0 —	0
Experiment 2									
1	+	5% AA	25	7 (28)	0.3±0.5	10 (40)	0.4±0.5	1 (4)	0.0±0.1
2	+	5% SE	23	23 (100)[c]	5.3±4.2[c]	19 (83)[c]	1.4±1.2[c]	15 (65)[c]	0.8±0.7[c]
3	+	—	24	8 (33)	0.5±1.0	5 (20)	0.2±0.4	1 (4)	0.0±0.2
4	–	5% AA	25	0 (0)	0	0 (0)	0	0 (0)	0
5	–	5% SE	25	0 (0)	0	0 (0)	0	0 (0)	0
Experiment 3									
1	+	5% EA	20	8 (40)	0.3±0.4	1 (5)	0.0±0.1	1 (5)	0.0±0.2
2	+	—	25	11 (44)	0.5±0.7	5 (20)	0.1±0.3	2 (8)	0.1±0.2
3	–	5% EA	20	0 (0)	0	0 (0)	0	0 (0)	0

a) Basement membrane. b) Mean±S.D. c) Significantly different from the control at $p<0.001$.

4 and 5) 0.05% BBN for 4 weeks and then for 32 weeks they were administered powdered basal diet (Oriental M) containing 5% AA (groups 1 and 4) or 5% SE (groups 2 and 5), or basal diet containing no added chemical (group 3) as a control. The total observation period in experiment 2 was 36 weeks.

Histological findings in the urinary bladders are summarized in Table 1. The incidences and numbers per 10 cm of basement membrane of PN hyperplasia and papilloma were significantly higher in groups given SE (group 2) than in controls (group 3). The incidence and number of carcinomas also increased significantly in group 2 treated with SE. However, treatment with AA was not associated with increase in the induction of any urinary bladder lesions in rats pretreated with BBN (group 1). In groups given the test chemicals alone without BBN (groups 4 and 5), AA and SE had no effect on the mucosa. These results show that SE promotes urinary bladder carcinogenesis, while AA does not.

In experiment 3, the same experimental schedule was applied. Male 6-week-old rats were given drinking water with (groups 1, 20 rats and 2, 25 rats) or without (group 3, 20 rats) 0.05% BBN supplement for 4 weeks. Groups 1 and 3 were then given powdered basal diet containing EA for 32 weeks prior to sacrifice. The incidences and numbers per 10 cm of basement membrane of PN hyperplasia, papilloma, carcinoma in group 1 did not differ from those in group 2 (Table 1). This result shows that EA does not exert promoting potential for urinary bladder carcinogenesis.

Urinalysis of Rats Treated with SA or AA

Analyses of the urine, including its electrolytes in rats treated with SA and AA were carried out as described previously (*14*). Male 6-week-old F344 rats were continuously administered powdered basal diet containing 5% SA and 5% AA or powdered basal diet with no added chemical as a control. At week 8, urinary pH was measured after forced urination. Electrolyte concentrations were determined in urine collected for 4 hr in the early morning. Apparent elevation in the pH of the urine was detected in the group given SA as shown in Table 2. SA also caused significant increase in the sodium ion concentration. At week 22, the urine of rats treated with SA or AA was collected for measurement of AA content. The values for total ascorbic acid and dehydroascorbic acid were significantly higher in rats treated with SA or AA than in controls.

TABLE 2. Urinary pH and Electrolyte Concentration in Rats Treated with SA or AA for 8 Weeks

Test chemical	pH	Electrolytes			
		Na (mEq/*l*)	K (mEq/*l*)	Ca (mEq/*l*)	Cl (mEq/*l*)
5%SA	7.60±0.27[a]	207±63[b]	107±35	6.3±2.0	60±28
5%AA	6.09±0.15	87±64	122±54	5.2±1.3	64±62
No chemical	6.31±0.18	93±47	153±36	3.5±0.9	117±66

Values are mean±S.D. [a] Significantly different from the control at $p<0.01$. [b] Significantly different from the control at $p<0.05$.

Modification of Urinary Bladder Carcinogenesis by a Combination of AA Plus $NaHCO_3$ or SA Plus NH_4Cl

Since urinary sodium ions appear to play a key role in the promoting activity of AA, and elevation of urinary pH by SA was related to promotion of urinary bladder carcinogenesis, we examined the effects of AA plus $NaHCO_3$ and SA plus NH_4Cl in urinary bladder carcinogenesis of rats (*17*). Male 6-week-old F344 rats were given drinking water containing 0.05% BBN and then for 32 weeks were given powdered basal diet (Oriental M) containing 5% AA, 3% $NaHCO_3$, 5% AA plus 3% $NaHCO_3$, 5% SA, 1% NH_4Cl, 5% SA plus 1% NH_4Cl, or no added chemical (control group). The total observation period was 36 weeks.

Macroscopically numbers of tumors of the urinary bladder were highest in groups treated with AA plus $NaHCO_3$ and SA. Histological findings of urinary bladder lesions in rats are summarized in Table 3. The incidences of PN hyperplasia in groups given $NaHCO_3$, AA plus $NaHCO_3$, SA, and SA plus NH_4Cl were significantly higher than in the control group. The incidences and numbers of carcinoma were significantly elevated in the groups given AA plus $NaHCO_3$, $NaHCO_3$, and SA. Moreover, the incidence in the group given AA plus $NaHCO_3$ was significantly different from those in the AA group. Treatment with AA plus $NaHCO_3$ resulted in much higher induction of carcinomas than treatment with $NaHCO_3$ alone. Treatment with SA plus NH_4Cl resulted in significantly fewer carcinomas than treatment with SA alone, the numbers of carcinomas in the group given SA plus NH_4Cl being similar to control group values. Combined treatment with AA and NH_4Cl alone also tended to decrease the induction of neoplastic lesions initiated by BBN.

Results of urinalysis showed that the urinary pH was increased in the groups administered $NaHCO_3$, AA plus $NaHCO_3$, and SA, but significantly decreased in groups given AA and NH_4Cl. The urinary pH in the SA plus NH_4Cl group was almost the same as that of controls (Table 4). The sodium ion content of the urine of rats in groups given $NaHCO_3$, AA plus $NaHCO_3$, SA, and SA plus NH_4Cl was elevated over control values (Table 4). However, that of the NH_4Cl group was lower

TABLE 3. Incidence of Urinary Bladder Lesions in Rats Treated with BBN Followed by Test Chemicals

Group	Test chemicals	No. of rats	PN hyperplasia (%)	Papilloma (%)	Carcinoma Incidence (%)	Carcinoma No./10 cm BM
1	AA	20	12 (60)	8 (40)	4 (20)	0.18±0.37[a]
2	$NaHCO_3$	20	20 (100)[g]	15 (75)	16 (80)[f]	1.03±0.70[e]
3	AA+$NaHCO_3$	20	20 (100)[b,i]	18 (90)[b,h]	19 (95)[c,j]	2.76±1.44[c,d,j]
4	SA	20	20 (100)[n]	19 (95)[k,m]	17 (85)[k,n]	1.93±1.18[l,n]
5	NH_4Cl	20	4 (20)	5 (25)	4 (20)	0.20±0.40
6	SA+NH_4Cl	20	19 (95)[o]	13 (65)	9 (45)	0.69±1.02
7	—	20	6 (30)	11 (55)	5 (25)	0.42±0.79

[a] Mean±S.D. Values for the following groups were significantly different: groups 1 and 3, [b] $p<0.01$, [c] $p<0.001$; groups 2 and 3, [d] $p<0.001$; groups 2 and 7, [e] $p<0.05$, [f] $p<0.01$, [g] $p<0.001$; groups 3 and 7, [h] $p<0.05$, [i] $p<0.01$, [j] $p<0.001$; groups 4 and 6, [k] $p<0.05$, [l] $p<0.01$; groups 4 and 7, [m] $p<0.01$, [n] $p<0.001$; groups 6 and 7, [o] $p<0.001$.

TABLE 4. Urinary pH, Sodium Ion Concentration, and Contents of Total Ascorbic Acid and Dehydroascorbic Acid in the Urine of Rats Treated with Test Chemicals

Test chemicals	pH	Na (mEq/l)	Total ascorbic acid (mg/100 ml)	Dehydroascorbic acid (mg/100 ml)
AA	6.12±0.11[a]	127± 22	96.4±19.3[b]	5.9±0.9[b]
$NaHCO_3$	7.89±0.51[a]	420± 56[a]	0.1± 0.2[c]	0.1±0.2[c]
AA+$NaHCO_3$	7.25±0.32[a]	471±116[a]	45.7± 5.8[b]	3.7±0.3[b]
SA	7.48±0.23[a]	316± 90[a]	66.6±10.7[b]	4.7±0.8[b]
NH_4Cl	5.78±0.07[a]	73± 27[c]	0.9± 0.1[a]	0.9±0.1[a]
SA+NH_4Cl	6.49±0.30	340± 57[a]	39.9±12.8[a]	4.8±0.9[b]
—	6.69±0.35	121± 36	0.5± 0.1	0.5±0.1

Values are means±S.D. Significantly different from control group at [a] $p<0.01$, [b] $p<0.001$, [c] $p<0.05$.

than the control. No significant increases or decreases in the levels of other electrolytes as compared to the control group were observed in the test groups. The values of total AA and dehydroascorbic acid were markedly higher in groups AA, AA plus $NaHCO_3$, SA, and SA plus NH_4Cl than in the control group (Table 4).

Thus, $NaHCO_3$ exhibited a promoting activity for urinary bladder carcinogenesis. The most interesting findings in the present study were that AA plus $NaHCO_3$ showed potent promoting activity in urinary bladder carcinogenesis like that of SA, whereas NH_4Cl inhibited the promoting activity of SA. Increases of sodium ion content and pH of the urine are apparently important factors for urinary bladder carcinogenesis. Moreover, high content of total AA or dehydroascorbic acid in the urine is associated with potent promoting potential under conditions of elevated pH and sodium ion concentration.

Modification by K, Ca, and Mg Ion Concentration of Urinary Bladder Carcinogenesis

Modifying effects of combinations of AA with K_2CO_3, $CaCO_3$, or $MgCO_3$ were investigated to determine the influence of urinary electrolytes other than sodium ions on promotion by AA. Male 6-week-old F344 rats were given drinking water containing 0.05% BBN and then administered powdered basal diet (Oriental M) containing 5% AA, 3% K_2CO_3, 5% AA plus 3% K_2CO_3, 3% $CaCO_3$, 5% AA plus 3% $CaCO_3$, 3% $MgCO_3$, 5% AA plus 3% $MgCO_3$, or no added chemical (control group) for 32 weeks.

Resultant incidences of urinary bladder carcinomas were 95% in the group given AA plus K_2CO_3, significantly elevated (the control group, 5%). However, treatment with K_2CO_3 alone did not increase the induction of carcinoma when compared to the control. Moreover, incidences of carcinoma in other test chemical-treated groups were not different from that of the control group. Thus the results suggested that elevation of urinary pH and increase of potassium ion concentration after ingestion of K_2CO_3 might be related to promotion of urinary bladder carcinogenesis by AA.

CONCLUSION

It was established that SA and SE exert full promotion potential for urinary bladder carcinogenesis, whereas AA and EA do not. $NaHCO_3$ also exhibited promoting activity in urinary bladder carcinogenesis. Sodium *o*-phenylphenate (*18*) but not *o*-phenylphenol has similar promoting activity, as does sodium saccharin but not saccharin acid (*6–9*). While raising the urinary pH and Na ion content of the urine, changes in those two parameters therefore appear to be important factors in the modification of urinary bladder carcinogenesis.

Urinary analyses showed differences in the contents of total AA and dehydroascorbic acid between the groups treated with AA plus $NaHCO_3$ and $NaHCO_3$ alone, but not in other parameters, such as the pH and electrolytes. However, the promoting activity for urinary bladder carcinogenesis of AA combined with $NaHCO_3$ was greater than that of $NaHCO_3$ itself, suggesting that AA exerts co-promoting potential.

The membrane potential of the epithelium in the early stage of urinary bladder carcinogenesis has been demonstrated to be significantly increased by BBN or sodium saccharin treatment (*19*). Since the apical membrane potential of the cell depends largely on the permeability of Na ion, it reflects the activity of the Na ion channel, which is essential for Na ion transport across the urinary bladder epithelium. Therefore, it seems likely that a high content of Na ion in the urine produces high levels of intracellular Na ion in the urinary bladder epithelium with consequent elevation of the intracellular pH. In general, a high intracellular concentration of Na ion is thought to be related to cellular proliferation (*20*, *21*). In addition, there is a good correlation between increase in the intracellular pH and DNA synthesis in cells (*22*). Increases in the Na ion concentration induce hyperplasia of urinary bladder epithelial cells (*23*, *24*) and renal pelvic epithelial cells (*25*). This phenomenon may be correlated with the promoting activity of Na ion in urinary bladder carcinogenesis.

Calcium ion is also known to be important in the proliferation of bladder epithelium cells (*26*). However, in the present study no promoting effect was observed after increasing calcium ion content of the urine. In contrast, increase of potassium ion content in the urine and AA did enhance urinary bladder carcinogenesis.

ACKNOWLEDGMENTS

This work was supported by Grants-in-Aid of Cancer Research from the Ministry of Education, Science and Culture of Japan and the Ministry of Health and Welfare of Japan and by a Grant-in-Aid from the Ministry of Health and Welfare of Japan for a Comprehensive 10 year Strategy for Cancer Control.

REFERENCES

1. Newberne, P. M. and McConnell, R. G. Nutrient deficiencies in cancer causation. J. Environ. Pathol. Toxicol., *3*: 323–356, 1980.
2. Young, V. R. and Newberne, P. M. Vitamins and cancer prevention: issues and dilemmas. Cancer (Phila.), *47*: 1226–1240, 1981.
3. Cameron, E., Pauling, L., and Leibovitz, B. Ascorbic acid and cancer: a review.

Cancer Res., *39*: 663–681, 1979.

4. Mirvish, S. S. N-Nitroso compounds: their chemical and *in vivo* formation and possible importance as environmental carcinogens. J. Toxicol. Environ. Health, *2*: 1267–1277, 1977.
5. Guttenplan, J. B. Mechanisms of inhibition by ascorbate of microbial mutagenesis induced by N-nitroso compounds. Cancer Res., *38*: 2018–2022, 1978.
6. Cohen, S. M., Arai, M., Jacobs, J. B., and Friedell, G. H. Promoting effect of saccharin and DL-tryptophan in urinary bladder carcinogenesis. Cancer Res., *39*: 1207–1217, 1979.
7. Fukushima, S., Friedell, G., Jacobs, J. B., and Cohen, S. M. Effect of L-tryptophan and sodium saccharin on urinary tract carcinogenesis initiated by N-[4-(5-nitro-2-futyl)-2-thiazolyl]formamide. Cancer Res., *41*: 3100–3013, 1981.
8. Hicks, R. M., Wakefield, J. St. J., and Chowaniec, J. Evaluation of a new model to detect bladder carcinogens or co-carcinogenesis; results obtained with saccharin, cyclamate and cyclophosphamide. Chem. Biol. Interact., *11*: 225–233, 1985.
9. Nakanishi, K., Hagiwara, A., Shibata, M., Imaida, K., Tatematsu, M., and Ito, N. Dose-response of saccharin in induction of urinary bladder hyperplasia in Fischer 344 rats pretreated with N-butyl-N-(4-hydroxybutyl)nitrosamine. J. Natl. Cancer Inst., *65*: 1005–1010, 1980.
10. Oyasu, R., Hirao, Y., and Izumi, K. Enhancement by urine of urinary bladder carcinogenesis. Cancer Res., *41*: 478–481, 1981.
11. Ito, N., Fukushima, S., Shirai, T., and Nakanishi, K. Effects of promoters on N-butyl-N-(4-hydroxybutyl)nitrosamine-induced urinary bladder carcinogenesis in the rat. Environ. Health Perspect., *50*: 61–69, 1983.
12. Fukushima, S., Hagiwara, A., Ogiso, T., Shibata, M., and Ito, N. Promoting effects of various chemicals in rat urinary bladder carcinogenesis initiated by N-nitroso-N-butyl-(4-hydroxybutyl)amine. Food Chem. Toxicol., *21*: 59–68, 1983.
13. Fukushima, S., Imaida, K., Sakata, T., Okamura, T., Shibata, M., and Ito, N. Promoting effects of sodium L-ascorbate on 2-stage urinary bladder carcinogenesis in rats. Cancer Res., *43*: 4454–4457, 1983.
14. Fukushima, S., Kurata, Y., Shibata, M., Ikawa, E., and Ito, N. Promotion by ascorbic acid, sodium erythorbate and ethoxyquin of neoplastic lesions in rats initiated with N-butyl-N-(4-hydroxybutyl)nitrosamine. Cancer Lett., *23*: 29–37, 1984.
15. Fukushima, S., Murasaki, G., Hirose, M., Nakanishi, K., Hasegawa, R., and Ito, N. Histopathological analysis of preneoplastic changes during N-butyl-N-(4-hydroxybutyl)nitrosamine induced urinary bladder carcinogenesis in rats. Acta Pathol. Jpn., *32*: 243–250, 1982.
16. Ito, N., Hiasa, Y., Tamai, A., Okajima, E., and Kitamura, H. Histogenesis of urinary bladder tumors induced by N-butyl-N-(4-hydroxybutyl)nitrosamine in rats. Gann, *60*: 401–410, 1969.
17. Shibata, M., Tamano, S., Miyata, Y., Ikawa, E., and Fukushima, S. Relationship between urinary Na^+ and pH and tumor development in 2-stage bladder carcinogenesis. Proc. Jpn. Cancer Assoc., *44*: 39, 1985.
18. Fukushima, S., Kurata, Y., Shibata, M., Ikawa, E., and Ito, N. Promoting effect of sodium *o*-phenylphenate and *o*-phenylphenol on two-stage urinary bladder carcinogenesis in rats. Gann, *74*: 625–632, 1983.
19. Imaida, K., Oshima, M., Fukushima, S., Ito, N., and Hotta, K. Membrane potentials of urinary bladder epithelium in F344 rats treated with N-butyl-N-(4-hydroxybutyl)-nitrosamine or sodium saccharin. Carcinogenesis, *4*: 659–661, 1983.

20. Cameron, I. L., Smith, N.K.R., Pool, T. B., and Sparks, R. L. Intracellular concentration of sodium and other elements as related to mitogenesis and oncogenesis *in vivo*. Cancer Res., *40*: 1493–1500, 1980.
21. Burns, C. P. and Rosenqurt, E. Extracellular Na^+ and initiation of DNA synthesis: role of intracellular pH and K^+. J. Cell Biol., *98*: 1082–1089, 1984.
22. Schuldiner, S. and Rozenqurt, E. Na^+/H^+ transport in Swiss 3T3 cells: mitogenic stimulation leads to cytoplasmic alkalinezation. Proc. Natl. Acad. Sci. U.S.A., *79*: 7778–7782, 1982.
23. Fukushima, S. and Cohen, S. M. Saccharin-induced hyperplasia of the rat urinary bladder. Cancer Res., *40*: 734–736, 1980.
24. Shibata, M.-A., Shibata, M., Tamano, S., and Fukushima, S. Sequential observation of alteration in the bladder epithelium and urine components in rats given various bladder promoters and their related compounds. J. Toxicol. Sci., *9*: 312, 1984.
25. Lalich, J. J., Paik, W.C.W., Pradhan, B., and Wis, M. Epithelial hyperplasia in the renal papilla of rats. Arch. Pathol., *97*: 29–32, 1974.
26. Reese, D. H. and Friedman, R. D. Suppression of dysplasia and hyperplasia by calcium in organ-cultured urinary bladder epithelium. Cancer Res., *38*: 586–592, 1978.

DIET, NUTRITION AND CANCER, Y. HAYASHI ET AL. (EDS.),
JAPAN SCI. SOC. PRESS, TOKYO/VNU SCI. PRESS, UTRECHT, PP. 169–182, 1986

Enhancing Effects of Dietary Salt on Both Initiation and Promotion Stages of Rat Gastric Carcinogenesis

Michihito TAKAHASHI and Ryohei HASEGAWA
Department of Pathology, National Institute of Hygienic Sciences, Tokyo 158, Japan

Abstract: Relatively short-term treatment (8 weeks) of rats with N-methyl-N′-nitro-N-nitrosoguanidine (MNNG) in the drinking water (100 mg/*l*) was shown to adequately initiate gastric carcinogenesis when 10% NaCl was simultaneously administered in the diet. Utilizing this MNNG plus high salt diet as the initiation stage of a two-step protocol, it was also established that subsequent dietary administration of NaCl (10% of the diet) for 32 weeks tended to enhance tumor development in the glandular stomach. Similar tumor promoting activity was demonstrated for other mucosal damaging agents, such as potassium metabisulfite and formaldehyde.

Biological changes of the gastric mucosa were examined after chronic administration or a single oral intubation of NaCl. Morphological lesions observed included diffuse mild erosions, atrophy of the glands, and hyperplasia of the foveolar epithelium when given 10% NaCl diet chronically. After a single oral intubation of NaCl, increased tritiated thymidine labeling index and ornithine decarboxylase (ODC) activity were observed in both pyloric and fundic mucosa. No remarkable effects of NaCl were observed on the forestomach or duodenal mucosa.

These results suggest that NaCl exerts an enhancing effect at both initiation and promotion steps within the two stage model system of the gastric carcinogenesis, and that these effects of NaCl are possibly related to its mucosal damaging activity.

Gastric cancer remains the commonest cause of cancer death in Japan (*1*). The gastric cancer rate is also elevated in Chile, Colombia, Iceland and Eastern Europe (*2*, *3*), China (*4*), and the colored population of South Africa (*5*). While racial differences cannot explain this distribution (*6*) a number of epidemiological studies have suggested that dietary and life style factors are very closely related to the incidence of this cancer (*7*, *8*). A striking report on this subject concerned a study of immigrants from Japan, a high-incidence area, to the United States, a low-incidence area, suggesting that differences in their life style, especially in the diet, were responsible for a decrease in the incidence of gastric cancer in the second generation (*9*). One proposed etiological agent is salt. Sato *et al.* (*10*) demonstrated a geographical correlation between gastric cancer mortality and the concentration of salt used in soybean paste in Japan. Hirayama (*11*) also found a close association of excessive intake of salted pickles with gastric cancer.

Experimentally, a large number of studies on animal models of gastric carcinogenesis have been performed especially since the introduction of N-methyl-N′-nitro-N-nitrosoguanidine (MNNG) (*12*). This direct acting carcinogen induces forestomach, glandular stomach, and duodenal tumors when administered to rodents in their drinking water or by intragastric intubation.

Neoplasms of the stomach induced by oral administration of MNNG are strikingly similar in most respects to gastric tumors in man (*13, 14*). Utilizing this compound as an initiator, we have established a two-stage gastric carcinogenesis model on the prevailing assumption that, in line with several other organs, the carcinogenic process consists of at least two stages, initiation and promotion. Using this two-stage model, we have examined the effects of NaCl administration on gastric carcinogenesis.

Effect of Salt as a Co-initiator of Gastric Carcinogenesis

Previously we demonstrated in rats that exceedingly high doses of NaCl given concurrently with carcinogen increased tumor incidences in the forestomach induced by 4-nitroquinoline 1-oxide and in the glandular stomach induced by MNNG (*15*). More recently we confirmed this co-initiation activity of NaCl in experimental gastric carcinogenesis (Experiment I) (*13*).

Male outbred Wistar rats received MNNG in the drinking water (100 mg/*l*) with or without simultaneous administration of 10% NaCl in the diet for 20 weeks. Sacrifice at week 40 revealed a clear increase in both the incidence and size of tumors induced in the pyloric mucosa in the group receiving the combined initiation treatment. On the other hand, administration of NaCl during the second 20 week period,

TABLE 1. Effect of High Salt Diet on Gastroduodenal Tumor Development in Male Wistar Rats Given MNNG in the Drinking Water (Experiment I)

Group	Treatment 20w	Treatment 20w	No. of effective animals	Fore-stomach: Papilloma	Glandular stomach, Fundus: Adeno-carcinoma	Glandular stomach, Pylorus: Adeno-carcinoma	Glandular stomach, Pylorus: Preneoplastic hyperplasia	Duodenum: Adeno-carcinoma
1	MNNG +NaCl	NaCl	19	0	3 (18)	12 (63)[a,b]	5	6 (32)[b]
2	MNNG +NaCl		20	0	0	16 (80)[c]	3	8 (40)[c]
3	MNNG	NaCl	19	0	0	4 (21)	8	5 (26)
4	MNNG		20	0	0	5 (25)	10	2 (10)
5	NaCl	NaCl	20	0	0	0	1	0

[a] Significantly different from group 3 at $p<0.05$ by Fisher exact test. [b] Significantly different from group 5 at $p<0.05$ by Fisher exact test. [c] Significantly different from group 4 at $p<0.05$ by Fisher exact test. Numbers in parentheses indicate percentage values.

after removal of the carcinogen, did not result in any significant enhancement of tumor development in the pyloric area. Although fundic adenocarcinomas were observed only in the group administered both carcinogen and NaCl, the low incidence precluded establishment of statistical significance (Table 1).

These data indicate that excess intake of NaCl enhances carcinogenesis only at the initiation stage. Shirai *et al.* (*16*) also reported an enhancing effect of a single oral intubation of 1 ml saturated NaCl solution 24 hr prior to a single oral dose of MNNG on the induction of adenocarcinoma in the glandular stomach. However, this administration of MNNG by gastric tube induced multiple tumors in the forestomach and only very few tumors in the glandular stomach (*16*, *17*).

Previously we demonstrated that concomitant administration of several surfactants with MNNG enhanced the resultant incidence of the more malignant and poorly differentiated type of adenocarcinoma (*18*, *19*). The presence of surfactant may be effective in facilitating absorption of the carcinogen and thereby enhancing the direct action of the carcinogen on the target cells. It was also shown that after induction of chronic ulcers by administration of iodoacetamide (*20*), by freezing (*21*), or by intramural injections of formalin (*22*), subsequent ingestion of MNNG produced cancers predominantly at the ulcer sites. As observed in other organs, cells undergoing replication are considered more susceptible to the carcinogenic or initiating action of chemical carcinogens.

Another factor of relevance to the carcinogenic action of MNNG is the gastric luminal pH and self-protective function of the gastric mucosa. Lowering of the pH by gastrin given during MNNG treatment was demonstrated to enhance tumor yield (*23*). NaCl and other mucosal damaging agents produces a gastric alkaline response (diffusion of HCO_3^-) possibly stimulating formation of prostaglandins by the gastric mucosa as a component of self-protection or repair processes (*24*). It is possible that chronic ingestion of NaCl changes this self-protective mechanism of the mucosa and affects the carcinogenic action of MNNG.

Promoting Effect of Salt in Two-step Gastric Carcinogenesis

It is widely accepted that the unfolding of the carcinogenic process can be accelerated subsequent to initiation by administration of promoting agents. While previous work (*13*, *16*) failed to demonstrate appreciable promoting effects of NaCl for gastric carcinogenesis, the question was raised whether this was due to an inappropriate experimental design (*21*). Therefore, utilizing the two-step carcinogenesis model with MNNG plus high salt diet as the initiator, we re-examined the promoting potential of NaCl for gastric carcinogenesis and compared the results with the actions of sodium saccharin, a known bladder promoter, phenobarbital, a hepatic promoter, and aspirin, a mucosal damaging agent (Experiment II) (*14*).

Male outbred Wistar rats were given MNNG in their drinking water (100 mg/*l*) for 8 weeks, and during this period were fed on diet supplemented with 10% NaCl. Thereafter, they were fed on the basal diet or diet supplemented with either 10% sodium chloride, 5% sodium saccharin, 0.05% phenobarbital, or 1% aspirin for 32 weeks. At final sacrifice (week 40) the incidence of adenocarcinoma

TABLE 2. Promoting Effects of Various Chemicals on Gastroduodenal Carcinogenesis Initiated by MNNG and NaCl in Male Wistar Rats

Group No.	Chemicals	Total No. of rats	No. of tumor bearing animals (%)	Number of animals (%) with tumor in					
				Forestomach	Glandular stomach				Duodenum
					Fundus		Pylorus		
				Papilloma	Adeno-carcinoma	Hyperplasia	Adeno-carcinoma	Preneoplastic hyperplasia	Adeno-carcinoma
Experiment II									
1	None	39	6 (15.4)	0	0	0	3 (7.7)	5 (12.8)	3 (7.7)
2	NaCl	20	7 (35.0)	0	0	0	4 (20.0)	8 (40.0)[a]	3 (15.0)
3	Saccharin	20	5 (25.0)	0	0	0	3 (15.0)	5 (25.0)	1 (5.0)
4	Pheno-barbital	19	2 (10.5)	0	0	0	1 (5.3)	4 (21.1)	1 (5.3)
5	Aspirin	20	1 (5.0)	0	0	0	0	2 (10.0)	1 (5.0)
Experiment III									
1	None	30	4 (13.3)	0	0	0	1 (3.3)	7 (23.3)	3 (10.0)
2	Ethanol	21	2 (9.5)	0	0	1 (4.8)	0	1 (4.8)	2 (9.5)
3	Potassium metabisulfite	19	6 (31.6)	2 (10.5)	0	1 (5.3)	5 (26.3)[a]	4 (21.1)	1 (5.3)
4	Form-aldehyde	17	5 (29.4)	15 (88.2)[b,c]	0	15 (88.2)[b]	4 (23.5)[a]	7 (41.2)	1 (5.9)
5	Hydrogen peroxide	21	2 (9.5)	21 (100)[b,c]	0	8 (38.1)[b]	2 (9.5)	6 (28.6)	0

[a,b] Significantly different from each control (None) at $p<0.05$ and $p<0.001$, respectively, by Fisher exact test. [c] Not significantly different from group treated with the chemical alone without the initiation procedure. Numbers in parentheses indicate percentage values.

was increased, but not significantly, in the group given NaCl following the initiation procedure as compared with the respective control group (Table 2). However, the incidence of preneoplastic hyperplasia was significantly increased. Sodium saccharin also enhanced the development of adenocarcinomas of the glandular stomach. The results indicated a tendency for dietary administration of both sodium chloride and sodium saccharin to promote tumor development. Phenobarbital did not enhance the tumor development and aspirin rather showed a tendency to decrease tumor incidence.

In a further study (Experiment III), we examined the promoting potentials of other mucosal damaging agents, ethanol, potassium metabisulfite, formaldehyde, and hydrogen peroxide using the two-step model described above. The results are shown in Table 2 (*25*). The incidence of adenocarcinoma in the pyloric mucosa was increased, but not significantly, in the groups given 1% potassium metabisulfite and 0.5% formalin (formaldehyde). Neither 10% ethanol nor 1% hydrogen peroxide showed any enhancement of tumor development. Papillomas in the forestomach were observed in formaldehyde- and hydrogen peroxide-treated rats, but the incidence of this lesion did not differ between the groups with or without the initiation treatment.

These results strongly suggest that NaCl, in addition to possessing co-initiator action, may also exert a promoting influence on gastric carcinogenesis. In contrast, ethanol and aspirin do not appear active in this respect. While it remains unclear why different mucosal damaging agents vary in their tumor promoting activity, one possibility is that the distribution of injuries within the gastric mucosa differs with individual agents, and another is that the effect on intramucosal prostaglandin, an important factor for mucosal self-protection (*26*), is variable.

Recently, Salmon *et al.* (*27*) and Kobori *et al.* (*28*) reported a significant enhancement of MNNG-initiated gastric tumorigenesis by dietary administration of taurocholic acid and sodium taurocholate, indicating that bile salts could also act as gastric promoters.

Effect of NaCl on the Gastric Mucosa

Male outbred Wistar rats were continuously administered 10% NaCl in the diet. Five rats each were sacrificed at weeks 5, 10, and 20, and 10 rats at week 30 for histopathological, autoradiographic, histochemical, and electron microscopic examination.

1. *Histopathological examination*

Histopathological changes were more remarkable in the fundic mucosa than in the pyloric area. At week 5, mild erosion of the surface and marked hyperplasia of the foveolar epithelium and slight dilatation of the pits were observed throughout the fundic mucosa. Lesions were most remarkable in the mucosa adjacent to the limiting ridge of both anterior and posterior walls. Cells with hyaline degeneration of the cytoplasm and pyknotic nuclei were prominent in this area, and superficial erosion was observed in 2 of 5 rats at this time. After 10 or more weeks of the treatment, foveolar hyperplasia and atrophy of the fundic glands were diffusely observed, and focal mild erosions were evident in the fundic area. In several animals, focal hyperplasia of foveolar epithelium accompanying atrophy of the surrounding fundic glands was apparent. No significant differences were observed between rats killed at weeks 10, 20, or 30.

In the pyloric mucosa, slight dilatation of the pits was the only change observed in rats given 10% NaCl in the diet, and erosion of the mucosa was rare. Pyloric metaplasia of the fundic mucosa was not evident in this experiment.

2. Autoradiography

After chronic ingestion of 10% NaCl, the number of cells labeled with ^{3}H-thymidine injected 1 hr before sacrifice was increased by 1.4- to 6.6-fold over control levels in both fundic and pyloric mucosa (Fig. 1). This regenerative response was observed in all regions of the gastric mucosa at each time point examined.

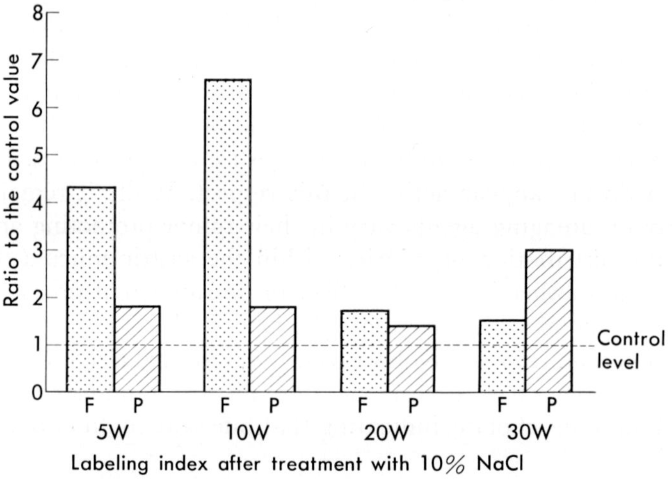

FIG. 1. Results of autoradiography employing ^{3}H-thymidine intraperitoneally injected 1 hr before sacrifice. All labeled cells on the basement membrane were counted per unit length of fundus (F) and pylorus (P). Vertical axis is ratio of the value in treated rats to the respective control value for each region. The numbers of labeled cells were increased by the treatment with NaCl in the diet in all regions.

3. Histochemical examination

Stomach samples fixed in 10% buffered formalin or ice-cold acetone solution were processed for mucin-histochemical investigation by periodic acid-Schiff (PAS) reaction, high-iron diamine-alcian blue (pH 2.5) staining (HID-AB), and paradoxical concanavalin A staining (Con A) for type III mucin, and for immunohistochemical localization of pepsinogen isozyme 1 (Pg 1).

In the control sections, the surface epithelium consisted entirely of PAS-positive cells which extended into the upper two-thirds of the gastric pits. The foveolar cells also contained acidic mucin. By the HID-AB sequence, sialomucin was predominant in upper foveolar surface mucous cells and sulfomucin was predominant in lower foveolar cells. While the upper region of the pyloric glands contained sulfomucin,

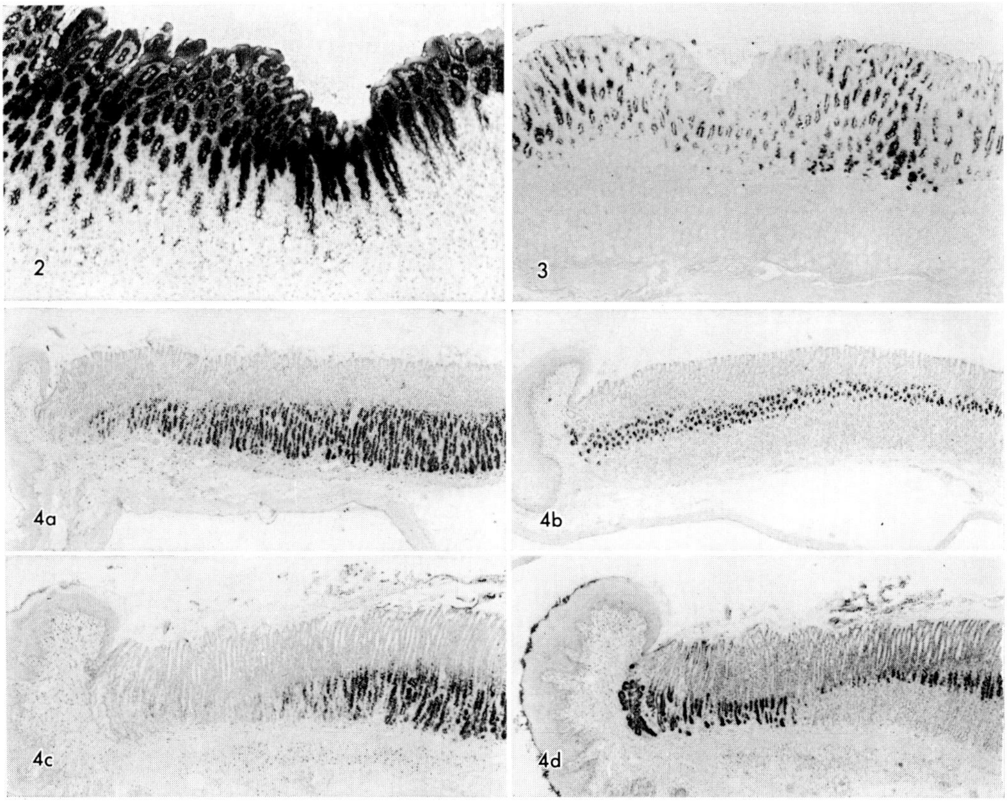

FIG. 2. PAS-AB positive mucin increased in the superficial layer after 5 weeks NaCl treatment. ×200.

FIG. 3. Increased acidic mucin content in the hyperplastic lesion. Black staining (looks darker) sulfomucin is predominant in a rat treated with NaCl for 10 weeks (HID-AB). ×100.

FIG. 4. Histochemistry of the fundic mucosa adjacent to the limiting ridge. ×80. a: Pg 1 in control rat, b: Con A type III mucin in control rat, c: Pg 1 in NaCl-treated rat after 5 weeks, d: Con A type III mucin in NaCl-treated rat after 5 weeks. Atrophy of the fundic glands and increased number of mucous neck cells (or cardiac gland cells) are evident.

the lower regions predominantly stained for sialomucin. The superficial surface mucous cells and mucous neck cells gave no reaction with the HID-AB stain. By the Con A method, stable class III mucin was found in mucous neck and pyloric gland cells. Immunohistochemically Pg 1 showed positive staining in the chief cells and pyloric gland cells. These findings are in agreement with previous reports (*29–33*).

After treatment with NaCl, PAS-positive mucin and both sialomucin and sulfomucin demonstrated by HID-AB staining were increased in the superficial layer of the gastric mucosa, especially in the hyperplastic and/or regenerating area (Figs. 2 and 3). The number of mucous neck cells and the width of their zone in fundic mucosa, observed as cells with clear cytoplasm by hematoxylin and eosin staining, was diffusely increased by NaCl administration, especially in the area close to the limiting ridge (Figs. 4 and 5). Similar changes were reported with MNNG treatment in the drinking water to rats (*29*). Fundic gland cells decreased in number which was

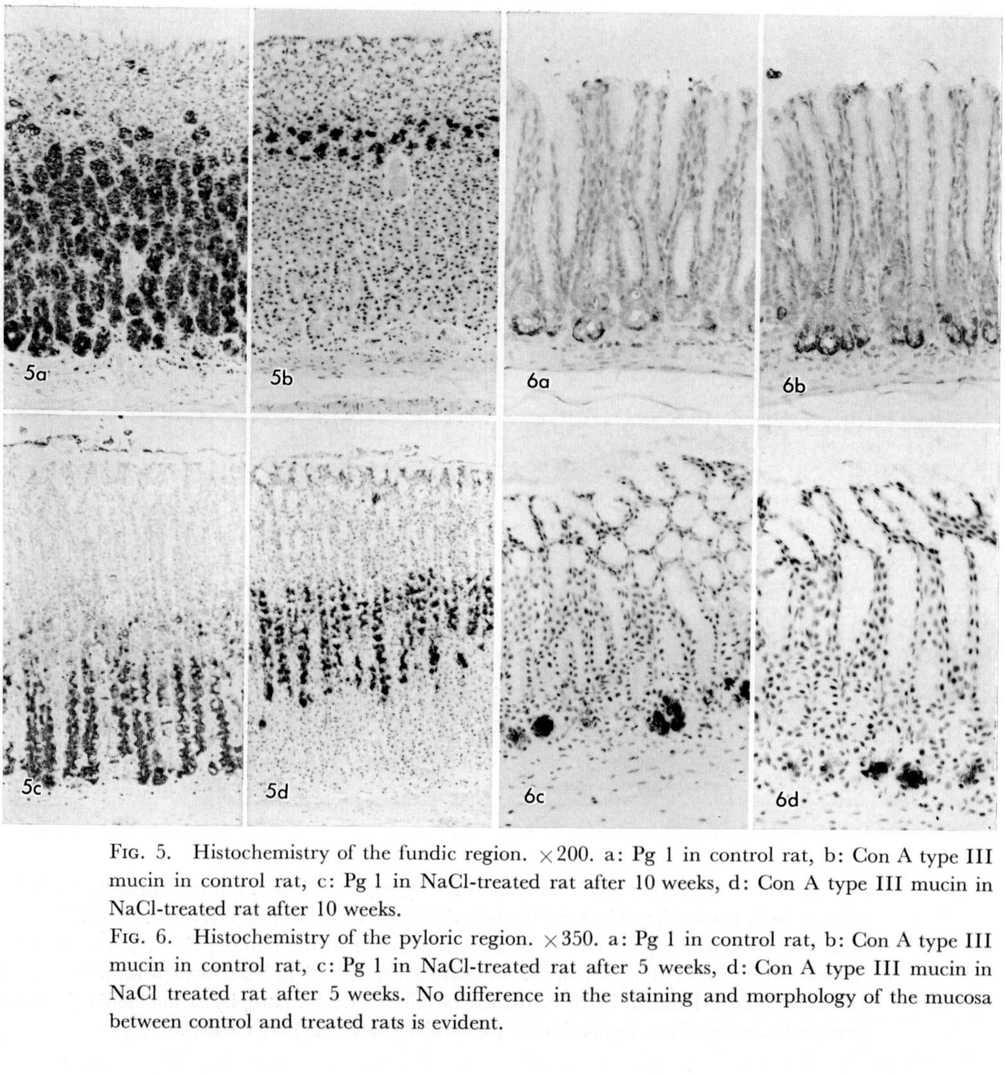

FIG. 5. Histochemistry of the fundic region. ×200. a: Pg 1 in control rat, b: Con A type III mucin in control rat, c: Pg 1 in NaCl-treated rat after 10 weeks, d: Con A type III mucin in NaCl-treated rat after 10 weeks.

FIG. 6. Histochemistry of the pyloric region. ×350. a: Pg 1 in control rat, b: Con A type III mucin in control rat, c: Pg 1 in NaCl-treated rat after 5 weeks, d: Con A type III mucin in NaCl treated rat after 5 weeks. No difference in the staining and morphology of the mucosa between control and treated rats is evident.

clear evidence of a decrease in the height of glands positive for Pg 1 immunohistochemistry (Fig. 5). In the pyloric mucosa, no remarkable changes due to the treatment were observed by mucin histochemical or Pg 1 immunohistochemical methods (Fig. 6), except for slightly increased staining for acidic mucins (sialomucin/sulfomucin) in the superficial epithelium of the pits. While the content of Pg 1 has been demonstrated to decrease in focal populations during experimental gastric carcino-

FIG. 7. Scanning electron micrographs of the fundic mucosa of a rat treated with NaCl for 30 weeks. a: Damage of the superficial cells is diffusely observed (×150), b: Higher magnification shows loosely attached epithelial cells and loss of cells, although denudation of the lamina propria is not evident (×700).

FIG. 8. Transmission electron micrograph of the fundic mucosa of a rat given 10% NaCl for 30 weeks. Two necrotic epithelial cells are protruding into the lumen. Many intracellular vacuoles and swelling of mitochondria are observed. ×3,000.

genesis (*34–36*), no remarkable change in pyloric gland staining pattern was evident in the present experiment.

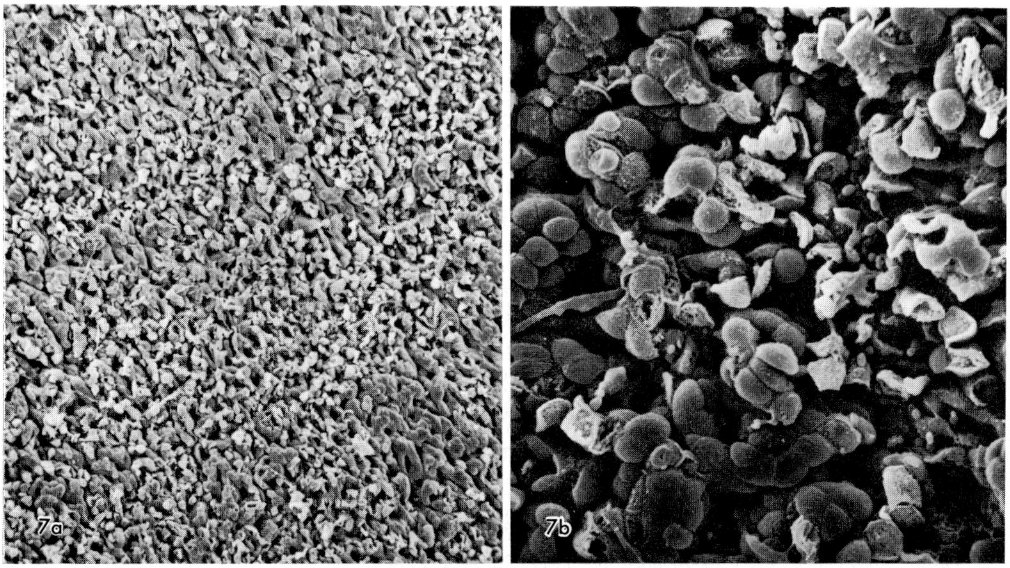

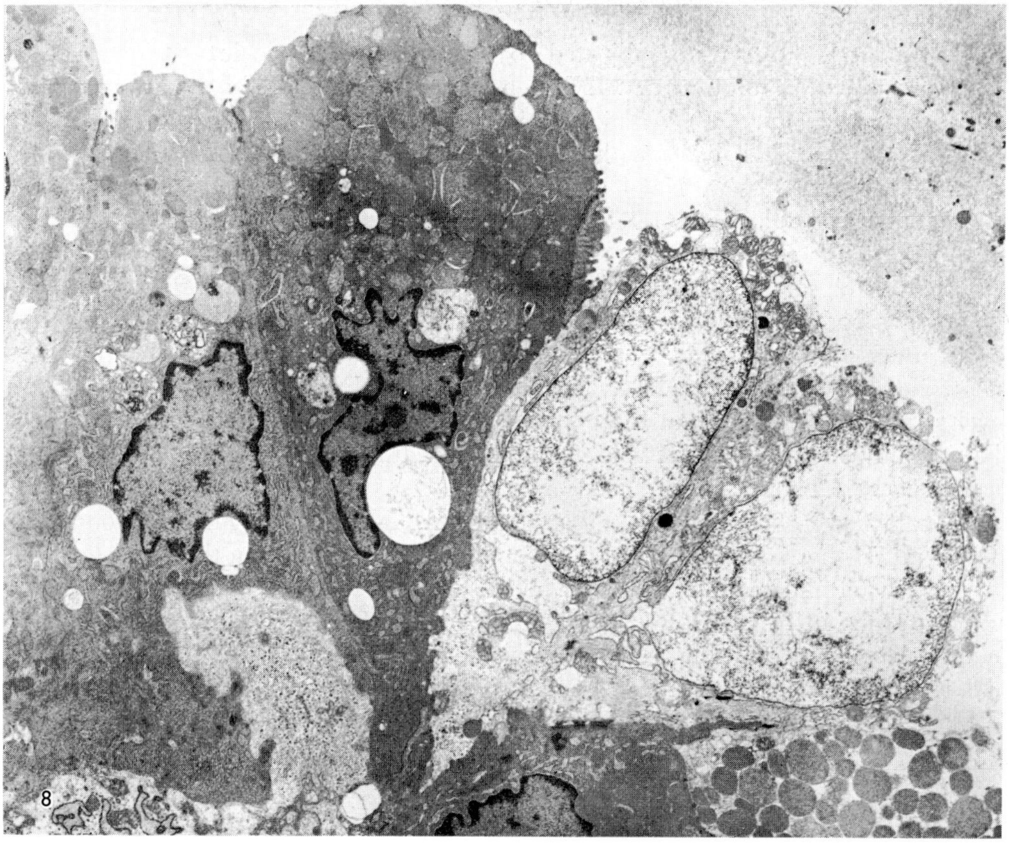

Although the enhancing effect of NaCl on MNNG-initiated gastric carcinogenesis was evident in the pyloric mucosal lesions, the histochemical studies revealed more prominent NaCl-dependent changes in the fundic mucosa.

4. Electron microscopic observations

Scanning electron microscopy of the gastric mucosal surface after treatment with 10% NaCl in the diet for 5–30 weeks revealed various degrees of focal exfoliation of the surface epithelial cells throughout the glandular stomach, but more prominently in the fundic region (Fig. 7). Denuded lamina propria was rarely observed. The changes were similar in rats killed after different periods of treatment.

By transmission electron microscopy, decrease in numbers of chief and parietal cells and increase in number of mucous neck cells were obvious. In the epithelial cells of the pits, various signs of degeneration were apparent in the cell cytoplasm not only in the upper regions but also in the deeper pits. Increased numbers of cells containing vacuoles and increased vacuoles in each cell were apparent and large lysosomes were occasionally observed in the cytoplasm. Protrusion and extrusion of the epithelial cells demonstrating extensive vacuolar changes and other signs of cellular degeneration were more frequently observed than in normal epithelium (Fig. 8).

5. Induction of ornithine decarboxylase (ODC) and DNA synthesis

Male Fischer rats were subjected to a single intragastric intubation of NaCl as 1 ml of aqueous solution or MNNG as 0.5 ml of solution in dimethylsulfoxide after overnight fasting (*37*). Administration of concentrated aqueous NaCl solution induced macroscopical ulcers within 3 hr in the glandular stomach mucosa which disappeared within 24 hr. NaCl at doses of 0.25 to 1.5 g/kg body weight caused dose-dependent induction of ODC in both pyloric and fundic mucosa. ODC increased from 3 to 16 hr with a peak of more than 200-fold control levels at 6 to 9 hr after treatment with saturated aqueous NaCl solution (1.5 g/kg body weight). MNNG (200 mg/kg body weight) also strongly induced ODC activity in the pyloric and fundic mucosa after between 16 and 48 hr with a maximum at 24 hr after the treatment. NaCl thus induced ODC earlier than did MNNG and the increased ODC level returned to the original level earlier than with MNNG. This study suggested that the induction of ODC is a useful marker of tumor promoting potential as has been suggested for the skin.

Small pieces of stomach pyloric mucosa were incubated in medium containing 10 μCi/ml ^{3}H-thymidine at 37°C for 2 hr and the DNA content in the extracted fraction was determined. DNA synthesis increased 5 to 7 times from 3 to 24 hr after administration of NaCl. Dose-dependent induction of DNA synthesis was observed in the pyloric mucosa after NaCl administration. DNA synthesis increased earlier and returned to the original level sooner than after MNNG administration. This DNA synthetic response is possibly due to a response to damage of the mucosa by NaCl.

CONCLUSION

Although understanding of experimental gastric carcinogenesis has recently progressed by utilization of different animal models, the results have yet to be translated into an effective means for controlling the human disease. However, the availability of experimental models like that used in the present series of experiments permits evaluation of risk factors, such as high salt consumption, and their relative importance in the initiation and promotion stages of multi-step carcinogenesis.

Co-initiating activity of NaCl may be related to both a disturbance of the mucous barrier, thereby enhancing the penetration of MNNG to the mucosa, and also repeated injury to the gastric mucosa resulting in increased cell proliferation. A similar co-carcinogenic effect of NaCl with MNNG was reported in mice (*38*), and similar effect of aspirin was reported in rats when co-administered with MNNG (*39*).

The present study provides evidence suggestive of a promotional role for dietary NaCl in gastric carcinogenesis. Although the mechanism remains unclear, one possibility, as mentioned above, is that high doses of NaCl have hyperplaseogenic potential. In the present model, the enhancement of tumor induction with NaCl was associated with diffuse proliferative changes in the superficial epithelium and increased generative zone of the glandular stomach. In contrast, phenobarbital and ethanol showed neither effect. Although erosive lesions could be seen with aspirin which showed no promoting effect, the lesions were focal in nature and no diffuse hyperplasia was apparent. Promotion potential was demonstrated only for potassium metabisulfite and formaldehyde in the mucosal damaging agents. The role of these mucosal damaging compounds in gastric carcinogenesis is of importance, since they may possess co-initiating or co-carcinogenic effects. Human exposure to N-nitroso compounds, which are widespread environmental chemicals, through food-stuffs, water, air, industrial and consumer products, and moreover, by their spontaneous formation in the stomach from nitrites and amides is important since we may take these with NaCl in daily life (*8*).

Induction of ODC activity and DNA synthesis in the glandular stomach mucosa of rats was observed with NaCl as well as with MNNG. This phenomenon is reminiscent of the changes in ODC and DNA synthesis observed after 12-O-tetradecanoylphorbol-13-acetate (TPA) or mezerein treatment of skin, although much higher concentrations of NaCl were necessary, both for significant induction and promoting activity in the gastric model.

ACKNOWLEDGMENTS

We gratefully acknowledge Dr. M. A. Moore for assistance in the preparation of this manuscript. This work was supported in part by the Princess Takamatsu Cancer Research Fund and by a Grant-in-Aid for Cancer Research from the Ministry of Education, Science and Culture of Japan.

REFERENCES

1. Kuroishi, T., Tominaga, S., Hirose, K., and Segi, M. Cancer mortality in Japan. GANN Monogr. Cancer Res., *26*: 1–91, 1981.

2. Gregor, O. Gastric cancer control. Neoplasma, *21*: 235–247, 1974.
3. American Cancer Society. Cancer facts and figures. Am. Cancer Soc., New York, 1982.
4. Armstrong, B. The epidemiology of cancer in the People's Republic of China. Int. J. Epidemiol., *9*: 305–315, 1980.
5. Harrington, J. S. Advances in cancer epidemiology in South Africa. South Afr. Cancer Bull., *25*: 9–18, 1981.
6. Charnley, G., Tannenbaum, S. R., and Correa, P. Gastric cancer: an etiologic model. *In;* P. N. Magee, (ed.), Nitrosamines and Human Cancer. Banbury Report 12, pp. 503–522, Cold Spring Harbor Laboratory, New York, 1982.
7. Coggon, D. and Acheson, E. D. The geography of cancer of the stomach. Br. Med. Bull., *40*: 335–341, 1984.
8. Mirvish, S. S. The etiology of gastric cancer: intragastric nitrosamide formation and other theories. J. Natl. Cancer Inst., *71*: 629–647, 1983.
9. Haenszel, W., Kurihara, M., and Segi, M. Stomach cancer among Japanese in Hawaii. J. Natl. Cancer Inst., *49*: 969–988, 1972.
10. Sato, T., Fukuyama, T., Suzuki, T., Murakami, T., Shiotsuki, N., Tanaka, R., and Tsuji, R. Studies of the causation of gastric cancer. 2. The relation between gastric cancer mortality rate and salted food intake in several places in Japan. Bull. Inst. Publ. Health, *8*: 187–198, 1959.
11. Hirayama, T. Epidemiology of stomach cancer. GANN Monogr. Cancer Res., *11*: 3–19, 1971.
12. Sugimura, T. and Fujimura, S. Tumour production in glandular stomach of rats by N-methyl-N′-nitro-N-nitrosoguanidine. Nature, *216*: 943–944, 1967.
13. Takahashi, M., Kokubo, T., Furukawa, F., Kurokawa, Y., Tatematsu, M., and Hayashi, Y. Effect of high salt diet on rat gastric carcinogenesis induced by N-methyl-N′-nitro-N-nitrosoguanidine. Gann, *74*: 28–34, 1983.
14. Takahashi, M., Kokubo, T., Furukawa, F., Kurokawa, Y., and Hayashi, Y. Effects of sodium chloride, saccharin, phenobarbital and aspirin on gastric carcinogenesis in rats after initiation with N-methyl-N′-nitro-N-nitrosoguanidine. Gann, *75*: 494–501, 1984.
15. Tatematsu, M., Takahashi, M., Fukushima, S., Hananouchi, M., and Shirai, T. Effects in rats of sodium chloride on experimental gastric cancers induced by N-methyl-N′-nitro-N-nitrosoguanidine or 4-nitroquinoline-1-oxide. J. Natl. Cancer Inst., *55*: 101–106, 1975.
16. Shirai, T., Imaida, K., Fukushima, S., Hasegawa, R., Tatematsu, M., and Ito, N. Effects of NaCl, Tween 60 and a low dose of N-ethyl-N′-nitro-N-nitrosoguanidine on gastric carcinogenesis of rats given a single dose of N-methyl-N′-nitro-N-nitrosoguanidine. Carcinogenesis, *3*: 1419–1422, 1982.
17. Hirono, I. and Shibuya, C. Induction of stomach cancer by a single dose of N-methyl-N′-nitro-N-nitrosoguanidine through a stomach tube. *In;* W. Nakahara, S. Takayama, T. Sugimura, and S. Odashima (eds.), Topics in Chemical Carcinogenesis, pp. 121–131, Japan Sci. Soc. Press, Tokyo, 1972.
18. Fukushima, S., Tatematsu, M., and Takahashi, M. Combined effect of various surfactants on gastric carcinogenesis in rats treated with N-methyl-N′-nitro-N-nitrosoguanidine. Gann, *65*: 371–376, 1974.
19. Takahashi, M., Fukushima, S., and Sato, H. Carcinogenic effect of N-methyl-N′-nitro-N-nitrosoguanidine with various kinds of surfactant in the glandular stomach of rats. Gann, *64*: 211–218, 1973.

20. Takahashi, M., Shirai, T., Fukushima, S., Hananouchi, M., Hirose, M., and Ito, N. Effect of fundic ulcers induced by iodoacetamide on development of gastric tumors in rats treated with N-methyl-N′-nitro-N-nitrosoguanidine. Gann, *67*: 47–54, 1976.
21. Takahashi, M., Shirai, T., Fukushima, S., Ito, N., Kokubo, T., and Kurata, Y. Ulcer formation and associated tumor production in multiple sites within stomach and duodenum of rats treated with N-methyl-N′-nitro-N-nitrosoguanidine. J. Natl. Cancer Inst., *67*: 473–479, 1981.
22. Saito, T., Sasaki, O., Matsukuchi, T., Iwamatsu, M., and Inokuchi, K. Experimental gastric cancer: pathogenesis and clinicohistopathologic correlation. *In;* C. H. Herfarth and P. Schlag (eds.), Gastric Cancer, pp. 22–31, Springer-Verlag, Berlin, 1979.
23. Tahara, E., Shimamoto, F., Taniyama, K., Ito, H., Kosako, Y., and Sumiyoshi, H. Enhanced effect of gastrin on rat stomach carcinogenesis induced by N-methyl-N′-nitro-N-nitrosoguanidine. Cancer Res., *42*: 1781–1787, 1982.
24. Nobuhara, Y. and Takeuchi, K. Possible role of endogenous prostaglandins in alkaline response in rat gastric mucosa damaged by hypertonic NaCl. Dig. Dis. Sci., *29*: 1142–1147, 1984.
25. Takahashi, M., Hasegawa, R., Furukawa, F., Toyoda, K., Sato, H., and Hayashi, Y. Effects of ethanol, potassium metabisulfite, formaldehyde and hydrogen peroxide on gastric carcinogenesis in rats after initiation with N-methyl-N′-nitro-N-nitrosoguanidine. Jpn. J. Cancer Res. (Gann), *77*: 118–124, 1986.
26. Robert, A. Cytoprotection by prostaglandins. Gastroenterology, *77*: 761–767, 1979.
27. Salmon, R. J., Laurene, M., and Thierry, J. P. Effect of taurocholic acid feeding on methyl-nitro-N-nitrosoguanidine induced gastric tumors. Cancer Lett., *22*: 315–320, 1984.
28. Kobori, O., Watanabe, J., Shimizu, T., Shoji, M., and Morioka, Y. Enhancing effect of sodium taurocholate on N-methyl-N′-nitro-N-nitrosoguanidine-induced stomach tumorigenesis in rats. Gann, *75*: 651–654, 1984.
29. Tatematsu, M., Katsuyama, T., Fukushima, S., Takahashi, M., Shirai, T., Ito, N., and Nasu, T. Mucin histochemistry by paradoxical concanavalin A staining in experimental gastric cancers induced in Wistar rats by N-methyl-N′-nitro-N-nitrosoguanidine or 4-nitroquinolin 1-oxide. J. Natl. Cancer Inst., *64*: 835–843, 1980.
30. Berrisford, R. G., Wells, M., and Dixon, M. F. Gastric epithelial mucus—a densitometric histochemical study of aspirin induced damage in the rat. Br. J. Exp. Pathol., *66*: 27–33, 1985.
31. Lev, R. The mucin histochemistry of normal and neoplastic mucosa. Lab. Invest., *14*: 2080–2099, 1965.
32. Goldman, H. and Ming, S. C. Mucins in normal and neoplastic gastrointestinal epithelium: histochemical distribution. Arch. Pathol., *85*: 580–586, 1968.
33. Katsuyama, T. and Spicer, S. S. Histochemical differentiation of complex carbohydrates with variants of the concanavalin A horseradish peroxidase method. J. Histochem. Cytochem., *26*: 233–250, 1978.
34. Furihata, C., Tatematsu, M., Shirai, T., Yokochi, K., Takahashi, M., and Sugimura, T. Pepsinogen and stomach cancer. *In;* E. Farber *et al.* (eds.), Pathophysiology of Carcinogenesis in Digestive Organs, pp. 49–63, Japan Sci. Soc. Press, Tokyo/Univ. Park Press, Baltimore, 1977.
35. Tatematsu, M., Furihata, C., Hirose, M., Shirai, T., Ito, N., Nakajima, Y., and Sugimura, T. Changes in pepsinogen isozymes in stomach cancers induced in Wistar rats by N-methyl-N′-nitro-N-nitrosoguanidine and in transplantable gastric carcinoma (SG2B). J. Natl. Cancer Inst., *58*: 1709–1716, 1977.

36. Tatematsu, M., Furihata, C., Masui, T., Shirai, T., Mera, Y., and Nakatsuka, T. Dose dependence and sequential nature of pyloric gland lesions demonstrating low Pg 1 activity in rats treated with MNNG. Proc. Jpn. Cancer Assoc., *44*: 48, 1985.
37. Furihata, C., Sato, Y., Hosaka, M., Matsushima, T., Furukawa, F., and Takahashi, M. NaCl induced ornithine decarboxylase and DNA synthesis in rat stomach mucosa. Biochem. Biophys. Res. Commun., *121*: 1027–1032, 1984.
38. Ohno, Y., Kurokawa, Y., Takahashi, M., Takamura, N., Imazawa, T., and Hayashi, Y. Effect of NaCl diet on jejunal carcinogenesis in mice given N-methyl-N′-nitro-N-nitrosoguanidine orally. Sci. Rep. Res. Inst. Tohoku Univ. C, *30*: 23–28, 1983.
39. Chang, T.-H., Lee, Y.-C., Lee, K.-Y., Sun, C.-H., and Chang, Y.-P. Cocarcinogenic action of aspirin on gastric tumors induced by N-nitroso-N-methylnitroguanidine in rats. J. Natl. Cancer Inst., *70*: 1067–1075, 1983.

DIET, NUTRITION AND CANCER, Y. HAYASHI ET AL. (EDS.),
JAPAN SCI. SOC. PRESS, TOKYO/VNU SCI. PRESS, UTRECHT, PP. 183–192, 1986

Non-Starch Polysaccharides as a Protective Factor in Human Large Bowel Cancer

Sheila A. Bingham

Medical Research Council and University of Cambridge, Dunn Clinical Nutrition Centre, Cambridge CB2 1QL, U.K.

Abstract: The hypothesis that lack of dietary fibre in the diet is responsible for a variety of large bowel problems, including cancer, has stimulated much discussion and research over the past 15 years. However, the epidemiological examination of this hypothesis has been hampered by the absence of data on the fibre content of most of the world's foods. In Scandinavia and Britain where the consumption of the major chemical fraction of dietary fibre, the non-starch polysaccharides has been measured using accurate methods, significant negative associations have been shown with large bowel cancer occurrence. These studies suggest that non-starch polysaccharides may be protective in populations at otherwise high risk of large bowel cancer from an excess of meat and fat. However, methodological problems in the assessment of non-starch polysaccharide consumption in individuals preclude the use of case control studies in verifying these associations within a single homogeneous population.

The hypothesis that dietary fibre is important in the aetiology of a number of non-infective large bowel disorders, including cancer and diverticular disease (*1*) has gained a measure of acceptance in recent years (*2*). However, a major difficulty with this hypothesis is that the mechanisms whereby fibre is thought to protect against these disorders are only suggestive. Fibre is known to decrease transit time, dilute intestinal contents, and increase faecal weight, partly because it is an energy substrate for bacteria in the colon leading to an increase in bacterial mass (*3*). In addition, the net effects of altered bacterial metabolism, such as reduced faecal pH, increased short chain fatty acid production and reduced faecal ammonia concentration, have all been suggested as protective factors in the aetiology of human large bowel cancer (*3*). Despite these suggested protective effects of dietary fibre however, the causative agent in human large bowel cancer is not established. Furthermore, other substrates for bacterial fermentation entering the colon may be equally important.

The major impetus for the fibre hypothesis was the comparative rarity of colon cancer and other large bowel disorders in rural Africa, where fibre intakes were thought to be high in comparison with westernised countries (*4*). This paper will concentrate on recent studies designed to assess whether or not a reduced intake of dietary fibre, or "deficiency" of it, is associated with a higher risk of large bowel cancer, both in populations and individuals.

Overview of Fibre Intakes and Large Bowel Cancer

The initial observations in Africa (*4*) were made using values for crude fibre, the deficiencies of which for human studies are well recognised (*5*). In addition, there are many other differences in diet and lifestyle between rural African and western populations, for example in starch, animal protein, and fat consumption, which have also been associated epidemiologically with large bowel cancer (*6*). Furthermore there is now uncertainty as to present day levels of fibre consumption, at least in South African black communities where large bowel disorders, including cancer, are rare (*7*). This uncertainty is due to the lack of values for the dietary fibre content of staple foods eaten in Africa, chiefly maize. This lack of analytical data is not confined to Africa and as we have pointed out elsewhere (*8*) the amount of fibre eaten in most countries of the world is unknown.

Despite the absence of firm data on fibre intakes worldwide, a number of investigators have attempted to correlate rates for large bowel cancer occurrence with national estimates of the consumption of fibre containing foods, or crude fibre intake values (*9–14*). In all of these international studies, no significant protective associations were found, or the associations became non-significant when controlling for meat or cholesterol consumption. The overriding factor emerging from these studies therefore is the positive association between meat consumption and colon cancer occurrence (*12*). In the most recent comparison, dietary fibre values were assigned to foods, using estimates from the British Food Tables (*15*) and a significant negative association ($r = -0.33$) did remain after controlling for meat consumption (*16*). However, the calculated intakes of fibre on which this comparison was based have been criticised (*17*).

The most recent and interesting data is from Japan, where despite the comparative rarity of large bowel cancer and diverticular disease (*18*, *19*), fibre intakes are no greater than in Britain where large bowel cancer is the second most common cancer (*20–23*). The reason for this is that the fibre content of rice is low when compared with other staple foods, such as white wheat bread (Table 1). As a consequence there have been no major changes in the Japanese fibre intake over the past 50

TABLE 1. Total Dietary Fibre (g per 100 g Fresh Weight)

Method	White rice	White bread	Cabbage
Hellendoorn *et al.* (*24*)	1.4	2.4	2.1
Southgate *et al.* (*25*)	2.4	2.7	3.4
Holloway *et al.* (*26*)	—	2.4	2.1
NDF (Van Soest (*27*))	—	1.5	1.1
Angus *et al.* (*28*)	1.7	2.0	1.0
Englyst *et al.* (*29*)	0.7	1.6	3.3

years (*21*). Fibre cannot therefore be the reason for the low rates of large bowel cancer and diverticular disease in Japan. This however does not exclude the possibility that fibre may be protective in populations otherwise assumed to be at risk from colon cancer from a westernised type of diet or lifestyle. The average consumption of meat in Japan for example is comparatively low, 70 g per day, compared with 160 g per day in Britain.

Our studies of dietary fibre intakes in relation to large bowel cancer have therefore been confined to western populations assumed to be at risk from a diet high in fat and meat. A critical problem however has been the choice of an accurate method for the analysis of dietary fibre in foods.

Definition and Analysis of Dietary Fibre

Food Table Values, based on well researched methods, are necessary for definitive epidemiological studies. At present however, there are a number of different methods for fibre analysis in use and, as can be seen from Table 1, all give different results for important staple foods such as white bread, potato, and rice.

Cummings has suggested (*30*) that analytical methods should confine themselves to measuring the chemically defined non-starch polysaccharides (NSP). The method of Englyst *et al.* (*29*) has therefore been developed to measure total NSP, uncontaminated with starch, and the cellulose, and the non-cellulosic polysaccharide fractions as the monosaccharides glucose, mannose, and galactose (the hexose sugars) and xylose and arabinose (the pentose sugars) and uronic acids. This method of analysis has been used in three types of epidemiological study.

Time Trends

A cornerstone of the hypothesis relating fibre intake to the aetiology of colon cancer was the suggested decline in intakes following the introduction of roller milling of wheat in Britain, Europe, and the U.S. around 1880. This allowed cheap white flour to be produced and distributed whereas previously it had been more expensive than brown. However, while it is clear that all types of bread were eaten throughout the 19th century, exactly how much was eaten by which social class is difficult to determine. In Britain, interest in the food supply of the general population and its nutritional value was only aroused at the beginning of the 20th century following the discovery of widespread and hitherto unrecognized malnutrition.

In countries with stable and well-documented population bases, per capita national statistics of food produced and imported with corrections for exported food are probably valid indicators of trends. In Britain, these statistics are available for selected years from 1909, and continuously from 1940. When interpreted to examine trends in fibre consumption (*31*). the most striking feature in Britain over the 20th century was a probable doubling of total dietary fibre intake due to an increase in the extraction rate of flour and increases in consumption of bread to conserve population food supplies during World War II. McMichael *et al.* (*32*) noted that time trends in colon cancer mortality rates could be associated in Britain and the U.S. with changes in crude fibre intakes, particularly in Britain, where there was a marked postwar fall in mortality. This analysis has recently been extended by Powles and Williams (*33*), using more accurate analyses for NSP consumption, based on the data of Englyst *et al.* (*29*), to include four other countries. The estimated war-time increases in NSP from flour were 15 g in Ireland, 12 g in England and Wales, 10 g in Switzerland, and 4 g in New Zealand, with virtually no change in Australia and the U.S. In these two latter countries the ratio of observed to expected mortaliy 11–15 years after the increase, extrapolating from prewar trends, was about 0.8, but only 0.6 in Ireland, the population that experienced the greatest increase in NSP consumption. Over the six populations studied, the correlation coefficient between these two variables was −0.88 (*33*). The most obvious problem with this type of analysis, however, is that changes in dietary habits rarely occur in isolation; in Britain, for example, fat consumption fell from 39% of total energy in the 1930s to 33% of total energy in 1947 (*34*).

Household Surveys

Food intakes assessed from household survey techniques on a national basis can provide information on regional dietary differences. In Britain, intakes of foodstuffs in the nine standard regions have been documented every year since 1940 by the British National Food Survey, under the auspices of the Ministry of Agriculture, Fisheries and Food (*35*). The housewife is asked to keep a record of her purchases for 1 week and from this the average quantity of food eaten per person per day in the household is calculated. Between 300 and 2400 households in nine standard regions of the U.K. take part every year, and the published reports show that there are consistent differences in nutrient and food intake between the regions. Butter and fat consumption is consistently higher in Wales for example, and beef in Scotland.

Average regional nutrient and food intake for the years 1969–73 were calculated and, in addition, regional intakes of NSP were estimated by direct analysis using the Englyst method (*36*) of food composites. Intakes were then compared with truncated

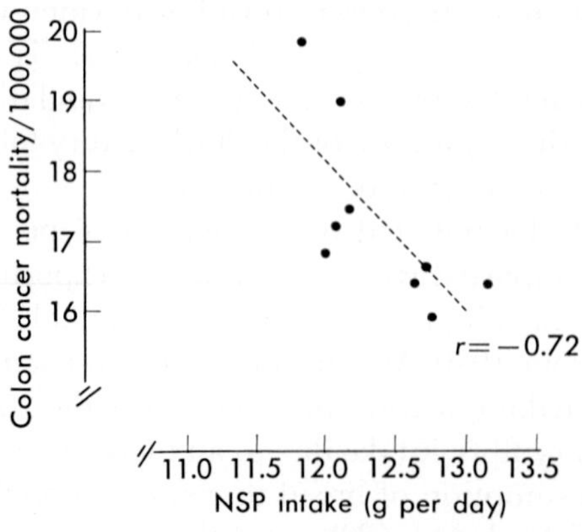

FIG. 1. Average regional NSP intakes in relation to age-truncated (35–64 years) average annual standardised colon cancer death rates, 1969–1973 in Great Britain (*22*).

age and sex-standardised mortality rates for colon and rectal cancer (*37*) in the same standard regions.

In summary, intakes of fat and animal protein were high and there was a significant negative association with vitamin C consumption ($r = -0.84$) (*37*). When the regional intakes of NSP were compared with death rates there were significant inverse correlations between colon cancer mortality and intakes of NSP ($r = -0.72$, $p < 0.05$) (Fig. 1) (*22*). We were however unable to confirm the strong inverse relationship between the pentose fraction of dietary fibre and colon cancer (*37*) although the association with vegetable consumption ($r = -0.94$) remained.

Studies of Representative Groups of Individuals

Established cancer registries in Scandinavia show a three- to four-fold range in colon cancer incidence within an area with a fairly homogeneous population. Age-standardized truncated colon cancer incidence rates per 100,000 men are lowest in rural Finland (6.7) and highest in Copenhagen (22.8). Rates in rural Denmark and Helsinki are intermediate (12.9 and 17.0). In collaboration with the International Agency for Research on Cancer, a study was undertaken to characterize the diets eaten by representative population samples in each area, with particular reference to the consumption of NSP using the method of Englyst (*38*). Thirty men aged 50–59 were randomly selected from population registers in each of the four areas: Parikkala in the Kymi region of southeast Finland; Them, in Jutland; Helsinki; and Copenhagen. Response rates among those approached were: Parikkala, 83.3%; Helsinki, 74.4%, Copenhagen, 75%, and Them, 62.5%.

Each of the 120 men was asked to keep a weighed food record for 4 days, in order to assess nutrient intake from local food tables. Because analyses for dietary fibre are generally not available in food tables, the amounts of NSP eaten in each of the four areas had to be established from direct chemical analysis of duplicate diets in Cambridge. However, it is notoriously difficult to collect complete duplicates of food eaten from free-living individuals, and a number of methodological checks were built into the protocol. On the final day of the survey, the men were asked to make a complete weighed duplicate of all food eaten, a 24 hr urine collection, and a 24 hr faecal collection. The weights of the food collected were compared with that recorded to have been eaten, and the chemically analysed content of fat, carbohydrate, and nitrogen (N) in the duplicate collections compared with calculated intakes for the same day, using food tables and the food records. In summary, as judged by the evidence available from weighing and analysis, the food samples were virtually complete duplicates. The overall correlation coefficient between protein intake calculated from food tables and that in the duplicates, for example, was 0.86, slope 0.99, intercept 2.3 g protein. The standard deviations of individual percentage differences between analysed and calculated intakes of protein ranged from 8 to 16% in the four areas, and those for fat from 7 to 21%. In addition, average dietary intake was equal to or related to average faecal and urine N excretion, with no evidence of systematic bias incurred as a consequence of attempting to obtain duplicate day's diets. There was also no evidence of a change in dietary habits during the survey, since a comparison of constituents of 24 hr urines collected on the day of the survey with those collected 3–6 weeks after the study showed no significant differences (*38*).

As in the British study, average intakes of fat and animal protein were high; fat ranging from 102 to 146 g/day, and meat consumption from 148 to 214 g/day. There were no significant associations with fat, cholesterol, or meat consumption and large bowel cancer incidence. The simple correlation coefficient between NSP consumption and large bowel cancer incidence was −0.776, although more sophisticated statistical techniques demonstrated a significant relationship (*38*).

Significance of Geographical Associations

Associations do not prove causation, but lack of association in this type of geographical comparative study would have been an important finding. In both Scandinavia and Great Britain, however, NSP consumption was negatively associated with large bowel cancer occurrence. Regression analysis has been conducted, and intercepts for zero colon and colorectal cancer incidence were obtained for 21 and 29 g NSP respectively (R. Doll, personal communication) in these areas of high meat and fat consumption.

The differences in average intake amongst the regions are small but should be viewed in the context of the likely distribution of individual intakes of NSP within the overall regional averages. With only a 1 g difference in the regional averages, 11% more of the population in Scotland would be eating less than the national average than in the Southeast area of England where rates for colon cancer are comparatively low. Hence, if risk from colonic cancer can be attributed to a low NSP diet, then there would be marked differences between the two populations in the numbers of individuals at risk from colon cancer. Figure 2 readily demonstrates the marked effect of an apparently small difference in the average between two groups (5 g NSP in this case) on the distribution of intakes of individuals within any one group. Whereas half the population in rural Finland were consuming more than 18 g NSP/day, only 10% in Copenhagen did so.

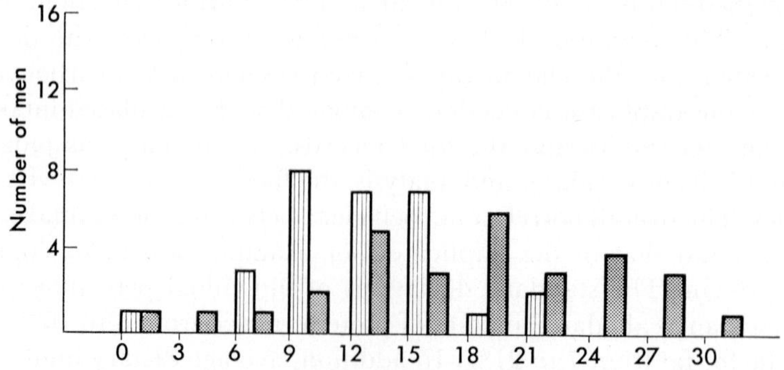

FIG. 2. Distribution of NSP intake (g/day) in two randomly selected groups of men aged 50–59 in Copenhagen (▥), and rural Finland (▩) (*38*).

Individual Studies

The epidemiologist looks for evidence of exposure from risk factors in relation to disease at the individual level, since in the group approach the data on exposure and outcome may relate to different groups in the population (*39*). The majority of case-control studies that have specifically looked at intakes of fibre or fibre-containing foods in large bowel cancer have not confirmed a role for dietary fibre in colon cancer. However, none have investigated NSP consumption.

Two well-controlled studies in Toronto and Adelaide (*40*, *41*) for example have recently been reported in which a quantitative estimate of fibre consumption was made using the British Food Tables. In both, the cases consumed more fat, but a protective effect of fibre was not seen, apart from slight weakening of the risk for males only consuming high-fat diets in Adelaide. In Toronto when cases and controls were subdivided into extremes of fibre consumption, there was no lessening of risk either when controlled or uncontrolled for fat consumption. This is despite the fact that individuals in the high extreme were eating more than 28 g of fibre per day, an amount greater than the average consumed by men in rural Finland, where the colon cancer risk is one of the lowest. However, numerical comparisons are difficult, because of the problems with systematic bias in methods of assessing food consumption and the analysis of dietary fibre. The particular method used in Toronto, for example, may overestimate crude fibre consumption by 56%, at least when compared with records of actual food eaten (*42*). Table 1 shows that the British Food Tables (Southgate analysis) may overestimate the NSP for common foods such as white bread by 50%. In addition, the methods used to assess diet in case control studies are prone to two other sources of error.

Retrospective Dietary Assessment

In making comparisons between patients and healthy controls, it is never known if the measured risk factor is a cause or consequence of the disease in question. This is particularly true of the study of diet. The most common symptoms in gastrointestinal disease for example are pain and altered bowel function (*42*) which the patient may attempt to ameliorate by a change in diet. Consequently the aim of dietary investigations in most case control studies is to ascertain past diet history, before the onset of symptoms. This procedure is impossible to validate with certainty, but circumstantial evidence suggest that recall of past diet is strongly influenced by present dietary habits (*43–48*).

Questionnaires

In addition, the dietary intakes of individuals in case control studies have been assessed by questionnaires or interviews, which are the least accurate methods of the assessment of present-day diet. Whilst it is generally assumed that these methods are capable of correctly classifying individuals into at least the correct third of a distribution in dietary intake, present evidence suggests the contrary. Validation studies in which the results of questionnaires have been compared with results from the same individuals who kept accurate records of actual foods consumed, consistently report correlation coefficients between the two sets of estimated nutrient intakes to be 0.4 or less (*43*, *49–51*). Where correlation coefficients are of the order of 0.2 to 0.4, only 4 to 20% of the variance of measured nutrient intake will be predicted by questionnaire, and less than half the population will be classified into the correct third of the distribution. Given these margins of misclassification in the measurement of diet, the inability of case control studies to confirm or refute international associations between diet and cancer occurrence is not surprising.

REFERENCES

1. Burkitt, D. P. Epidemiology of cancer of the colon and rectum. Cancer, *28*: 3–13, 1971.
2. Royal College of Physicians. Medical Aspects of Dietary Fibre. Pitman Medical, London, 1980.
3. Cummings, J. H. Cancer of the large bowel. *In;* H. Trowell, D. Burkitt, and K. Heaton (eds.), Dietary Fibre, Fibre-Depleted Foods and Disease, pp. 161–189, Academic Press Inc., London, 1985.
4. Trowell, H. Definition of dietary fibre and hypotheses that it is a protective factor in certain diseases. Am. J. Clin. Nutr., *29*: 417–427, 1976.
5. Van Soest, P. J. and McQueen, R. W. The chemistry and estimation of fibre. Proc. Nutr. Soc., *32*: 123–130, 1973.
6. Gregor, O., Toman, R., and Prusova, F. Gastrointestinal cancer and nutrition. Gut, *10*: 1031–1034, 1969.
7. Segal, I. and Walker, A.R.P. Falling fibre yet low fat intakes compatible with rarity of non-infective bowel diseases in blacks in Soweto, Johannesburg. Nutr. Cancer, 1986 (in press).
8. Bingham, S. Dietary fibre intakes. *In;* H. Trowell, D. Burkitt, and K. Heaton (eds.), Dietary Fibre, Fibre-Depleted Foods and Disease, pp. 77–104, Academic Press Inc., London, 1985.
9. Drasar, B. S. and Irving, D. Environmental factors and cancer of the colon and breast. Br. J. Cancer, *27*: 167–172, 1972.
10. Irving, D. and Drasar, B. S. Fibre and cancer of the colon. Br. J. Cancer, *28*: 462–463, 1973.
11. Howell, M. A. Diet as an etiological factor in development of cancer of the colon and rectum. J. Chron. Dis., *28*: 67–80, 1975.
12. Armstrong, B. and Doll, R. Environmental factors and cancer incidence in different countries, with special reference to dietary practices. Int. J. Cancer, *15*: 617–631, 1975.
13. Schrauzer, G. H. Cancer mortality correlation studies. *Med. Hypoth.*, *2*: 39–49, 1976.
14. Liu, K., Stamler, J., Moss, D., Garside, D., Persky, U., and Soltero, I. Dietary cholesterol, fat and fibre and colon-cancer mortaility. Lancet, *2*: 782–785, 1979.
15. Paul, A. A. and Southgate, D.A.T. (eds.) McCance and Widdowson's The Composition of Foods, 4th ed. of MRC Spec. Rep. 297, HMSO, London, 1978.
16. McKeown-Eyssen, G. and Bright-See, E. Dietary factors in colon cancer. Nutr. Cancer, *6*: 160–170, 1984.

17. Rutishauser, I.H.E. Estimation of dietary fibre supply. Am. J. Clin. Nutr., *41*: 824–826, 1985.
18. Waterhouse, J., Muir, C., Shanmugaratnam, K., and Powell, J. Cancer incidence in 5 continents, vol. IV, IARC Scientific Publications No. 42, IARC, Lyon, 1982.
19. Inoue, M. Diverticular disease of the colon in Japan. Stomach and Intestine, *15*: 807–815, 1980 (in Japanese).
20. Bingham, S., Cummings, J. H., and McNeil, N. I. Intakes and sources of dietary fiber in the British population. Am. J. Clin. Nutr., *32*: 1313–1319, 1979.
21. Minowa, M., Bingham, S., and Cummings, J. H. Dietary fibre intake in Japan. Hum. Nutr., *37A*: 113–119, 1983.
22. Bingham, S., Williams, D.R.R., and Cummings, J. H. Dietary fibre consumption in Britain; new estimates and their relation to large bowel cancer mortality. Br. J. Cancer, *52*: 399–402.
23. Kuratsune, M., Honda, T., Englyst, H. N., and Cummings, J. H. Dietary fibre in the Japanese diet. This volume, pp. 247–253, 1986.
24. Hellendoorn, E. W., Noordhoff, M. G., and Slagman, J. Enzymatic determination of the indigestible residue content of human food. J. Sci. Food Agric., *26*: 1461–1468, 1975.
25. Southgate, D.A.T., Bailey, B., Collinson, E., and Walker, A. F. A guide to calculating intakes of dietary fibre. J. Hum. Nutr., *30*: 303–313, 1976.
26. Holloway, W. D., Tasman-Jones, C. and Maher, D. Towards an accurate measurement of dietary fibre. N.Z. Med. J., *85*: 420–423, 1977.
27. Van Soest, P. J. Fibre analysis tables. Am. J. Clin. Nutr., Suppl. *31*: S. 284, 1978.
28. Angus, R., Sutherland, T. M., and Farrel, D. J. Insoluble dietary fibre content of some local foods. Proc. Nutr. Soc. Aust., *6*: 161, 1981.
29. Englyst, H., Wiggins, H. S., and Cummings, J. H. Determination of the NSP in plant foods by GLC of constituent sugars as alditol acetates. Analyst, *107*: 307–318, 1982.
30. Cummings, J. H. Some aspects of dietary fibre metabolism in the human gut. *In;* G. G. Birch and K. J. Parker (eds.), Food and Health—Science and Technology, pp. 441–458, Applied Science Publishers, London, 1980.
31. Southgate, D.A.T., Bingham, S., and Robertson, J. Dietary fibre in British diet. Nature, *274*, 51–52, 1978.
32. McMichael, A. J., Potter, J. D., and Hetzel, B. S. Time trends in colo-rectal cancer mortality in relation to food and alcohol consumption. Int. J. Epidemiol., *8*: 295–303, 1979.
33. Powles, J. and Williams, D.R.R. Trends in bowel cancer in selected countries in relation to war-time changes in flour milling. Nutr. Cancer, *6*: 40–48, 1984.
34. Greaves, J. P. and Hollingsworth, D. F. Trends in food consumption in the United Kingdom. World Rev. Nutr. Diet., *6*: 34–89, 1966.
35. Ministry of Agriculture, Fisheries and Food. Household Food Consumption and Expenditure 1969–1973. Annual Reports of the National Food Survey Committee. HMSO, London, 1971–75.
36. Englyst, H. N. and Cummings, J. H. Simplified method for the measurement of total NSP by GLC of constituent sugars as alditol acetates. Analyst, *109*: 937–942, 1984.
37. Bingham, S. A., Williams, D.R.R., Cole, T. J., and James, W.P.T. Dietary fibre consumption and regional large bowel cancer mortality in Britain. Br. J. Cancer, *40*: 456–463, 1979.
38. International Agency for Research on Cancer. Large Bowel Cancer Group. Nutr. Cancer, *4*: 3–79, 1982.

39. National Research Council. Diet, nutrition and cancer. Report of the National Academy of Sciences, Assembly of Life Sciences, National Research Council, Washington, D.C., National Academy Press, Washington, 1982.
40. Miller, A. B., Howe, G. R., Jain, M., Craib, K.J.P., and Harrison, L. Food items and food groups as risk factors in a case control study of diet and colorectal cancer. Int. J. Cancer, *32*: 155–161, 1983.
41. Potter, J. D., McMichael, A. J., and Bonnett, A. J. A case control study of colorectal cancer in South Australia. *In;* Fibre in Human and Animal Nutrition (abstr.), p. 35, Bulletin 20, Roy. Soc. New Zealand, Wellington, New Zealand, 1983.
42. Cummings, J. H. Dietary fibre and large bowel cancer. Proc. Nutr. Soc., *40*: 7–14, 1981.
43. Jain, M., Howe, G. R., Johnson, K. C., and Miller, A. B. Evaluation of a diet history questionnaire for epidemiologic studies. Am. J. Epidemiol., *111*: 212–219, 1980.
44. Jensen, O. M., Wahrendorf, J., Rosenquist, A., and Geser, A. The reliability of questionnaire-derived historic dietary information and temporal stability of food habits in individuals. Am. J. Epidemiol., *120*: 281–290, 1984.
45. Garland, B., Ibrahim, M., and Grimson, R. Assessment of past diet in cancer epidemiology. Am. J. Epidemiol., *166*: 577, 1982.
46. Byers, T. E., Rosenthal, R. I., Marshall, J. R., Rzepka, T. F., Cummings, M., and Graham, S. Dietary history from the distant past: a methodological study. Nutr. Cancer, *5*: 69–77, 1983.
47. Rohan, T. E. and Potter, J. D. Retrospective assessment of dietary intake. Am. J. Epidemiol., *120*: 876–877, 1984.
48. Acheson, E. D. and Doll, R. Dietary factors in carcinoma of the stomach: a study of 100 cases and 200 controls. Gut, *5*: 126–131.
49. Hankin, J. H., Rawlings, V., and Nomura, A. Assessment of a short dietary method for a prospective study on cancer. Am. J. Clin. Nutr., *31*: 355–359, 1978.
50. Stuff, J. E., Garza, C., O'Brien-Smith, E., Nichols, B. L., and Montandon, C. M. A comparison of dietary methods in nutritional studies. Am. J. Clin. Nutr., *37*: 300–306, 1983.
51. Yarnell, J.W.G., Fehilly, A. M., Milbank, J. E., Sweetnam, P. M., and Walker, C. L. A short dietary questionnaire for use in an epidemiological study. Hum. Nutr., *37A*: 103–112, 1983.

DIET, NUTRITION AND CANCER, Y. HAYASHI ET AL. (EDS.),
JAPAN SCI. SOC. PRESS, TOKYO/VNU SCI. PRESS, UTRECHT, PP. 193–203, 1986

Inhibition of Carcinogenesis by Some Minor Dietary Constituents

Lee W. WATTENBERG,[*1] A. Bryan HANLEY,[*2] George BARANY,[*3] Velta L. SPARNINS,[*1] Luke K. T. LAM,[*1] and G. Roger FENWICK[*2]

Department of Laboratory Medicine and Pathology, University of Minnesota, Minneapolis, Minnesota 55455, U.S.A.,[*1] *AFRC, Food Research Institute, Norwich NR7UA, England,*[*2] *and Department of Chemistry, University of Minnesota, Minneapolis, Minnesota 55455, U.S.A.*[*3]

Abstract: Previous work has shown that food contains a large number of minor dietary constituents that can inhibit the occurrence of cancer. Additional inhibitors from four different natural sources will be the subject of this presentation. 1. *Citrus fruit oils.* Orange, tangerine, lemon, and grapefruit oils given p.o. induce increased glutathione (GSH) S-transferase activity in tissues of the mouse. When fed in the diet prior to and during the course of administration of benzo(a)pyrene (BP), the four citrus fruit oils inhibit formation of tumors of both the forestomach and lungs of mice. When fed either before or after the administration of 7,12-dimethylbenz(a)anthracene (DMBA) orange oil inhibits mammary tumor formation. 2. *Garlic oil.* Allyl methyl trisulfide (AMT), a constituent of garlic oil, has been synthesized recently. When given p.o. 96 and 48 hr prior to BP, AMT inhibits the occurrence of forestomach tumors in mice. 3. *Green coffee beans.* Two diterpene esters, kahweol palmitate and cafestol palmitate, which are potent inducers of GSH S-transferase activity have been isolated from coffee beans. When administered p.o. prior to DMBA the two diterpene esters inhibit mammary tumor formation. 4. *Cruciferous vegetables.* Several glucosinolates occur in cruciferous vegetables. Efficient procedures for the isolation of these compounds have been developed recently. The inhibitory effects of three of these *i.e.* glucobrassicin, glucotropaeolin, and glucosinalbin were studied in several animal models. Glucobrassicin caused substantial inhibition of BP-induced neoplasia of the lung and forestomach of the mouse and DMBA-induced neoplasia of the breast in rats. Glucosinalbin and glucotropaeolin are less active in these systems. In addition to protective effects, indoles derived from the hydrolysis of glucobrassicin have potential harmful properties. The implications of multiple properties and factors which may determine their consequence will be discussed.

There is a growing body of evidence showing that the diet contains compounds that can cause cancer and compounds that have the capacity to protect against this entity. Attention in this paper will be focussed on compounds that offer protection. There is a surprising variety of such compounds. They include phenols, indoles, aromatic isothiocyanates, methylated flavones, coumarins, terpenes, dithiolthiones, plant sterols, protease inhibitors, selenium salts, ascorbic acid, tocopherols, and retinol (*1, 2*). Not only is the number of inhibitors substantial but they are highly diverse chemically. This diversity, coupled with the widespread occurrence of these compounds in food, make it virtually impossible to consume a diet that does not contain inhibitory substances. One of the implications of the wide variation in structure amongst inhibitors is that others not yet identified almost certainly exist. They should be identified in order to fully evaluate the impact of diet on occurrence of cancer. This area of research is in its infancy. We lack a full knowledge of the range of inhibitors present in food and have incomplete information as to the mechanism of action of many of the inhibitory compounds already identified. One means of classifying inhibitors is on the basis of the time in the carcinogenic process at which they exert their inhibitory effects. Some prevent formation of carcinogens. Others, termed "blocking agents," prevent carcinogens from reaching or reacting with critical target sites; thus they exert a barrier function. A third group called "suppressing agents" are effective when given subsequent to administration of carcinogens. They prevent evolution of the neoplastic process in cells that would otherwise become malignant (*1, 2*). Some compounds are capable of inhibition at more than one time point.

The major emphasis in this paper is on blocking agents, in particular those that act by enhancing host detoxification systems. Many blocking agents produce a coordinated enhancement of multiple enzymes that can detoxify carcinogens. Two distinctive patterns of enzyme induction termed type A and type B have been identified (*1, 2*). One enzyme system commonly induced by blocking agents is glutathione (GSH) S-transferase (*3–7*). On the basis of this information, induction of GSH S-transferase activity may be used to detect the presence of blocking agents in complex natural products. The inhibitory effects of four different natural food sources will be the subject of this presentation.

Citrus Fruit Oils

In preliminary work, a number of natural materials were studied for their capacity to induce increased GSH S-transferase activity. One of the natural products included in the study was orange oil. This oil induced a marked increase in GSH S-transferase activity. Subsequently, in more complete investigations, orange, lemon, and grapefruit oils were added to a semi-purified diet fed to mice for a period of 10 days. The results of this study are shown in Table 1. It will be seen that all four oils produced considerable increases in GSH S-transferase activity. In addition to studying the effects of the citrus fruit oils added directly to the diet, other work was done in which the citrus fruit oils were given by oral intubation for three days prior to termination of the experiment. Under these conditions, a marked induction of increased GSH S-transferase activity was again found. The major component of citrus fruit oils is D-limonene. This compound accounts for much but not all of the inducing activity of the citrus fruit oils. Other constituents of orange oil including β-caryophyllene and valencene also show inducing activity (*8*).

On the basis of the results obtained in the study of induction of increased GSH

TABLE 1. Effect of Feeding Citrus Oils on the Glutathione S-transferase Activity in the Cytosol of Liver and Small Bowel Mucosa in Mice

Test compound[a)]	Amount (mg/g diet)	GSH S-transferase activity[b)]	
		Liver (μmol/min/mg protein)	Small bowel mucosa (μmol/min/mg protein)
None	—	1.63	0.44
Orange oil	100	8.71	2.80
Tangerine oil	100	8.62	3.04
Grapefruit oil	100	8.51	2.99
Lemon oil	100	8.14	2.28

[a)] Female ICR/Ha mice 8 weeks of age were randomized by weight and divided into groups of five animals. The mice were fed a semi-purified diet consisting of 27% vitamin-free casein, 59% starch, 10% corn oil, 4% salt mix (U.S.P. XIV), and a complete mixture of vitamins (Teklad Inc., Madison, WI). [b)] Glutathione S-transferase activity was assayed according to the method of Habig *et al.* (J. Biol. Chem., *249*: 7130–7139, 1974) using 1-chloro-2,4-dinitrobenzene as substrate.

TABLE 2. Inhibition by Citrus Fruit Oils of BP-induced Neoplasia

Additions to the diet[a)]	No. of mice at risk	Weight gain[b)] (g)	Pulmonary Adenomas		Forestomach Tumors	
			No. of tumors/mouse[c)]	Ratio: test/control	No. of tumors/mouse[c)]	Ratio: test/control
None	31	13.3	19.5±2.1[d)]	—	4.3±0.6[d)]	—
Orange oil	20	13.8	10.8±2.8[e)]	0.56	1.8±0.3[f)]	0.42
Tangerine oil	19	12.9	7.4±1.8[f)]	0.38	2.0±0.4[f)]	0.44
Lemon oil	17	11.9	10.7±2.4[e)]	0.55	2.8±0.6[g)]	0.65
Grapefruit oil	20	12.3	8.1±1.9[f)]	0.42	3.1±0.4	0.72

[a)] Female ICR/Ha mice at 8 weeks of age were randomized by weight and divided into groups. The mice were fed a semi-purified diet consisting of 27% vitamin-free casein, 59% starch, 10% corn oil, 4% salt mix (U.S.P. XIV), and a complete mixture of vitamins (Teklad Inc., Madison, WI). The diets of experimental groups were supplemented as indicated. After 10 days the first of 8 doses of BP were given by oral intubation (1.5 mg, 2× a week for 4 weeks). The experimental diets were continued until 3 days after the last dose of BP. The animals were sacrificed 20 weeks after the first dose of BP. [b)] From 8 to 29 weeks of age. [c)] Number of tumors in the entire group divided by the number of mice in the group. [d)] Mean±S.E. [e)] $p<0.01$. [f)] $p<0.005$. [g)] $p<0.05$.

S-transferase activity, an experiment was carried out to determine the capacity of the citrus fruit oils to inhibit carcinogen-induced neoplasia of the forestomach and lungs of mice having received benzo(a)pyrene (BP) by oral intubation. The results of this study are shown in Table 2. It will be seen that all four citrus fruit oils inhibited the occurrence of neoplasia in both sites. In addition, orange oil fed in the diet for 8 days prior to administration of 7,12-dimethylbenz(a)anthracene (DMBA) to female Sprague Dawley rats inhibited mammary tumor formation (unpublished).

In an additional study, the effect of feeding orange oils subsequent to administration of DMBA to Sprague Dawley rats was investigated. The results of this study are reported elsewhere (*2*). They showed that orange oil inhibited the occurrence of mammary tumor formation. Studies by Elegbede *et al.* (*9*) have shown that D-limonene fed in the diet inhibits mammary tumors in the same experimental model. In further work it has been found that high doses of D-limonene can cause regression of mammary tumors that have already reached a size that can be palpated grossly.

Garlic Oil

Allyl methyl trisulfide (AMT) is a constituent of garlic oil which has recently been synthesized in pure form free of disproportionation by-products (*10*). A number of sulfur-containing compounds have been shown to induce increased GSH S-transferase activity. However, studies of the effects on GSH S-transferase activity of compounds containing a trisulfide moiety have not been reported. Work was initiated to determine the effects of AMT on GSH S-transferase activity in tissues of the mouse. It was found that AMT induced increased GSH S-transferase activity in forestomach, small bowel mucosa, liver, and lung of female A/J mice. The forestomach and small bowel mucosa responded to a single low dose of AMT (3.0 μmol) given by oral intubation whereas liver and lung were less reactive. A dose schedule of 2 administrations of 15 μmol AMT given 48 hr apart gave close to maximum induction in all four tissues and was chosen for investigation of its inhibitory effects against the occurrence of neoplasia (unpublished). With this dose schedule, AMT showed inhibition of BP-induced neoplasia of the forestomach but not the lung (Table 3). Thus AMT is a member of a new class of naturally-occurring chemicals that have the capacity to inhibit chemical carcinogenesis.

TABLE 3. Effects of Oral Administration of AMT on BP-induced Neoplasia in A/J Mice

Material administered[a]	No. of mice	Forestomach tumors		Pulmonary adenomas	
		No. of mice with tumors	No. of tumors per mouse[b]	No. of mice with tumors	No. of tumors per mouse[b]
None	15	15	2.9±0.5[c]	15	24±1.8[c]
Cottonseed oil	14	14	3.7±0.4	14	18±2.0
AMT	15	13	1.0±0.1[d]	15	19±1.8

[a] 15 μmol of AMT in 0.2 ml cottonseed oil, cottonseed oil or nothing was given by oral intubation 96 and 48 hr prior to oral administration of 2 mg of BP in 0.2 ml cottonseed oil. This sequence was repeated at 2 week intervals twice. [b] Number of tumors in the entire group divided by the number of mice in the group. [c] Mean ±S.E. [d] AMT *vs.* none, $p<0.005$; AMT *vs.* cottonseed oil, $p<0.005$.

Green Coffee Beans

During studies aimed at identifying constituents of natural products that might protect against chemical carcinogens, the consumption of diets containing powdered green coffee beans was found to produce a marked enhancement of GSH S-transferase activity in the liver and mucosa of the small bowel of the mouse. Green coffee beans were employed in order to determine if the most natural available form of the bean would have the biological attributes being tested for. This experiment obviously does not reflect normal coffee consumption but was aimed at providing favorable experimental conditions for determining whether or not the beans contained an inducer. The coffee beans used in the original study were from Guatemala. In subsequent work, coffee beans from Colombia, Brazil, El Salvador, Mexico, and Peru were all found to have similar enhancing effects on GSH S-transferase activity. Studies then were carried out to identify the inducing material in green coffee beans. Most of the activity was extracted into petroleum ether and two inducers of increased GSH S-transferase have been identified thus far. They are kahweol palmitate and cafestol palmitate. These two compounds account for approximately 40 percent of the inducing activity occurring in the petroleum ether extract (*11*).

Following identification of kahweol palmitate and cafestol palmitate as inducers of increased GSH S-transferase activity, studies were conducted to determine if they would inhibit chemical carcinogenesis. This work has been summarized previously (*12*). When a dosage regime entailing administration of the two diterpenes 3, 2, and 1 day before the carcinogen was used, kahweol palmitate significantly inhibited the occurrence of DMBA-induced mammary tumor formation in the rat. Cafestol plamitate has a marginal inhibitory effect under these conditions. A second experiment was performed using a different administration regime. In this instance kahweol palmitate was given by oral administration as a single dose 4 hr prior to the carcinogen. Again a significant inhibition was obtained. This administration schedule is based on observations with other blocking agents which show inhibition if administered a relatively short time prior to the carcinogen (*1*).

Glucosinolates

In previous studies three indoles derived from a glucosinolate occurring in cruciferous vegetables were found to inhibit the occurrence of carcinogen-induced neoplasia. The three compounds, indole-3-acetonitrile, 3,3′-diindolylmethane and indole-3-carbinol, were investigated for their capacity to inhibit the formation of neoplasms in 2 experimental models. In the first, indole-3-carbinol and 3, 3′-diindolylmethane given by oral intubation to Sprague Dawley rats 20 hr prior to DMBA inhibited the occurrence of mammary tumors. Indole-3-acetonitrile was ineffective under these conditions. In the second experimental model, inhibitory effects of the indoles on forestomach tumor formation resulting from BP administration to female ICR/Ha mice was studied. The compounds were added to a diet which was fed for 10 days prior to the first of 8 administrations of BP (twice weekly for 4 weeks) and continued until the last administration of BP. All three indoles inhibited the occurrence of forestomach tumors (*13*).

The three indoles that have been studied are derived from the hydrolysis of a precursor compound, glucobrassicin. High concentrations of glucobrassicin exist in some cruciferous vegetables (*14–16*). The compound is compartmentalized in the plant cells and separated from a hydrolyzing enzyme, myrosinase. When the cells are damaged, glucobrassicin undergoes hydrolysis to the three indoles (*14*, *17*), the exact proportions of which depend upon the conditions of the hydrolysis. An additional aspect of the chemistry of the breakdown of glucobrassicin is a consequence of the high concentrations of ascorbic acid in cruciferous vegetables. A reaction between ascorbic acid and indole-3-carbinol can occur leading to the formation of a product called ascorbigen.

At the time of the original studies of the effect of indoles on carcinogen-induced neoplasia, methods for isolating glucobrassicin were crude. Adequate amounts of pure glucobrassicin were not available for carrying out carcinogen inhibition studies with this compound. In recent years, there has been a marked improvement in the techniques for isolating relatively large quantities of pure glucobrassicin and other glucosinolates (*18*). Accordingly, it became possible to study the effects of such compounds on the occurrence of carcinogen-induced neoplasia. Experiments of this nature have been performed in the mouse and the rat. In Table 4, the results of two

TABLE 4. Effects of Glucotropaeolin, Glucosinalbin, and Glucobrassicin on BP-induced Neoplasia of the Forestomach and Lungs of Female ICR/Ha Mice

Experiment number	Material administered by oral intubation[a]	No. of mice at risk	Weight gain (g)	Forestomach tumors: % mice with tumors	Forestomach tumors: Tumors per mouse[c]	Pulmonary adenomas: % mice with tumors	Pulmonary adenomas: Tumors per mouse
1	None	17	10	100	8.6±0.9[d]	94	6.1±1.3[d]
	H_2O	20	10	100	7.9±0.8	90	6.5±1.0
	Glucotropaeolin-2.2 mg	19	11	100	6.9±0.7	95	7.1±1.0
	Glucotropaeolin-13.0 mg	19	13	100	6.5±0.9[e]	95	7.5±2.1
	Glucosinalbin-2.2 mg	20	10	100	6.4±1.0[e]	100	6.4±0.8
	Glucosinalbin-13.0 mg	17	11	100	6.2±0.7[e]	100	7.4±1.6
2	None	18	12	100	6.1±0.7	100	8.9±1.7
	H_2O	18	12	100	6.9±0.7	94	5.5±1.3
	Glucobrassicin-2.0 mg	19	10	100	4.5±0.6[f]	100	5.1±0.9
	Glucobrassicin-12.0 mg	19	10	100	4.8±0.7[f]	74	3.7±0.8[e]
	None[g]	19	12	0	0	37	0.6±0.26

[a] Test compound in 0.2 ml H_2O or H_2O was administered to female ICR/Ha mice by oral intubation 72, 48, and 24 hr prior to 3 mg of BP also given by oral intubation. This sequence was repeated at two week intervals 4 times. The experiment was terminated 24 weeks after the first dose of BP. [b] From 8 to 39 weeks of age. [c] Number of tumors occurring in all mice of the group divided by the number of mice at risk. [d] Mean ±S.E. [e] Test compounds *vs.* "H_2O" control—not significant but test compound *vs.* combined data from both controls ("None" and "H_2O"), $p<0.05$. [f] Test compound *vs.* H_2O control, $p<0.01$. [g] No carcinogen control.

TABLE 5. Effects of Glucotropaeolin, Glucosinalbin, and Glucobrassicin on BP-induced Neoplasia of the Forestomach and Lungs of Female A/J Mice

Material administered by oral intubation[a]	No. of mice at risk	Weight gain[b] (g)	Forestomach tumors: % mice with tumors	Forestomach tumors: Tumors per mouse[c]	Pulmonary adenomas: % mice with tumors	Pulmonary adenomas: Tumors per mouse[c]
H_2O	19	9.6	100	6.3±0.5[d]	100	24.4±2.2[d]
Glucotropaeolin-13 mg	20	9.0	100	7.7±0.8	100	16.5±1.8[e]
Glucobrassicin-12 mg	20	8.2	100	5.8±0.6	100	7.3±0.6[e]
Glucosinalbin-13 mg	20	7.4	100	7.3±0.8	100	13.3±1.4[e]

[a] Test compound in 0.2 ml H_2O or H_2O was administered to female A/J mice by oral intubation 4 hr prior to 3 mg of BP also given by oral intubation. This sequence was repeated at 2 week intervals for a total number of 3 administrations. [b] Mice from 7 to 35 weeks of age. [c] Number of tumors occurring in all mice of the group divided by the number of mice at risk. [d] Mean±S.E. [e] Test compounds *vs.* "H_2O" control, $p<0.005$.

TABLE 6. Effects of Glucobrassicin and Glucotropaeolin on DMBA-Induced Mammary Tumor Formation in Female Sprague Dawley Rats

Material administered by oral intubation[a]	No. of rats at risk	Weight gain[b] (g)	Mammary tumors	
			% rats with tumors	Tumors per rat[c]
H_2O	16	161	75	1.25±0.28[d]
Glucobrassicin-60 mg	16	143	25[e]	0.50±0.26[f]
Glucotropaeolin-60 mg	16	165	38[e]	0.69±0.28[f]

[a] Test compound in 1 ml H_2O or H_2O were administered to 7 week old female Sprague Dawley rats by oral intubation 4 hr prior to administration of 12 mg of DMBA in 1 ml olive oil also by oral intubation. [b] Rats from 6 to 24 weeks of age. [c] Number of tumors occurring in all rats divided by the number of rats at risk [d] Mean ±S.E. [e] Test compound *vs.* H_2O control, $p<0.05$. [f] Test compound *vs.* H_2O control, $p<0.01$.

experiments are presented in which three glucosinolates were given by oral intubation 3, 2, and 1 days prior to administration of BP also by oral intubation. Glucobrassicin inhibited the occurrence of forestomach tumors and pulmonary adenomas. Glucotropaeolin and glucosinalbin inhibited forestomach tumor formation but not the occurrence of pulmonary adenomas. In Table 5 the effects of administering the glucosinolates by oral intubation 4 hr prior to BP are shown. All three glucosinolates inhibit pulmonary adenoma formation, but the most striking inhibitory effect was with glucobrassicin. An experiment has been performed in the rat using the four hr time interval between administration of glucobrassicin or glucotropaeolin and DMBA. The results of this study are shown in Table 6. As in the mouse experiment in which a four hour time interval between administration of the test compound and carcinogen was employed, both glucosinolates exerted inhibitory effects. The number of rats bearing mammary tumors and the number of tumors per rat were decreased by administration either of glucobrassicin or glucotropaeolin, but the former was again more effective. Further studies of the relationships between time of administration of glucosinolates and carcinogen on inhibition of carcinogenesis are in progress. Should inhibition occur in intervals of the order of minutes, consumption of cruciferous vegetables could have an impact on exposure to a carcinogen consumed during the same meal.

Increasing attention has been paid to the possible effects of indoles on the occurrence of neoplasia. Two lines of experimentation have recently suggested possible adverse consequences that might follow consumption of these compounds. The first are studies showing that the administration of indole-3-carbinol in the diet of rainbow trout subsequent to carcinogen administration results in an enhanced carcinogenic response (*19*). Indole-3-carbinol is a moderately potent inducer of increased microsomal mixed function oxidase activity (*20*). Compounds that have this biological property have been shown to be promoters of carcinogenesis in some tissues. The experiments of Bailey *et al.* (*19*) have demonstrated that addition of indole-3-carbinol to the diet prior to and during exposure of rainbow trout to aflatoxin B_1 results in inhibition of carcinogenesis. However, if the indole-3-carbinol is fed in the diet subsequent to the aflatoxin B_1 administration, there is an increase in tumor response. Whereas these studies indicate circumstances under which indole-3-carbinol can produce an increased carcinogenic response, the experimental conditions are such that they appear unlikely to occur in humans except under most unusual circumstances. The tumor promotion studies with indole-3-carbinol entail continuous daily consumption of a dose of indole-3-carbinol of 2 mg/gm of diet. In an adult trout, the intake of this indole would be about 30 mg/kg body weight. The average daily consumption per person of all glucosinolates in the United Kingdom, which has an unusually high intake of cruciferous vegetables, is approximately 50 mg. Glucobrassicin accounts for considerably less than half of the glucosinolate content of these vegetables. About 30% of the molecular weight of glucobrassicin is due to the indole moiety, thus an estimate of the average daily consumption of indoles in the United Kingdom would be less than 10 mg/day. On a body weight basis this would be less than 1% of that consumed by the trout. A further consideration that would make the risk less is that indole-3-carbinol, which is the most potent inducer of mixed function oxidase activity of the indoles resulting from hydrolysis of glucobrassicin, accounts for only a fraction (one half or less) of the indoles produced on hydrolysis of the parent compound. In addition to the quantitative factors, it should be emphasized that studies of tumor promotion, as that reported for the trout, entail continuous daily administration over very long time intervals. Tumor promotion is reversible so that interruptions of consumption of a promoting agent are likely to result in loss of promotional effects.

A second possible hazard from indoles in the diet is that they may react with nitrites to form nitroso compounds. Research work on reactions of this type *in vitro* has been presented elsewhere in this Symposium by Nagao *et al.* Uncertainties that exist pertaining to this possible adverse effect reside in two areas. The first is whether such reactions actually can occur *in vivo*, resulting in the formation of a carcinogenic compound or compounds. The second uncertainty relates to the inhibiting effects that the high ascorbic acid content of cruciferous vegetables would have on such nitrosation. If the ascorbic acid is not destroyed by procedures in food handling or cooking, it would react with nitrite and reduce or eliminate its availability for reaction with indoles.

DISCUSSION

Inhibitors of carcinogenesis from four different natural sources have been discussed. Three were found on the basis of their capacity to induce increased GSH S-transferase activity and demonstrate the usefulness of the assay for identifying blocking agents. In the case of the citrus fruit oils, the induction of increased GSH S-transferase activity focussed attention on the likelihood that these oils might contain a compound or compounds that could act as blocking agents. The subsequent study of inhibition of carcinogens by the citrus fruit oils demonstrated the predictive value of the assay. Further work remains to be performed on identifying the particular constituent or constituents of the citrus fruit oils responsible for the inhibitory effects.

The example of the use of the predictive attributes of the assay in the studies of AMT is of a somewhat different nature. In this instance a compound, which on a theoretical basis might be thought to be an inhibitor of carcinogenesis, became available for testing. The finding that this compound, in fact, did induce increased GSH S-transferase activity made it worthwhile to carry out a full carcinogen inhibition study. Thus in this instance the screening assay was applied originally to a defined compound rather than a natural product. Garlic and onion oils contain a series of related sulfur containing compounds. The usefulness of the GSH S-transferase assay in identifying the inhibitory capacities of one of their constituents suggests the likelihood that it will also be applicable to studies of related compounds in these two oils.

In the case of coffee diterpenes, a full sequence of procedures was employed. The inductive capacity of green coffee beans was the initial observation. Subsequently, using GSH S-transferase induction as an assay procedure for isolation studies, the two diterpene esters, kahweol palmitate and cafestol palmitate were isolated and identified. Although induction of increased GSH S-transferase activity is useful in identifying new blocking agents, it is imperfect as a screening procedure. Instances exist in which blocking agents do not enhance GSH S-transferase activity as might be anticipated for compounds that inhibit by other mechanisms. We have also encountered compounds that induce increased GSH S-transferase activity but do not inhibit carcinogen-induced neoplasia. Thus the procedure is useful but it is important to recognize and accept its limitations.

Finally, we would again like to stress that with continued examination of natural products and naturally-occurring compounds present in foods, further increases in the number and variety of inhibitors of carcinogenesis are very likely to be found. Indeed this is evident from the studies presented above. Until the full range of inhibitors occurring in food are known, studies of the impact of diet on the occurrence of cancer will be hampered. The identification of a compound as an inhibitor of carcinogenesis should be understood to be only a beginning step for evaluating its potential role in modulating carcinogenesis. Further work would then be required to assess potency, to examine the range and variety of carcinogenic agents inhibited, to define the conditions under which inhibition occurs and to demonstrate any possible adverse effects. Only when such a range of studies is performed will potential negative attributes become apparent. The task of evaluating the role that minor dietary constituents of food play on the occurrence of cancer is clearly a very large one. Yet until this evaluation is accomplished, not only will there be a serious void in our ability to understand these dietary effects but opportunities for optimizing the role of diet in preventing cancer may also be lost.

ACKNOWLEDGMENTS

This work was supported by American Cancer Society Grant SIG-5 and USPHS Grant CA 37797 from the National Cancer Institute. The support of the glucosinolate studies by Prof. R. F. Curtis, Director, Food Research Institute, Norwich, is gratefully acknowledged.

REFERENCES

1. Wattenberg, L. W. Chemoprevention of cancer. Cancer Res., *48*: 1–8, 1985.
2. Wattenberg, L. W. Inhibition of neoplasia by minor dietary constituents. Cancer Res.,

43 (Suppl.): 2448s–2453s, 1983.
3. Benson, A. M., Batzinger, R. P., Ou, S. L., Bueding, E., Cha, Y. N., and Talalay, P. Elevation of hepatic glutathione S-transferase activities and protection against mutagenic metabolites by dietary antioxidants. Cancer Res., *38*: 4486–4495, 1978.
4. Benson, A. M., Cha, Y. N., Bueding, E., Heine, H. S., and Talalay, P. Elevation of extrahepatic glutathione S-transferase and epoxide hydratase activities by 2(3)-*tert*-butyl-4-hydroxyanisole. Cancer Res., *39*: 2971–2977, 1979.
5. Chasseaud, L. F. The role of glutathione and glutathione S-transferase in the metabolism of chemical carcinogens and other electrophilic agents. Adv. Cancer Res., *29*: 175–274, 1979.
6. Sparnins, V. L., Venegas, P. L., and Wattenberg, L. W. Glutathione S-transferase activity: enhancement by compounds inhibiting chemical carcinogenesis and by dietary constituents. J. Natl. Cancer Inst., *68*: 493–496, 1982.
7. Sparnins, V. L. and Wattenberg, L. W. Enhancement of glutathione S-transferase activity of the mouse forestomach by inhibitors of benzo(a)pyrene-induced neoplasia of this anatomic site. J. Natl. Cancer Inst., *66*: 769–771, 1981.
8. Sparnins, V. L. and Wattenberg, L. W. Effects of citrus fruit oils on glutathione S-transferase activity and benzo(a)pyrene-induced neoplasia. Proc. Am. Assoc. Cancer Res., *26*: 123, 1985.
9. Elegbede, J. A., Elson, C. E., Qureshi, A., and Gould, M. N. Inhibition of DMBA-induced mammary cancer by the monoterpene d-limonene. Carcinogenesis, *5*: 661–664, 1984.
10. Mott, A. W. and Barany, G. A new method for the synthesis of unsymmetrical trisulfanes. Synthesis, 657–660, 1984.
11. Lam, L.K.T., Sparnins, V. L., and Wattenberg, L. W. Isolation and identification of kahweol palmitate and cafestol palmitate as active constituents of green coffee beans that enhance glutathione S-transferase activity. Cancer Res., *42*: 1193–1198, 1982.
12. Wattenberg, L. W. and Lam, L.K.T. Protective effects of coffee constituents on carcinogensis in experimental animals. *In;* Banbury Report 17: Coffee and Health, pp. 137–145, Cold Spring Harbor Press, New York, 1984.
13. Wattenberg, L. W. and Loub, W. D. Inhibition of polycyclic aromatic hydrocarbon-induced neoplasia by naturally occurring indoles. Cancer Res., *38*: 1410–1413, 1978.
14. Fenwick, G. R. and Heaney, R. K. Glucosinolates and their breakdown products in cruciferous crops, foods and feeding stuffs. Food Chem., *11*: 249–271, 1983.
15. McGregor, D. I., Mullin, W. I., and Fenwick, G. R. Analytical methodology for determining glucosinolate composition and content. J. Assoc. Off. Anal. Chem., *66*: 825–849, 1983.
16. Sones, K., Heaney, R. K., and Fenwick, G. R. An estimate of the average daily intake of glucosinolates *via* cruciferous vegetables in the UK. J. Sci. Food Agric., *35*: 712–720, 1984.
17. Fenwick, G. R., Heaney, R. K., and Mullin, W. J. Glucosinolates and their breakdown products in food and food plants. CRC Crit. Rev. Food. Soci. Nutr., *18*: 123–201, 1983.
18. Hanley, A. B., Heaney, R. K., and Fenwick, G. R. Improved isolation of glucobrassicin and other glucosinolates. J. Sci. Food Agric., *34*: 869–873, 1983.
19. Bailey, G., Goeger, D., Hendricks, J., Nixon, J., and Pawlowski, N. Indole-3-carbinol promotion and inhibition of aflatoxin B_1 carcinogenesis in rainbow trout. Proc. Am. Assoc. Cancer. Res., *26*: 115, 1985.

20. Loub, W. D., Wattenberg, L. W., and Davis, D. W. Aryl hydrocarbon hydroxylase induction in rat tissues by naturally-occurring indoles in cruciferous plants. J. Natl. Cancer Inst., *54*: 985–988, 1975.

DIET, NUTRITION AND CANCER, Y. HAYASHI ET AL. (EDS.),
JAPAN SCI. SOC. PRESS, TOKYO/VNU SCI. PRESS, UTRECHT, PP. 205–222, 1986

The Role of Nutrients in Cancer Causation

P. M. NEWBERNE and A. E. ROGERS

Department of Applied Biological Sciences, Massachusetts Institute of Technology, Cambridge, Massachusetts 02139 and Department of Pathology, Boston University School of Medicine, Boston, Massachusetts 02118, U.S.A.

Abstract: The role of nutrients in cancer causation has been a subject of considerable interest, research, and public discussion in recent years. Results from epidemiologic, clinical, and animal studies have suggested that: 1) a reduction in total calories decreases risk for a number of tumor types; 2) dietary protein is directly correlated with liver, prostate, and colon cancer, among others, with increasing dietary protein increasing the risk; 3) increased dietary fat is correlated with increased risk for breast cancer; the evidence for an effect of fat on colon cancer is equivocal in human and animal studies; 4) a deficiency of vitamin A may enhance lung and colon tumors in animal experiments but in human this is equivocal. Increasing vitamin A above normal levels, as an anticarcinogenic effect, has not been satisfactorily demonstrated in animal models. The synthetic retinoid, 13-*cis* retinoic acid, inhibits both colon and lung cancer in animal models; 5) zinc deficiency is associated with enhanced esophageal cancer in humans and markedly enhances animal tumors; selenium inhibits this form of neoplasia in animals, 6) diets low in lipotropes enhance liver cancer induced by a variety of hepatocarcinogens.

Our data from studies in animal models agree in some cases with epidemiological observations, but disagree with others, particularly fat and colon cancer. Overall, some forms of cancer are enhanced by excessive calories, increased dietary protein and fat, and by deficiencies of vitamin A, selenium, zinc, and lipotropes. Decreasing total intake of calories, protein, and fat, and ensuring adequate dietary levels of vitamin A, selenium, zinc, and lipotropes decreases risk for some forms of cancer.

There has been much discussion and debate in the scientific literature, in public meetings and in the press in recent years, relative to nutritional factors which influence cancer incidence. Some of the discussions have been based on epidemiologic studies, some on laboratory animal studies, and some on personal beliefs. The public is now asking about the causes of cancer which are not associated with smoking and they wish to know what these causes are and how they can be avoided. Of particular interest is the influence of diet and nutrition on risk for cancer (*1*). There can no longer be any doubt about the enormous influence that the nutritive status of an individual has on the capacity to respond to environmental stress, including infectious disease agents, toxins, and carcinogens. The wide variation in incidence of some types of cancer in various populations of ethnic groups living in different geographic locations and the change in risk of migrant populations as they relocate and assume dietary habits and nutritive status of low or high risk groups, strongly support this concept (*2–4*).

Beginning with the publication in the United States of the Surgeon General's Report on smoking and health in 1979 (*5*) the public appears to have accepted the fact that cigarettes are a prime factor in the occurrence of lung cancer in human populations. While the amount of tobacco consumed remains relatively constant and the number of people who smoke remains about the same, the population using cigarettes continues to shift, generally to a younger age group, predicting even more tobacco-related disease in the future. The numbers of cancer cases is increasing steadily as the population grows but the age adjusted total incidence in mortality rates for sites other than the respiratory tract, the pancreas and stomach have remained rather stable over the past 30 to 40 years. Cancer of the lung and of the pancreas have increased; stomach cancer has steadily decreased.

The association of nutritional deficiencies or imbalances and food contaminants, or additives, with a high incidence of some forms of cancer and the supporting experimental data indicate strongly that a relationship exists between our diet and some forms of cancer.

In a general way the causes of cancer may be categorized into two major areas, namely, exogenous and endogenous factors (Table 1). These classifications do not identify any single factor or condition that may be causally related to human cancer but they do provide an overall framework for approaching an assessment of cancer cause and prevention.

TABLE 1. Factors Contributing to Chronic Disease in Human Populations

A. Exogenous factors
   1. Industrial exposures
   2. Air pollution
   3. Food and water contaminants
   4. Lifestyle
      (a) Diet
      (b) Tobacco
      (c) Alcohol
      (d) Sexual behavior
      (e) Infection
      (f) Medication
      (g) UV light
      (h) Radiation
B. Endogenous Factors
   1. Internal chemicals
   2. Hormonal factors
   3. Metabolic deficiencies
   4. Dietary deficiencies

As shown in Table 1, lifestyle stands out as one of the primary areas for potential growth of knowledge regarding cancer cause and prevention. High on the list of lifestyle factors, over which we exercise considerable control, is diet. This complex aspect of human habits, referred to generically as diet, affords an excellent opportunity to explore in a systematic way, the probable causes or conditions, related to food intake and various forms of neoplasia that afflict humankind as well as lower animals. With respect to diet, two subpopulations within the United States have provided much valuable information on tumor incidence and lifestyle; these are the Mormons and the Seventh Day Adventists (SDA) (*6*, *7*) whose lifestyle and personal habits differ from those of non-Mormons and non-SDAs. Diet, alcohol use, and smoking habits are important differences.

Genetics and familial influences on certain types of tumors have been known for a long period of time and these will not be considered here (*8*). Rather, those factors which may be modulators of cancer risk will be considered in as much detail as space permits and those which are associated with substantial negative or positive risk will be given priority. In regard to genetics, it would appear that some factors in our lifestyle may overide genetic control of some aspects of cancer risk. If we abuse tobacco and alcohol, and do not exercise, as only three examples, we can expect to be more susceptible to degenerative and proliferative diseases.

Some of the better known nutritional factors and conditions, developed from epidemiological studies in human populations, along with modulation of carcinogenesis in experimental animals, will be briefly alluded to in this presentation. The tumors will be considered according to the nutrient or nutrients involved.

Calories

The international distribution of hormone dependent cancers has generated suspicion that these types of tumors may be related to affluence and an increased overall total caloric intake. Berg (*9*) has suggested that diets consumed by affluent populations, if ingested from early life on through old age, may overstimulate the endocrine system and in this manner lead to metabolic aberrations resulting in increased cancer induction.

Hill *et al.* (*10*) showed that an affluent group in Hong Kong had more than twice the mortality from colo-rectal cancer, compared to the poorest group. The relative proportions of nutrients in their diets were similar but the estimated daily caloric intake was 2,700 in the lowest socioeconomic group and 3,900 in the highest. Gregor *et al.* (*11*), Doll and associates (*2*, *12*), and Gaskill *et al.* (*13*) among many others, have observed similar associations.

The American Cancer Society has, on the other hand (*14*), in long-term prospective studies, shown that cancer mortality was significantly elevated in both sexes only among those that were 40% or more overweight.

In animal studies, Tannenbaum (*15*, *16*) showed that tumors induced by benzo(a)pyrene in mice were inhibited by caloric restriction. The level of dietary fat had a profound effect on the growth of skin tumors or of spontaneous and chemically induced breast tumors. Mice, with a daily intake of 11.7 calories had 25% more

TABLE 2. Liver Tumor Incidence in Control *Ad Libitum* and 75% Restricted Groups Exposed to Aflatoxin B_1

Treatment	No./%					
	12-months		24-months		Total	
	M	F	M	F	M	F
Control, *ad libitum*	0/80	0/77	0/48	0/54	0/80	0/77
Control *ad libitum* + Aflatoxin B_1	3/79	1/78	41/53[a] (77.3)	28/60[a] (46.6)	44/79[a] (55.7)	29/78[a] (37.2)
75% restricted	0/80	0/80	0/64	0/72	0/80	0/80
75% restricted + Aflatoxin B_1	0/80	0/79	19/60[a] (31.7)	12/67[a] (17.9)	19/80[a] (23.8)	12/79[a] (15.2)

[a] Statistically significant ($p = < 0.01$).

spontaneous mammary tumors than mice with a daily caloric intake of 9.6 calories. Furthermore, it appeared that dietary fat had more of an enhancing effect if the calories from fat were 18% compared to 2%. This seemed to indicate to Tannenbaum that dietary fat exerted a specific influence beyond its caloric contribution. Lavik and Bauman (*17*) reported similar results.

We have recently completed studies designed to determine the effect on tumor incidence of reducing overall caloric intake to about seventy-five percent of the *ad libitum* group of rats. The various groups were started on the study at post-weaning, about four weeks of age. They were administered aflatoxin B_1 25 μg per day by gavage for 15 daily doses, or a total amount of 375 μg. The food intake of the *ad libitum* groups was measured daily; on the subsequent day the paired mates were given 75% of that consumed by the *ad libitum* groups allowed to consume all they wished of a control, semisynthetic diet (*18*) with vitamins and minerals adjusted upward to that the intake of all animals were comparable in the latter nutrients. The rats were observed daily, weighed weekly and all animals that died or that were moribund were sacrificed and necropsied. At the end of twenty four months on study, the survivors were sacrificed and, along with the others, tissues prepared for histologic examined. The results of tumor induction in that study are listed in Table 2.

Survival was significantly increased in all groups that were restricted, whether or not they received the carcinogen, and in both sexes. Liver tumor incidence was also reduced significantly, in both sexes. Moreover, when caloric intake was decreased, by only 25%, pituitary, adrenal, and mammary tumors were reduced by 70 to 90% of the *ad libitum* groups. In addition renal lesions were markedly reduced. These data will be published in more complete form in another communication. These few examples are representative of many studies which provide evidence that caloric intake alone influences significantly, risk for cancer (*19*).

Protein

Dietary protein has been associated with cancer of the kidney, pancreas, colorectum, prostate, endometrium, and breast in human populations (*20*). Kolonel *et al.* (*20*) analyzed the diet histories of more than 4,000 subjects and observed a correlation of 0.7 between total protein and total fat consumption. In a larger study the incidence rates for 27 types of cancer in 23 countries and the mortality rates for 14 types of cancer in 32 countries were examined by Armstrong and Doll (*12*) who reported a relationship between many variables including the correlations of total rpotein and total fat: these correlations turned out to be 0.7 and 0.93, respectively.

In some of the studies there have been strong correlations between international mortality rates for colon cancer and meat consumption, particularly beef (*21–23*). Others have recorded the strongest correlations between cancer of the large intestine with the consumption of eggs: this was followed by beef, sugar, and pork, but time trend data for per capita beef intake and colorectal cancer incidence showed no clear association in studies in the United States (*7*).

However, Haenszel *et al.* (*22*) found an association between cancer of the colon and consumption of legumes, starches, and meats with the strongest association for beef.

In a Japanese study Ishii *et al.* (*24*) found an association of pancreatic cancer with the consumption of high meat diets by men. A relative risk of 2.5 for daily meat intake in pancreatic cancer incidence in Japan in a cohort of 265,000 subjects followed prospectively has been reported by Hirayama (*25*).

Evidence for a correlation between cancer and protein intake has also been derived from studies in animals (*26*). The work of Ross and Bras (*27, 28*) has served clearly to point out the effects of protein on the development of spontaneous tumors in experimental animals. These workers have shown that the total incidence of various types of tumors was related directly to caloric intake and tumors appeared sooner if the caloric intake was high. Moreover, the highest number of any tumor type for any one diet occurred as fibrosarcomas in animals fed a 30% casein diet.

Madhavan and Gopalan (*29*) observed 11/30 tumors in rats fed a high protein

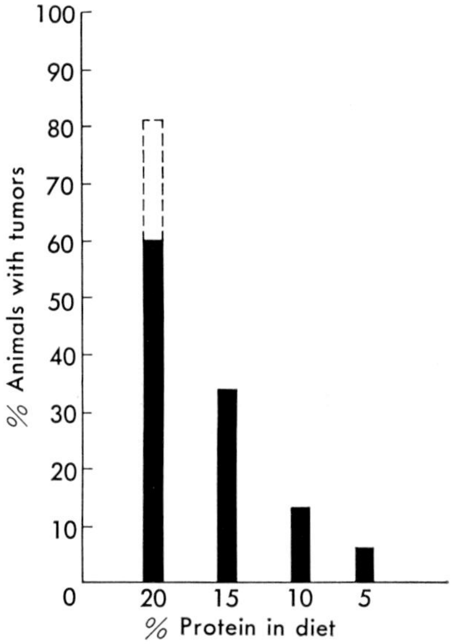

FIG. 1. Influence of dietary protein concentration and aflatoxin B_1 induced liver cancer. As the protein level was decreased, there was a linear decrease in the incidence of liver tumors.

TABLE 3. Dietary Protein, Aflatoxin B_1 Adducts Covalently Bound

Treatment dietary protein +aflatoxin B_1	Total aflatoxin B_1 in liver (mg/g)	Covalent aflatoxin adducts (ng/mg macromolecule)	
		DNA	RNA
20%	973±35	20.5±1.5	51.4±3.0
10%	780±65	14.3±1.1	39.0±2.5

Ten male rats assayed in duplicate. ^{3}H-aflatoxin B_1+cold aflatoxin B_1 to specific activity 7μCi/μmol at dose of 1 mg/kg, killed 6 hr later.

diet and exposed to aflatoxin, compared to 0/12 in the low protein group. The studies of Wells *et al.* (*30*) and Temchroen *et al.* (*31*) essentially confirmed the work of Madhavan and Gopalan.

In recent studies with rats, using pair-feeding we have found an interesting linear decrease in aflatoxin-induced liver cancer as the dietary protein was reduced from 20 to 10%. Figure 1 illustrates our findings which clearly indicate the effect of dietary protein; the higher the protein level, the greater the tumor incidence. Table 3 suggests that, mechanistically, the increased tumor incidence with higher protein levels is related to the higher DNA adducts in the higher protein groups, in agreement with the work of Mgbodile and Campbell (*32*).

Carbohydrates

When fiber is excluded and considered separately from other carbohydrates there is very little in the literature about the effects of carbohydrates on cancer. Armstrong and Doll (*12*) found a direct correlation between sugar intake and pancreatic cancer mortality in women and a weak association between liver cancer incidence and the intake of potatoes. Hems (*33*) reported that a high intake of refined sugar was associated with an increased incidence of breast cancer. In a study involving data from 37 countries, Drasar and Irving (*34*) found a direct correlation between the intake of several sugars and breast cancer. In other studies, starchy foods including bread and potatoes have been reported to correlate with both esophageal and gastric cancer (*35, 36*). Experimental animal studies with carbohydrates have been essentially negative (*37–39*).

Fiber

Fibers are not all the same; they differ in digestibility and in the chemical constituency, *i.e.*, cellulose, lignin, hemacellulose, pectins, and gums. A major characteristic common to all of these is the capacity to form bulk in the gastrointestinal tract.

Populations in the Western world, during the past several decades, have decreased their consumption of fiber (*40*). On the basis of this knowledge and in considering other populations, particularly Africans, where the food fiber intake is high, Burkitt and Trowell (*40*) suggested that chronic diseases of the bowel, including colon cancer, are associated with a low intake of dietary fiber. It has been suggested (*41*) that the differences in colon cancer incidence among East Indian populations in the North of India and those in the South of India might be explained by the high levels of roughage in the North Indian diet, compared to the very low levels in the diet of Indians living in the south of country. The Punjabis in the north, who consume a high level of vegetable fiber have a very low incidence of large bowel cancer. This compares to a higher incidence in the south Indians where vegetable fibers are rare or absent in the diets.

Comparing adult males from Denmark, who are at a high risk for colon cancer, a group from Finland, a low-risk population, it was found that the Danes consumed less fiber and stool weights were much less than those of the Finns (*42*) a finding similar to that in the Indian study.

There have been reports of studies which have yielded results in opposition to those recorded about (*43*, *44*). Considering the data from the relatively large number of studies in human pouplations, however, the effects of fiber on human colon cancer is still an open question.

Animal models with colon tumors have been chemically induced by dimethylhydrazine (DMH), azoxymethane (AOM), methylazoxymethanol acetate dimethylaminobiphenyl (DMB), and nitrosomethylurea (NMU). Bran has been re ported to have some protective effect in rats against DMH induced colon cancer (*45*, *46*). It is of particular interest however that some investigators have demonstrated that some types of fiber result in a denuding of the epithelium and an increase in cell turnover in the colon which exposes more of the cells to the effects of carcinogens (*47*).

Dietary Lipids

The influence of quality and quantity of dietary fats have been studied from both epidemiological and experimental standpoints, with regard to specific tumor types, probably more than any other single dietary ingredient. There is more convincing evidence regarding fat and its relationship to some forms of cancer than for any other macronutrient. However, it is sometimes difficult to interpret the findings because protein is almost always an accompanying nutrient.

A number of epidemiologic studies have shown direct associations between the per capita intake of fat and breast cancer incidence or mortality (*12*, *13*, *20*). In most of these studies the correlations have been higher for total fat than for other dietary factors considered at the time, including animal protein, meat, or specific fat components in oils.

Both intracountry and intercountry studies have been used to compare dietary fat intake with breast cancer. A large study by Gaskill *et al.* (*13*) found that there was a significant correlation within the United States of fat intake and breast cancer mortality rates when all states were combined. Other factors, however, including age at first marriage and geographic location, seemed to nullify some of these correlations. Kolonel *et al.* (*20*) correlated the consumption of fat with breast cancer incidence in Hawaii and found that there were significant associations between breast cancer and total fat consumption, with animal fat, and with both saturated and unsaturated fats. In other studies, the consumption of fat has been associated with increased risk for breast cancer (*48*, *49*).

In addition to breast cancer there is also epidemiological evidence for an association between dietary fat and prostate cancer, supported by the observations of Kolonel *et al.* (*20*); Armstrong and Doll (*12*); and Hirayama (*25*). These epidemiologic observations support the suggestion of many investigators that dietary fat is associated in some as yet unclear manner with increased risk for prostate cancer in men.

Perhaps the most studied aspect of fat in cancer interrelationships, aside from the breast, is that of the gastrointestinal tract. In these investigations dietary fat has been associated with cancer at a number of sites along the gastrointestinal tract, including the stomach and large bowel. Major emphasis however has been placed on large bowel cancer and the relation of dietary fat to neoplasms at this site.

A strong correlation exists between mortality and cancer of the large intestine and per capita fat intake but the data of Enstron (*7*) is probably as correct as any in pointing out that trends in beef intake (and thus fat intake) in the United States do not correlate with trends in the incidence of colorectal cancer, either from incidence or from mortality.

The experimental evidence for an effect of dietary fat on mammary carcinogenesis was first reported by Tannenbaum (*16*). Dietary fat enhanced the development of mammary tumors in mice regardless of whether they were spontaneous or induced by chemicals. Tannenbaum also showed that fat, rather than calories *per se*, was responsible for enhancing tumorigenesis by feeding mice isocaloric high and low fat diets (*16*).

The quality as well as the quantity of fat was shown by Carroll and Khor to be as important a factor in the induction of breast cancer as was the carcinogen and its dose (*50*). The incidence of mammary tumors was uniformly high with all dietary

TABLE 4. Influence of Dietary Fat on Mammary Tumorigenesis in Sprague-Dawley Rats

Fat	% (weight)	DMBA (mg)	Tumors	
			Incidence (%)	Latency (days)
Corn oil	5	2.5	68	110
Corn oil	20	”	83	96
Lard +	4	”	57	117
corn oil	1			
Lard +	19	”	71	89
corn oil	1			
Beef tallow +	4	2.5	83	154
corn oil	1			
Beef tallow +	19	”	80	130
corn oil	1			
Rapeseed oil +	4	”	71	114
corn oil	1			
Rapeseed oil +	19	”	66	120
corn oil	1			

DMBA, 7,12-dimethylbenz (a) anthracene.

fats, when they were fed at a level of 20% in the diet, but the number of tumors per group was greater in the rats fed unsaturated fats. These investigators concluded that dietary fat exerts its effect during the promotional stage of carcinogenesis (*51*). Recent studies (*52*) in our own laboratory suggest that it is a function of the type of fat that determines whether or not it influences carcinogenesis during exposure to the chemical or afterwards (initiation or promotion stages of tumorigenesis). Table 4 illustrates some of our findings. In most of the studies reported to date it appears that the amount of unsaturated fatty acids in the fat under consideration has a significant effect on response of animals to chemical carcinogenesis.

There has been an increased interest in the effects of dietary fat on experimental carcinogenesis of the large bowel in animals. Nigro *et al.* (*53*) demonstrated that rats treated with AOM developed more intestinal tumors when fed a diet containing 35% beef fat, compared to those fed a regular chow type control diet that contained considerably less fat. These data are difficult to interpret because the laboratory chow is an ill-defined diet and one cannot effectively sort out the effect of calories from that of fat on tumorigenesis.

Reddy *et al.* (*54*) have reported that low fat corn oil diets resulted in more tumors than low fat lard diets; however, rats fed 20% fat, whether it was saturated or unsaturated, had about the same tumor incidence. It appeared from these studies that the animals ate the same quantity of diet but since the caloric density of low and high fat diet differed, those eating the high fat diets received more calories. This has been associated with an increased tumor risk.

Experiments in our laboratory using both DMH (which requires metabolic activation) and NMU (a direct acting carcinogen) to induce colon tumors in rats have failed to support the hypothesis that quality and quantity of fat in experimental systems do influence carcinogenesis (*18*); we have observed no effect on tumor incidence by either the quality (corn oil, beef fat, or Crisco®) or the quantity (5% or 20% or 24%) of fat on tumor induction (Table 5). These have been rigidly controlled studies with nutrient to calories ratios carefully controlled. We have conducted six long-term studies with various permutations and have found no effect on tumor incidence of either quality or quantity of fat. It is out conclusion that the effect of fat alone is at best minimal and that the etiology is more complex; animal studies, when critically reviewed, do not indicate an effect of fat quality or quantity, *per se*, on colon cancer.

TABLE 5. Quality and Quantity of Dietary Fat and Colon Carcinogenesis with Two Colon Carcinogens

Dietary fat % (weight)	Type	Carcinogen/% DMH	Colon tumors NMU
5	Mixed	77	55
24	Beef tallow	68	63
24	Corn oil	63	55
24	Crisco®	55	30

From ref. *18*, abridged. Each group comprised of 40 rats each.

Vitamins

A number of epidemiological investigations have indicated an inverse relationship between "vitamin A" and a variety of cancers of various organ sites, primarily those with epithelial surfaces which require vitamin A for their integrity. In the epidemiologic studies the estimates of vitamin A intake have been based primarily on the frequency of ingestion of the group of foods known to be rich in vitamin A, in beta carotene, and in certain other members of this family of nutrients.

One of the first investigators to report the correlation between vitamin A and lung cancer was Bjelke (*55*). This investigator studied Norwegian men and observed low values for lung cancer cases in nonsmokers. Others (*56*, *57*) have also pointed out that lung cancer incidence has varied inversely with carotene intake and with foods that contain higher concentrations of vitamin A.

A relationship between vitamin A consumption and urinary bladder cancer, cancer of the larynx, the esophagus, and the stomach has been identified (*58–60*).

A number of studies regarding vitamin A and, in particular the synthetic retinoids, have been reported by Sporn and Roberts (*61*). This area of research has a great deal of promise because many of the synthetic retinoids have unquestioned inhibitory effects on experimentally induced tumors of many sites in animal models. We have observed (*62*) a protective effect on lung tumors in hamsters by 13-*cis* retinoic acid (Table 6). It is still too early to know whether or not this can be extrapolated to human populations; some very interesting clinical studies are now in progress which should provide some answers within a few years, relative to the inhibitory effects of retinoids on cancer in subsets of human populations as high risk.

TABLE 6. Vitamin A, 13-*cis*-retinoic Acid and Lung Cancer in the Syrian Golden Hamster

Treatment (diet)	Malignant tumors of respiratory tract	
	Number	%
Control, 2 μg/g retinyl acetate, BP	46/89	51.7
Low vitamin A, 0.3 μg/g retinyl acetate, BP	102/127	80.3
High vitamin A, 30 μg/g retinyl acetate, BP	40/88	45.4
Control, 2 μg/g retinyl acetate + 13-*cis*-retinoic acid during dosing BP	38/83	45.8
Control, 2 μg/g retinyl acetate + 13-*cis*-retinoic acid during and after dosing BP	4/91	4.4
Control, 2 μg/g retinyl acetate + 13-*cis*-retinoic acid after dosing BP	11/84	13.1

BP, benzo(a) pyrene. From ref. *62*, abridged, by permission.

Vitamin C (Ascorbic Acid)

Early in the 1960s reports appeared in the literature noting that consuming foods rich in ascorbic acid was inversely related to the appearance of certain types of cancer (*59*). Shortly thereafter there were other reports that suggested that vitamin C protected against gastric cancer, perhaps by blocking the reaction of secondary amines with nitrite to form N-nitroso compounds, some of which are gastric carcinogens (*63*).

Choline, B Vitamins, and Methionine (Lipotropes)

The complex interrelationships between the B vitamins, choline, folic acid, and B_{12}, and methionine, have been examined in detail in our laboratory with reference to effects on tumor induction by N-nitroso compounds, mycotoxins, and polycyclic aromatic hydrocarbons. The most consistent results obtained with animals fed diets low in the lipotropic factors choline and methionine are an enhancement of cancer of the liver, the pancreas and, to a lesser extent, the colon. The diets used by Rogers and Newberne (*64*) do not have a consistent effect on tumor induction in target organs other than the liver. The effects of lipotropic agents on carcinogenesis appear to be mediated through effects on metabolism but other possible mechanisms are under investigation. Table 7 summarizes findings, typical of numerous studies.

Shamberger and colleagues (*65*) and Schrauzer *et al.* (*66*) have correlated selenium status with cancer mortality. These investigators found an inverse relationship between selenium intake and leukemia as well as cancers of other sites including the colon, breast, ovary, prostate, bladder, and lung.

Studies that were conducted at the FDA during the 1940s indicated that high levels of selenium induced or enhanced tumor formation (*67*). In addition, results of studies from Russia (*68*) reported two decades later, suggested that animals given high selenium supplements in the diet developed liver cirrhosis as well as hepatocellular carcinoma. These studies were defective however and there is now an accumulation of data which clearly indicates a protective role, in animal models (*69*), for selenium against certain types of tumors (*70*) and, more recently, for humans in studies in China. All of these suggestions are now under intense investigation.

Studies in our laboratory (*71, 72*) have shown that selenium, interacting with various fats, does indeed have an effect on biological response but that this may not be through its influence on modulating tissue peroxidation. These data point to a

TABLE 7. Chemical Carcinogenesis in Rats Fed a Control Diet or a Low Lipotrope Diet

Carcinogenic treatment	Tumor site	Tumor incidence (%)	
		Control	Low lipotrope
Aflatoxin B_1, 350–375 μg, total, intragastrically (3 experiments)	Liver	6–15	22–87
DENA, 40 mg/kg, in the diet, 18 weeks	Liver	70	88
12 weeks	Liver	24–43	60–86
DBN, 3.8 g/kg, total	Liver	24	64
DMN, 100 mg/kg	Liver	28	27
25, 50 mg/kg	Liver	74	50
AAF, 0.0125%	Liver	19–61	41–91

DENA, diethylnitrosamine; DBN, dibutylnitrosamine; DMN, dimethylnitrosamine; AAF, acetylaminofluorene.

potentially very important aspect of the biological activity of selenium but clarification of these issues await further studies.

Zinc is essential to more than 100 enzyme systems; furthermore, it has been demonstrated to be effective in reversing certain types of diseases in human populations. Its role in animal nutrition has been established for many decades. It is also accepted that severe zinc deficiency in humans is not uncommon (*73*). Schrauzer *et al.* (*66*) observed that the mean zinc concentration in food and in blood correlated inversely with mortality rates from cancer of the large bowel, breast, ovary, lung, bladder, and oral cavity. Strain *et al.* (*74*) reported that the level of zinc in the plasma of bronchogenic carcinoma patients is lower than that of other patients with other types of cancer and also, lower than normal laboratory values.

We have observed that the levels of zinc in serum, hair, and from the diseased but noncancerous esophageal tissue from esophageal cancer patients were significantly lower than levels from patients with other types of cancer, patients with other types of disease, or in normal subjects. The concentration of zinc in serum and hair was lower in all cancer patients than in normal subjects (Table 8).

In animal studies we (*75*) have consistently demonstrated an enhancing effect

TABLE 8. Zinc and Copper Levels in Serum and Hair of Normal Humans and Those With Disorders

Specimen	Zn	Cu
Serum (μg%)		
Normal subjects	102.4±18.5[a)] (15)	133.4±30.5[a)] (15)
Patients with esophageal cancer	78.0±14.9[b)] (20)	159.4±44.4[a)] (20)
Patients with other types of cancer	114.4±31.8[a)] (6)	148.6±10.2[a)] (6)
Patients with other disorders	96.2±15.0[b)] (9)	146.7±39.4[a)] (9)
Hair (ppm)		
Normal subjects	195.0±29.0[a)] (16)	
Patients with esophageal cancer	162.0±33.0[b)] (19)	
Patients with other types of cancer	169.0±37.0[b)] (4)	
Patients with other disorders	212.0±48.0[a)] (8)	

[a,b)] Values in each group with different superscripts (a or b) are significantly different. Mean±S.D. numbers in parentheses are the number of subjects.

TABLE 9. Tumor Incidence Induced by MBN in Rats Deficient in Zinc and Given Ethyl Alcohol and 13-*cis*-retinoic Acid

Treatment, zinc content	MBN	4% alcohol in drinking water	13-*cis*-retinoic acid	No. rats with tumors (%)	
Control, 60 ppm	−	−	−	0/12	0
Control, 60 ppm	+	−	−	14/35	40.0
Deficient, 7 ppm	+	−	−	25/33	75.7
60 ppm control deficient 7 ppm to post dosing	+	−	−	18/35	51.4
Deficient	+	+	−	29/34	85.3
Deficient	+	+	+	33/35	94.3

From ref. *75*, abridged. MBN, methylbenzylnitrosamine.

TABLE 10. Summary of Nutrient Effects on Cancer

Nutrient	Dietary level ↑	Dietary level ↓	Agent	Target organ	Tumor induction ↑	Tumor induction ↓	Tumor induction Variable	Tumor induction None
Total calories		×	Aflatoxin B_1	Liver		×		
		×	None	Endocrine		×		
Protein	×		Aflatoxin B_1	Liver	×			
		×	Aflatoxin B_1	Liver		×		
	×		DMH	Colon	×			
Fat	×		Aflatoxin B_1	Liver	×			
	×		DMH	Colon				×
	×		NMU	Colon				×
	×		DMBA	Breast	×			
Vitamins								
Carotene	×		BP	Lung		×		
			BP	Lung		×		
			DMH	Colon		×		
Vitamin A		×	Aflatoxin B_1	Liver		×		
,, A		×	Aflatoxin B_1	Colon	×			
,, A		×	DMH	Colon	×			
,, A	×		BP	Lung				×
,, A	×		BP	Stomach		×		
		×	BP	Lung	×			
Vitamin C	×		MNNG	Stomach		×		
Vitamin B		×	MBN	Esophagus	×			
Minerals								
Zn		×	MBN	Esophagus	×			
Se	×		MBN	Esophagus		×		
Lipotropes								
		×	DENA	Liver	×			
		×	DENA	Esophagus	×			
		×	DBN	Liver	×			
		×	DBN	Bladder				×
Lipotropes								
		×	DMN	Liver				×
		×	DMN	kidney		×		
		×	AAF	Liver	×			
		×	DMBA	Mammary			×	
		×	DMH	Colon	×			
		×	DDCP	Liver	×			
		×	MNNG	Forestomach				×
Other								
Salt	×		MNNG	Glandular Stomach	×			
Aspirin	×		MNNG	,,	×			
Bile	×		MNNG	,,	×			

MNNG, methyl-N-nitrosoguanine; DDCP, 3,3-diphenyl-3-dimethylcarbamoyl-1-propyne.

of zinc deficiency on nitrosamine-induced esophageal cancer and an inhibitory effect when supplements were administered at levels of nutritional requirements (Table 9). In our model the enhancing effect by zinc deficiency is further elevated by administration of ethanol. Supplementation with 13-*cis*-retionic acid did not protect and may have advanced the neoplasia.

Nutrients in our diet may provide a major tool for prevention of cancer. Migrant populations exhibit changes in incidence and frequency of certain types of cancer indicating external causes for cancer with diets high on the list of suspects. The foods we eat are probably the most important risk factors that human populations encounter and they offer a major focus for hope of preventing cancer; of all lifestyle factors associated with cancer, diet is one over which we exercise control. Epidemiologic observations, supported by controlled animal model research on carcinogenesis strongly suggest that through dietary control, some types of cancer are preventable. Table 10 summarizes current evidence for nutrient influences on cancer.

ACKNOWLEDGMENTS

These studies were supported in part by the following grants or contracts: 1 CP33238; CA08870; AM11158; CA20079; HE10112; CA26731; CA26917; CA21401; ES00616; ES00597; ES00183 by contract from DHHS; contract 43–62–468; 69–2083; and by grants from Hoffman-LaRoche; Merrell Dow and Eli Lilly Companies.

REFERENCES

1. National Academy of Sciences/National Research Council Diet, Nutrition and Cancer. National Academy of Sciences Press, p. 477, Washington, DC, 1982.
2. Doll, R. and Peto, R. The causes of cancer, a quantitative estimate of avoidable risks of cancer in the United States today. J. Natl. Cancer Inst., *66*: 1191–000, 1981.
3. Haenszel, H. M. and Kuribari, M. Studies of Japanese migrants. I. Mortality from cancer and other diseases among Japanese of the United States. J. Natl. Cancer Inst., *40*: 43–68, 1968.
4. Haenszel, N., Berg, J. W., Segi, M., Kurihara, M., and Locke, F. B. Large bowel cancer in Hawaiian Japanese. J. Natl. Cancer Inst., *51*: 1765–1779, 1973.
5. Surgeon General Report, Chap. 14. Constituents of tobacco smoke. Natl. Cancer Inst., 1979.
6. Phillips, R. L. Role of lifestyle and dietary habits in risk of cancer among Seventh-Day Adventists. Cancer Res., *35*: 3513–3522, 1975.
7. Enstrom, J. E. Health and dietary practices in cancer mortality among California Mormons. *In*: J. Cairns, J. L. Lyon, and M. Skolnick (eds.), Cancer Incidence in Defined Populations, pp. 69–80, Cold Spring Harbor Laboratory, New York, 1980.
8. Lynch, H. T., Guirgis, H. A., Lynch, P. M., and Lynch, J. F. The role of genetics and host factors in cancer susceptibility and cancer resistance. Cancer Detect. Prev., *1*: 175–190, 1976.
9. Berg, J. W. Can nutrition explain the pattern of international epidemiology of hormone-dependent cancer. Cancer Res., *35*: 3345–3355, 1975.
10. Hill, M., MacLennan, R., and Newcombe, K. Letter to the editor: diet and large-bowel cancer in three socioeconomic groups in Hong Kong. Lancet, *1*: 436–444, 1979.

11. Gregor, O., Toman, R., and Prusova, F. Gastrointestinal cancer and nutrition. Gut, *10*: 1031–1040, 1969.
12. Armstrong, B. and Doll, R. Environmental factors and cancer incidence and mortality in different countries, with special reference to dietary practices. Int. J. Cancer, *15*: 617–631, 1975.
13. Gaskill, S. P., McGuire, W. L., Osborne, C. K., and Stern, M. P. Breast cancer mortality and diet in the United States. Cancer Res., *39*: 3628–3635, 1979.
14. Lew, E. A. and Garfinkel, L. Variations in mortality by weight among 750,000 men and women. J. Chronic Dis., *32*: 563–572, 1979.
15. Tannenbaum, A. The genesis and growth of tumors. II. Effects of caloric restriction *per se*. Cancer Res., *2*: 460–467, 1942.
16. Tannenbaum, A. The genesis and growth of tumors. III. Effects of a high-fat diet. Cancer Res., *2*: 468–475, 1942.
17. Lavik, P. S. and Baumann, C. A. Further studies on the tumor promoting action of fat. Cancer Res., *3*: 749–756, 1943.
18. Nauss, K. M., Locniskar, M., and Newberne, P. M. Effect of alterations in the quality and quantity of dietary fat on 1,2-dimethylhydrazine-induced colon tumorigenesis in rats. Cancer Res., *43*: 4083–4090, 1983.
19. Newberne, P. M. The influence of nutrition, immunologic status and other factors on the development of cancer. *In;* D. Clayson, R. Krewski, and I. Munro, (eds.), Toxicological Risk Assessment, vol. II, pp. 43–87, CRC Press, Boca Raton, Florida, 1985.
20. Kolonel, L. N., Hankin, J. H., Lee, J., Chu, S. Y., Nomura, A.M.Y., and Hinds, M. W. Nutrient intakes in relation to cancer incidence in Hawaii. Br. J. Cancer, *44*: 332–339, 1981.
21. Gray, G. E., Pike, M. C., and Henderson, B. E. Breast-cancer incidence and mortality rates in different countries in relation to known risk factors and dietary practices. Br. J. Cancer, *39*: 1–20, 1979.
22. Haenszel, W., Locke, F. B., and Segi, M. A case-control study of large bowel cancer in Japan. J. Natl. Cancer Inst., *64*: 17–22, 1980.
23. Hirayama, T. A large-scale cohort study on the relationship between diet and selected cancers of the digestive organs. *In;* W. R. Bruce, P. Correa, M. Lipkin, and S. R. Tannenbaum (eds.), Gastrointestinal Cancer, Endogenous Factors: Banbury Report 7, p. 409–429, Cold Spring Harbor Laboratory, New York, 1981.
24. Ishii, K., Nakamura, K., Ozaki, H., Yamada, N., and Takeuchi, T. Epidemiological problems of pancreas cancer. Jpn. J. Clin. Med., *26*: 1839–1846, 1968.
25. Hirayama, T. Changing patterns of cancer in Japan with special reference to the decrease in stomach cancer mortality. *In;* H. H. Hiatt, J. D. Watson, and J. A. Winsten, (eds.), Origins of Human Cancer. Book A: Incidence of Cancer in Humans, pp. 55–80, Cold Spring Harbor Laboratory, New York, 1977.
26. Voegtlin, C. and Maver, M. E. Lysine and malignant growth. II. The effect on malignant growth of a gliadin diet. Publ. Health Rep., *51*: 1436, 1936.
27. Ross, M. H. and Bras, G. Tumor incidence patterns and nutrition in the rat. J. Nutr., *87*: 245–253, 1965.
28. Ross, M. H. and Bras, G. Influence of protein under- and overnutrition on spontaneous tumor prevalence in the rat. J. Nutr., *103*: 944–960, 1973.
29. Madhavan, T. V. and Gopalan, C. The effect of dietary protein on carcinogenesis of aflatoxin. Arch. Pathol., *85*: 133–141, 1968.
30. Wells, P., Alftergood, L., and Alfin-Slater, R. B. Effect of varying levels of dietary

protein on tumor development and lipid metabolism in rats exposed to aflatoxin. J. Am. Oil Chem. Soc., *53*: 559, 1976.
31. Temcharoen, P., Anukarahanonta, T., and Bhamarapravati, N. Influence of dietary protein and vitamin B_{12} on the toxicity and carcinogenicity of aflatoxins in rat liver. Cancer Res., *38*: 2185–2192, 1978.
32. Mgbodile, M.U.K. and Campbell, T. C. Effect of protein deprivation of male weanling rats on the kinetics of hepatic microsomal enzyme activity. J. Nutr., *102*: 53–60, 1972.
33. Hems, G. Associations between breast-cancer mortality rates, child-bearing and diet in the United Kingdom. Br. J. Cancer, *41*: 429–438, 1980.
34. Drasar, B. and Irving, D. Environmental factors and cancers of the colon and breast. Br. J. Cancer, *27*: 167–176, 1973.
35. Modan, B., Lubin, F., Barrell, V., Greenberg, R. A., Modan, M., and Graham, S. The role of starches in the etiology of gastric cancer. Cancer, *34*: 2087–2092, 1974.
36. deJong, U.W., Breslow, N., Hong, J.G.E., Sridharan, M., and Shanmugaratnam, K. Aetiological factors in oesophageal cancer in Singapore Chinese. Int. J. Cancer, *13*: 291–298, 1974.
37. Roe, F.J.C., Levy, L. S., and Carter, R. L. Feeding studies on sodium cyclamate, saccharin and sucrose for carcinogenic and tumour-promoting activity. Food Cosmet. Toxicol., *8*: 135–148, 1970.
38. Friedman, L., Richardson, H. L., Richardson, M. E., Lethco, E. J., Wallace, W. C., and Sauro, F. M. Toxic response of rats to cyclamates in chow and semisynthetic diets. J. Natl. Cancer Inst., *49*: 751–763, 1972.
39. Hunter, B., Graham, C., Heywood, R., Prentice, D. E., Roe, F.J.C., and Noakes, D. N. Tumorigenicity and carcinogenicity study with xylitol in long-term dietary administration. *In;* Xylitol, vol. 20–23, F. Hoffman N LaRoche Company, Ltd., Basel, Switzerland, 1978.
40. Burkitt, D. P. and Trowell, H. C. Refined carbohydrate foods and disease. *In;* Some Implications of Dietary Fibre, Academic Press, New York, 1975.
41. Malhotra, S. L. Dietary factors in a study of cancer colon from cancer registry, with special reference to the role of saliva, milk and fermented milk products and vegetable fibre. Med. Hypotheses, *3*: 122–130, 1977.
42. MacLennan, R., Jensen, O. M., Mosbech, J., and Vuori, H. Diet, transit time, stool weight, and colon cancer in two Scandinavian populations. Am. J. Clin. Nutr., *31*: S239–S242, 1978.
43. Liu, K., Stamler, J., Moss, D., Garside, D., Persky, V., and Soltero, I. Dietary cholesterol, fat and fibre and colon cancer mortality. Lancet, *2*: 782, 1979.
44. Lyon, J. L. and Sorenson, A. W. Colon cancer in a low-risk population. Am. J. Clin. Nutr., *31*: S227–S236, 1978.
45. Barbolt, T. A. and Abraham, R. The effect of bran on dimethylhydrazine-induced colon carcinogenesis in the rat. Proc. Soc. Exp. Biol. Med., *157*: 656, 1978.
46. Wilson, R. B., Hutcheson, D. P., and Wideman, L. Dimethylhydrazine-induced colon tumors in rats fed diets containing beef fat or corn oil, with and without wheat bran. Am. J. Clin. Nutr., *30*: 176–183, 1977.
47. Cassidy, M. M., Lightfoot, F. G., Grau, L. E., Story, J. A., Kritchevsky, D., and Vahouny, G. V. Effect of chronic intake of dietary fibers on the ultrastructural topography of rat jejunum and colon: a scanning electron microscopy study. Am. J. Clin. Nutr., *34*: 218–225, 1981.
48. Lubin, J. H., Burns, P. E., Blot, W. J., Ziegler, R. G., Less, A. W. ,and Fraumeni, J. F., Jr. Dietary factors and breast cancer risk. Int. J. Cancer, *28*: 685–694, 1981.

49. Nomura, A., Henderson, B. E., and Lee, J. Breast cancer and diet among the Japanese in Hawaii. Am. J. Clin. Nutr., *31*: 2020–2038, 1978.
50. Carroll, K. K. and Khor, H. T. Effects of level and type of dietary fat on incidence of mammary tumors induced in female Sprague-Dawley rats by 7,12-dimethylbenz(a)anthracene. Lipids, *6*: 415–423, 1971.
51. Carroll, K. K. Lipids and carcinogenesis. J. Environ. Pathol. Toxicol., *3*: 253–260, 1980.
52. Lee, S. Y. and Rogers, A. E. Dimethylbenzanthracene mammary tumorigenesis in rats fed beef tallow or rapeseed oil. Nutr. Res., *3*: 361–371, 1983.
53. Nigro, N. O., Singh, D. V., Campbell, R. L., and Pak, M.S. Effect of dietary beef fat on intestinal tumor formation by azoxymethane in rats. J. Natl. Cancer Inst., *54*: 439–447, 1975.
54. Reddy, B. S., Watanabe, K., and Weisburger, J. H. Effect of high-fat diet on colon carcinogenesis in F344 rats treated with 1,2-dimethylhydrazine, methylazoxymethanol acetate, or methylnitrosourea. Cancer Res., *37*: 4156–4163, 1977.
55. Bjelke, E. Dietary vitamin A and human lung cancer. Int. J. Cancer, *15*: 561–570, 1975.
56. Gregor, A., Lee, P. N., Roe, F.J.C., Wilson, M. J., and Melton, A. Comparison of dietary histories in lung cancer cases and controls with special reference to vitamin A. Nutr. Cancer, *2*: 93–100, 1980.
57. Smith, P. G., and Jick, H. Cancers among users of preparations containing vitamin A. Cancer, *42*: 808–817, 1978.
58. Mettlin, C., Graham, S., and Swanson, M. Vitamin A and lung cancer. J. Natl. Cancer Inst., *62*: 1435–1444, 1979.
59. Graham, S., Mettlin, C., Marshall, J., Priore, R., Rzepka, T., and Shedd, D. Dietary factors in the epidemiology of cancer of the larynx. Am. J. Epidemiol., *113*: 675–684, 1981.
60. Mettlin, C., Graham, S., Priore, R., Marshall, J., and Swanson, M. Diet and cancer of the esophagus. Nutr. Cancer, *2*: 143–147, 1981.
61. Sporn, M. B. and Roberts, A. B. Suppression of carcinogenesis by retinoids: interactions with peptide growth factors and their receptors as a key mechanism. This volume, pp. 149–158, 1986.
62. Newberne, P. M. and McConnell, R. G. Nutrient deficiencies in cancer causation. J. Environ. Pathol. Toxicol., *3*: 323–334, 1980.
63. Mirvish, S. S. Inhibition of the formation of carcinogenic N-nitroso compounds by ascorbic acid and other compounds. *In;* J. H. Burchenal, and H. F. Oettgen, (eds.), Cancer: Achievements, Challenges, and Prospects for the 1980s, vol. I, p. 557–571, Grune and Stratton, New York, 1981.
64. Rogers, A. E. and Newberne, P. M. Lipotrope deficiency in experimental carcinogenesis. Nutr. Cancer, *2*: 104–112, 1960.
65. Shamberger, R. J., Tytko, S. A., and Willis, C. E. Antioxidants and cancer. VI. Selenium and age-adjusted human cancer mortality. Arch. Environ. Health, *31*: 231–239, 1976.
66. Schrauzer, G. N., White, D. A., and Schneider, C. J. Cancer mortality correlation studies. IV. Associations with dietary intakes and blood levels of certain trace elements, notably Se-antagonists. Bioinorg. Chem., *7*: 35–42, 1977.
67. Nelson, A. A., Fitzhugh, O. G., and Calvery, H. O. Liver tumors following cirrhosis caused by selenium in rats. Cancer Res., *3*: 230–238, 1943.

68. Volgarev, M. N. and Tscherkes, L. A. Further studies in tissue changes associated with sodium selenate. *In;* Symp. Selenium in Biomedicine, 1st Int. Symp., pp. 179–186, AVI Publishing Co., Westport, Conn., 1967.
69. Harr, J. R., Exon, J. H., Whanger, P. D., and Weswig, P. H. Effect of dietary selenium on N-2-fluorenyl-acetamide (FAA)-induced cancer in vitamin E supplemented, selenium depleted rats. Clin. Toxicol., *5*: 187–194, 1972.
70. Medina, D. and Shepherd, F. Selenium-mediated inhibition of mouse mammary tumorigenesis. Cancer Lett., *8*: 241–245, 1980.
71. Newberne, P. M. and McConnell, R. G. Dietary nutrients and contaminants in laboratory animal experimentation. Proc. 4th Int. Symp. Lab Animals, Boston, MA, October 28–31. J. Environ. Pathol. Toxicol., *4*: 105–121, 1980.
72. Newberne, P. M. Influence of selenium on response to toxins. Toxicology, *3*: 140, (Abstr.) 1983.
73. Prasad, A. S. Trace Elements and Iron in Human Metabolism. Plenum Publishing Corp., New York, 1978.
74. Strain, W. H., Mansour, E. G., Flynn, A., Pories, W. J., Tomaro, A. J., and Hill, O. A., Jr. Letter to the editor: plasma-zinc concentration in patients with bronchogenic cancer. Lancet, *1*: 1021–1028, 1972.
75. Gabrial, G., Schrager, T., and Newberne, P. M. Zinc deficiency, alcohol and a retinoid: association with esophageal cancer in rats. J. Natl. Cancer Inst., *68*: 785–794, 1982.

CANCER EPIDEMIOLOGY OF MUTAGENS/CARCINOGENS IN FOODS

DIET, NUTRITION AND CANCER, Y. HAYASHI ET AL. (EDS.),
JAPAN SCI. SOC. PRESS, TOKYO/VNU SCI. PRESS, UTRECHT, PP. 225–235, 1986

Measurement of Individual Aflatoxin Exposure among People Having Different Risk to Primary Hepatocellular Carcinoma

Tsung-tang SUN,*[1] Shao-ming WU,*[1] Yu-ying WU,*[1] and Yuan-rong CHU*[2]

Cancer Institute, Chinese Academy of Medical Sciences, Beijing[1] *and Qidong Liver Cancer Institute, Jiangsu,*[2] *China*

Abstract: The establishment of the carcinogenic role of aflatoxins has been impeded by the lack of suitable tests to measure individual exposure for long-term studies. It was shown that the use of immuno-concentration followed by high pressure liquid chromatography (HPLC) or immunoassay can regularly detect aflatoxins down to pg/ml in fluids incluing urine. Aflatoxin M_1 (AFM_1) in free form was identified as the major metabolite in urine suitable for use as an approximate dosimetric indicator of recent exposure to aflatoxin B_1 (AFB_1). A panel of monoclonal antibodies of IgG class were developed in the murine system against AFB_1 and/or AFM_1. Their affinity constants are at the level of 10^8–10^9 L/mol, and they are suitable for use in either effective immuno-concentration or in high sensitivity immunoassays. It was shown in a high risk area of liver cancer (Qidong of China) that 10% or more of the local inhabitants had a urinary output of AFM_1 one or two orders of magnitude higher than that of Beijing people, especially during the wet seasons. The ingestion of AFB_1 among these local people was estimated to exceed 1 mg/year. The major source came from contaminated corn and rice, but local alcoholic beverages also contributed. The increased excretion of AFM_1 was more pronounced in patients with chronic active hepatitis. This observation of very low AFM_1 content in urine of a significant percentage of local people implies their food storage practices may offer a feasible method for such prevention. The value and possible problems in conducting long-term studies on primary prevention are discussed.

Liver cancer is one of the commonest forms of cancer in the world (*1*) and hepatitis B virus has been considered to be the major causative factor (*2, 3*). This is based mainly on the observations in several independent studies which reached similar conclusions (*4–6*). However, the accumulated epidemiological data are consistent with its multi-factorial etiology (*3, 7*). The additional risk factors are most probably mycotoxins, among which aflatoxin seems to be the major candidate.

Aflatoxin B_1 (AFB_1), the chief member of the family, has been shown to induce hepatic injury and hepatoma in trout, salmon, ducklings, mice, rats, shrews, and also in non-human primates, such as rhesus monkeys and mamosets, though with varying susceptibility. Its wide spectrum of animal hepato-carcinogenicity indicates that human might not be an exception. Its possible role as an environmental risk factor was suggested by the close geographical correlation found between liver cancer incidence and aflatoxin contamination of major food items (*8, 9*). Existence of quantitative association was claimed between aflatoxin exposure measured in local food ready for ingestion and liver cancer incidence, based on limited local registered data in several early field studies (*10*). However, only long-term studies of individuals at high risk of hepato-cellular carcinoma (PHC) will conclusively establish aflatoxin as a human carcinogen. Such studies have long been impeded by the lack of suitable tests which permit the measurement of individual exposure to aflatoxins (*11*).

Techniques which can regularly detect aflatoxin exposure on an individual basis have recently been developed using mainly immunologic approaches. Individual exposure to aflatoxins was found to be much increased among people in areas of high incidence of PHC. A feasible approach to reduce aflatoxin contamination in high risk areas has been suggested. These aspects will be summarized and updated in the present article.

New Approach to Measurement of Individual Aflatoxin Exposure

Ingested AFB_1 is quickly metabolized in liver and its various forms of metabolites excreted mainly in urine. Even through the use of small Sep-Pak columns to pre-clean the urine samples, the concentrated materials are usually dark in color and have strong interference when measured by sensitive methods such as high pressure liquid chromatography (HPLC) or immunoassay. In order to increase the signal to noise ratio, the recognition capacity of antibodies to aflatoxins was explored. It was found that immuno-concentration by immuno-precipitation in case of small samples or by mini-affinity columns for general purpose enabled the detection of aflatoxins down to the level of pg/ml of urine (*12, 13*). The results are quantitative and reproducible when HPLC and/or radioimmunoassay are used for measurement. For preparation of such columns, either 1 ml of polyclonal antibodies or 10 ml of monoclonal culture fluid, after concentration by ammonium sulfate precipitation, can be added to 1 ml of Sepharose 4B which can then bind around 1 μg of the relevant aflatoxin. Antibodies with an affinity constant of 10^8 L/mol or higher were found to be effective for concentration purpose. Usually, a small column of 0.2 ml bed size is sufficient for concentrating aflatoxin M_1 (AFM_1) from several hundred human urine specimens or for AFB_1 from a similar amount of diluted beverage when antibodies with required reactivity have been added to the matrix. The bound aflatoxins can be easily recovered with methanol or ethanol which may be quickly evaporated off. The subsequent measurement can be done by immunoassay, either radioimmunoassay or enzyme linked immunosorbent assay (ELISA) technique. Such a two-step, immunologic procedure has been found suitable for both laboratory and field uses. Whenever necessary, part of the concentrated materials can be confirmed by HPLC analysis.

AFB_1 ingested is metabolized in the body to form various metabolites such as

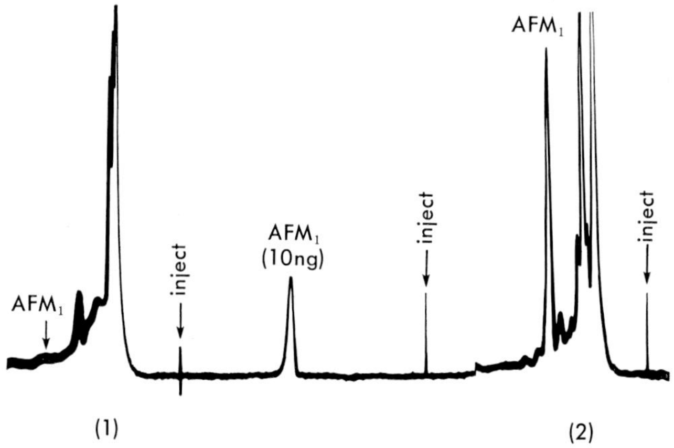

FIG. 1. HPLC profile of AFM_1 in 24-hour urine from a normal person after immuno-concentration. (1) during ingestion of rice contaminated with 10 ppb AFB_1. (2) 3 days after change of the rice. Column, C18 μ-Bondapak; mobile phase, 50% methanol in H_2O, flow rate, 0.8 ml/min; fluorescence detector, Spectro-Glo, Gilson. In the middle, standard AFM_1.

AFM_1, aflatoxin P_1 (AFP_1), aflatoxin Q_1 (AFQ_1), and others. These metabolites may be excreted in either free or conjugated state. The question is, can we identify the major one in human urine which can quantitatively reflect the amount of AFB_1 ingested? Through the use of affinity columns possessing a relatively broad spectrum of reactivity toward various aflatoxins including their metabolites known in the experimental systems, AFM_1 in free form was found to be an important metabolite stably excreted in the urine following recent exposure to aflatoxins. Figure 1 shows such an example. HPLC analysis of the immuno-concentrated 24 hr urine from an individual demonstrated the prominent peak of AFM_1 on a relatively clean background. This was confirmed by two dimensional thin layer chromatography done on the eluted peak and also quantitatively correlated to RIA of the peak collected for inhibitors. The rice was found to be contaminated by AFB_1 at 10 ppb level. On the third day following the change of rice, urinary AFM_1 dropped to a very low level, around 1 ng per 24 hr urine. Such phenomena were regularly observed in peoples having increased exposure from main food or from contaminated alcoholic beverages in areas of PHC prevalence. In the latter case, the amount of ingestion of AFB_1 could be more accurately measured, and 1.5 to 5% of ingested aflatoxins was found to be excreted in the free form of AFM_1 in the first 24 hr urine from volunteers (*14*). On the basis of these observations, it has been concluded that unconjugated AFM_1 may serve as a dosimetric indicator which reflects in quantitative terms recent individual exposure to ingested AFB_1 (*14–16*). It should be noted that 95% or more of the ingested AFB_1 has been excreted in other forms which escape the immunologic procedures used. However, the techniques described above enable the measurement of individual aflatoxin exposure down to a level equivalent to the ingestion of "clean" food having aflatoxin content below 0.1 ppb. Such sensitivity is probably sufficient for monitoring purposes in conducting critical long-term studies to assess the real hepato-carcinogenic role of aflatoxins.

A Panel of Monoclonal Antibodies Having Adequate Affinities toward Aflatoxins

Since AFB_1 is the chief constituent of the contaminated food source and AFM_1 is an indicator of individual exposure, two antibodies having sufficient reactivity against each of them will be enough for the assessment of individual exposure through an immunologic approach. Such requirements can be fulfilled without much difficulty by raising polyclonal antibodies in rabbits or in sheep according to standard procedures (*11*).

However, long-term studies require the measurement of exposure over a large number of samples for a relatively long period of time. In addition, such measurement may also be useful in other areas of health concern, including the food industry. A panel of monoclonal antibodies directed against required specificities with adequate affinities is therefore clearly desirable. This may provide a stable and virtually unlimited supply of these useful immunochemicals of consistent quality, including their affinity constants. Such a panel has been achieved in the authors' laboratory for AFB_1 and AFM_1. Antibodies were produced in the murine system where NS-1 myetoma cells were fused with sensitized lymphocytes following standard protocols. Radioimmunoassay was used for screening. Hybridomas selected were stabilized through several cycles of recloning and have been maintained in a stable state for 1 to 3 years (*13*, *17*).

Table 1 shows some characteristics of three representatives of a panel of seven monoclonals mainly directed against AFB_1, AFM_1, or both. AW1 was derived from lymphocytes sensitized by B2A-BSA conjugate, and AW11 as well as AW13 were derived following immunization with AFM_1-oxime-BSA. All three are IgG but belong to different subclasses. They grow fast in culture fluids where the concentration of these monoclonals usually exceeds 30 μg/ml. An important parameter is their affinity constants toward defined specificities. This constant is calculated from data obtained in competitive radioimmunoassay following the method of Muller (*18*). AW13 was found to have a high affinity toward AFM_1 and AFB_1 at a level of 1×10^9 L/mol, being 3-fold higher to AFM_1. This is correlated with the fact that AW13 is highly effective in concentrating a minute amount of AFM_1 from a large volume of human urine, and is thus highly sensitive when used in immuno-assay systems (*17*). The crossreactivity of AW13 does not cause problems in practical use because the main constituent of contaminants in food and beverages is AFB_1, whereas the presence of AFM_1 is usually very low or even below the level of detectability. The opposite situa-

TABLE 1. Specificity and Affinity Constant of Monoclonal Antibody to Aflatoxins

Monoclonal antibody	Immunogen used	Ig class	Major specificity	Affinity constant (L/mol)
AW1	B2a-BSA	IgG1	AFB_1	1×10^8
AW11	AFM1-oxime-BSA	IgG1	AFM_1	$\simeq 1 \times 10^8$
AW13	AFM1-oxime-BSA	IgG2b	AFM_1	3×10^9
			AFB_1	0.9×10^9

Affinity constants were calculated according to R. Muller's method using immunoassay. This is a summary of data to be published (*17*).

tion occurs in the case of urine. Therefore, AW13 was found to be a very useful reagent in the measurement of individual aflatoxin exposure. Where a higher distinction is required, either AW1 or AW11 can be selected for immuno-concentration or immunoassay. It should be noted that AW11 has affinity toward AFM_1 over ten times lower than that of AW13, however, its capacity to immuno-concentrate AFM_1 from urine was close to the latter in affinity chromatography.

Hydrophobic interaction appears to play an important role in the binding of the antibodies with their relevant haptens, the various members of the aflatoxin family. Once bound, it cannot be eluted by high salt solution or by lowering the pH. Nonetheless, the bound aflatoxins can be readily separated by elution with methanol, ethanol, acetone, and other organic solvents. The availability of a panel of monoclonals possessing different spectra of reactivity toward various members of the aflatoxin family also offers research opportunities for study of the mechanism immunologic recognition of haptens by combining the sites of their relevant monoclonal antibodies (*17*).

Increased Individual Exposure to Aflatoxin among People at High Risk of PHC

People of Qidong County on the north shore of the Yangtze valley opposite Shanghai in China are at high risk to hepatoma. The average incidence rate of this one-million-population county over a 10-year period is around $50/10^5$ individuals per annum (*19*). In addition to its prevalence in hepatitis B virus infection, aflatoxin contamination was commonly observed from randomly selected samples of corn, the major food of the local inhabitants. Long-term studies to assess the hepatocarcinogenic role of aflatoxins in this area has been strongly indicated.

Shortly after the development of an immunologic technique for the measurement of individual exposure, 24-hour urine specimens were collected from male individuals randomly selected from Qidong and also from the Beijing area. The incidence rate of liver cancer in Beijing was reported to be $8/10^5$ per year, about 6 times lower than that of Qidong. In these earlier studies, the immuno-concentrated samples were first analyzed by HPLC which provided more definitive results. Portions of these samples were parallelly detected by two dimensional thin layer chromatography and/or radioimmunoassay. The initial observations (*15*) which were extended and confirmed by subsequent studies (*17*, *20*) yielded three meaningful aspects of information. First, measurements showed that aflatoxin exposure on an individual basis was indeed much higher among Qidong inhabitants. As shown in Fig. 2, among 24 Beijing normal males tested, none had a urinary output of AFM_1 exceeding 10 ng per day. Elevations beyond this level were occasionally seen in an extended series of observations, however, they were of a transient nature. The major source came from contaminated rice in some lots of the food supply. In contrast, among 40 normal males of Qidong County tested, 18% (7/40) of them demonstrated AFM_1 presence above a 10 ng level in 24-hour urine. Since in rural areas people generally ingested corn or rice from a large batch kept in the home, the increased exposure was expected to be persistent and took the time course in plateaus, the height of which depended mainly on the level of contamination of the corn, the major source of aflatoxins in

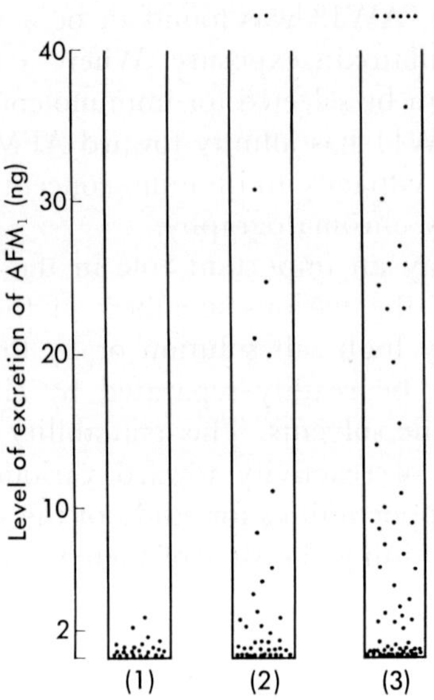

FIG. 2. Scattergram of AFM_1 content in 24-hour urine of people at different risk to PHC. (1) Beijing normal male adults. (2) Qidong normal male adults. (3) Qidong chronic hepatitis male patients. Urine samples were immuno-concentrated and then quantitated for AFM_1 by HPLC.

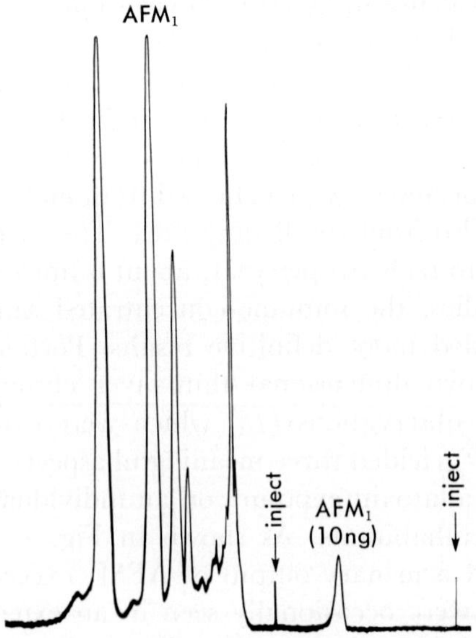

FIG. 3. HPLC profile of AFM_1 immuno-concentrated from 250 ml of urine from a Qidong normal during ingestion of AFB_1 contaminated corn in wet season. Experimental conditions: same as in Fig. 1. Adapted from ref. *15*.

the area studied. In a season causing heavy mold, the exposure reached a high level. Figure 3 shows the HPLC profile of an immuno-concentrated urine sample of a Qidong inhabitant who ingested moldy corn from his own store. The AFM_1 in 250 ml of urine was 115 ng. If his 24-hour urine was 1 *l* and 3% was taken as the conversion ratio of AFB_1 into AFM_1, then the amount of AFB_1 ingested by this individual the preceding day was estimated to be 15 μg. This is consistent with the observations that over 20% of corn samples randomly drawn from farmers' personal stores were shown to contain AFB_1 above the 20 ppb level during wet seasons (Z. Chen, personal communications). Previous years' approximations on the basis of limited measurements of urinary AFM_1 indicated that the integrated dose of individual aflatoxin exposure in a pattern of plateaus having different height and width was around or exceeded 1 mg/year (*15*, *16*, *20*). This might give a cumulative dose of AFB_1 ingestion of over 30 mg in a 30-year period among 10% or more of the local inhabitants in Qidong County. Such exposure might have carcinogenic potential among those with persistent liver cell hyperplasia associated with chronic infection by hepatitis B virus. Since young and middle-aged males generally eat and usually also drink more after strenuous physical labor, this fact might explain the epidemiological features in areas of prevalence, such as higher incidence, male preponderance, and shift of incidence age toward the younger side with a peak starting at the fourth decade (*3*). In previous studies, contaminated corn was identified as the major source of aflatoxin ingested; moldy rice was second. Local alcoholic beverages might also contribute to some extent. Since people may drink large quantities of such beverages, the potential effect of alcohol plus co-existing mycotoxins on hepatic cells with hyperplasia should also be considered. Since fungal contamination may yield other mycotoxins and promote the formation of nitrosoamines, potential aflatoxin exposure may indicate the existence of a group of fungus-related toxic products. Such consideration has been substantiated by the fact that experimental animals (ducks and rats) fed moldy corn containing a known amount of aflatoxin were found to have significantly more liver cancers than those fed with "clean" corn with a similar amount of purified aflatoxin added (*21*). These points should be considered in assessing the additional risk factors involved in human hepato-carcinogenesis.

The second aspect of possible interest is the observation that urinary excretion of AFM_1 was found to be significantly elevated among Qidong people whose serum alpha-fetoprotein was at a low but fluctuating level (see Fig. 2). These people belonged to the chronic hepatitis group who had serological expression of persistent liver cell hyperplasia. They were closely associated with hepatitis B virus infection but did not necessarily demonstrate obvious clinical and serum enzyme manifestations. They had been identified by several independent follow-up studies as extreme risks for subsequent development of PHC (*22*, *23*). The increased urinary output of AFM_1 observed may have been due to the change of metabolic pattern of AFB_1 in hepatocytes generally involved in chronic hepatitis, so that more aflatoxin metabolite was excreted in the form of free AFM_1, or it may have been due to increased exposure; the actual cause is not yet known. If increased urinary excretion of AFM_1 among chronic liver disease patients could be confirmed in subsequent studies and traced to increased exposure, it might be of value in assessing the broader significance of the pathogenic role of aflatoxins.

The third aspect of possible significance lies in the fact that a significant proportion of inhabitants in areas of greater exposure to aflatoxins excreted a very low level of AFM_1 in their urine. Because of its pragmatic implications, this point will be treated in some detail in the following section.

Feasible Approach to Reduction of Aflatoxin Exposure

In formulating a strategy to prevent liver cancer by reducing aflatoxin exposure, two issues deserve serious consideration. The first is an assessment of aflatoxin's carcinogenic role. This provides the rationale for such a prevention approach and constitutes the central theme of this paper. The other issue is the feasibility of implementing a prevention program once it has been indicated on a firm scientific basis.

Data shown in the second column of Fig. 2 indicate the possibility of finding a more feasible preventive approach in developing countries where control of aflatoxin exposure is strongly indicated. Forty-five percent (18/40) of normal individuals in Qidong had very low content of AFM_1 in their urine, at or below 1 ng per 24-hour sample. This reflected that their major food, rice or corn, had an aflatoxin content below 0.1 ppb which should be considered uncontaminated according to the standard of food safety regulations in various countries of the world. The secret of these farmers' food hygiene lies mainly in their way of storing foodstuffs. Therefore, in order to reduce aflatoxin exposure in endemic areas, infroduction of these local methods rather than a complete change of food source might be preferably accepted. This approach may further be facilitated by selective monitoring of AFM_1 in urine using the techniques described. In this way, the efficacy of reducing aflatoxin exposure might be achieved without a great economic investment.

DISCUSSION

Following the development of techniques for measuring aflatoxin exposure among individuals and the accumulation of epidemiologic data on such measurement, long-term studies including intervention trials should be conducted to assess the material's hepato-carcinogenic role. Research along this direction has been designed and carried out by several groups in endemic areas. If a significant decline of liver cancer incidence can be observed in close association with the reduction of integrated doses of aflatoxin ingestion on an individual basis, the carcinogenic role of this chemically well-defined environmental mycotoxin can be conclusively established. Furthermore, this may also provide a new approach to the reduction of liver cancer incidence among high risk groups. Prevention may then begin to be apparent within a several-year period, much sooner than the effect of the universal program of newborn immunizations to preventing this disease will be known (*16*). However, it should also be recognized that such an expensive long term study, even though well designed and carefully conducted, may not lead to a significant reduction of liver cancer incidence among high risk groups. This is true because aflatoxin exposure during the study period may be inadequate to confer the final "hits" to complete the process of hepatocytic malignancy. The relative insensitivity of the hepatocytes of non-human primates to aflatoxin, the carcinogenic dose of which is about one order of magnitude higher than the estimated human exposure mentioned in the previous section, also indicates such possibility (*24*). Alternatively, and more likely, it may be due to the initiative effect which occurs early in the course of aflatoxin exposure. Therefore, the "risk" of running a long, laborious and expensive trial without reaching a desirable "positive" endpoint should also be considered.

Vaccination projects against hepatitis B virus are planned or are under way in regions of the world where hepatitis B virus infection is endemic. Although its effect can only be clarified 30 or 40 years hence, on the basis of existing knowledge such projects offer major promise of control of PHC and its associated hepatitis. In such large scale intervention programs with hepatitis B virus vaccine, concurrent efforts should be devoted to better defining the possibly associated role of aflatoxin as a human hepato-carcinogen by monitoring human aflatoxin exposure over a long period (*11*). This offers the unique opportunity to dissect the carcinogenic role of multiple etiologic factors in the development of a major human cancer.

Multi-factorial interaction seems to play a key role in human hepato-carcinogenesis (*3*, *7*). This is further supported by a recent study on hepatitis B virus infection in young PHC patients in a high risk area where aflatoxin exposure was known to be high (*25*). Occurrence of PHC at a young age may be explained by excessive ingestion of such carcinogens during early years. However, immuno-histological analysis of the liver tissues of a consecutive series of 55 pathologically-proven PHC patients below 30 years of age demonstrated that 100% of them had chronic hepatitis and 96% were clearly hepatitis B virus related (*25*). This supports the notion that the probability of malignant change in hepatocytes will be greatly enhanced when two factors, one viral and one chemical, act together and synergistically. A similar situation exists in the duck hepatoma which frequently occurs in areas where human PHC is prevalent and is probably related to multi-factorial interaction (*3*, *23*). In this context. PHC may serve as an appropriate system for carcinogenesis studies, because its multi-factorial etiology and multi-staged development have been determined and might be conclusively defined through long-term studies along two directions to achieve etiologic prevention.

ACKNOWLEDGMENTS

This work was supported by Grants from the Ministry of Public Health of the People's Republic of China, and from the Cancer Research Institute/Roger E. Green Memorial Fund, New York, N.Y.

REFERENCES

1. Parkin, D. M., Stjernsward, J., and Muir, C. S. Estimates of the worldwide frequency of twelve major cancers. Bull. WHO, *62* (2): 163–182, 1984.
2. World Health Organization. Technical Report Series 691. Prevention of Liver Cancer, World Health Organization, Geneva, 1983.
3. Sun, T. and Chu, Y. Carcinogenesis and prevention strategy of liver cancer in areas of prevalence. J. Cell. Physiol., *3* (Suppl.): 39–44, 1984.
4. Beasley, R. P., Hwang, L. Y., Lin, C. C., and Chien, C. S. Hepatocellular carcinoma and HBV: a prospective study of 22,707 men in Taiwan. Lancet, *2*: 1129–1133, 1981.
5. Lu, J., Li, W., Jiang, Z., Hwang, F., and Ni, C. P. Matched prospective study on

chronic carriers of HBsAg and primary hepatocellular cancer. Chin. J. Oncology, *5*: 406–409, 1983 (in Chinese).
6. Sakuma, K., Takahara, T., Okuda, K., Tsuda, F., and Mayumi, M. Prognosis of hepatitis B virus surface antigen carriers in relation to routine liver function tests: a prospective study. Gastroenterology, *83*: 114–117, 1982.
7. Harris C. C. and Sun T. Multifactorial etiology of human liver cancer. Carcinogenesis, *5*: 697–701, 1984.
8. Wogan, C. N. Aflatoxins and their relationship to hepatocellular carcinoma. *In;* K. Okuda and R. L. Peters (eds.), Hepatocellular Carcinoma, pp. 25–42, John Wiley and Sons, Inc., New York, 1976.
9. Wang, Y., Yeh, P., Li, W., and Liu, Y. Correlation between geographical distribution of liver cancer and aflatoxin B1 climate conditions. Acta Sinica, Series B, 431–437, 1983 (in Chinese).
10. Linsell, C. A. and Peers, F. G. Field studies on liver cell cancer. *In;* H. H. Hiatt, J. D. Watson, and J. A. Winsten (eds.), Origins of Human Cancer, Book A, pp. 549–556, Cold Spring Harbor Laboratory, New York, 1976.
11. IARC Meeting Report, Monitoring of aflatoxins in human body fluids and application to field studies. Cancer Res., *45*: 922–928, 1985.
12. Wu, S., Yang, G., and Sun, T. Studies on immuno-concentration and immunoassay of aflatoxins. Chin. J. Oncol., *5*: 81–84, 1983 (in Chinese).
13. Sun, T., Wu, T., and Wu, S. Monoclonal antibody against aflatoxin B1 and its potential applications. Chin. J. Oncol., *5*: 401–405, 1983 (in Chinese).
14. Sun Z(T)., Wu, S., and Wu, Y. Prospect of using monoclonal antibodies against aflatoxins in studying human liver carcinogenesis. Acta Acad. Med. Sinicae, *5*: 344, 1983 (in Chinese).
15. Wu, S., Wei, Y., Ku, G., Liu, K., and Sun, Z(T). Study on urinary excretion of AFM1 in Beijing and Qidong people. Chin. J. Oncol., *6*: 163–167, 1984 (in Chinese).
16. Sun, T., Chu, Y., Hsia, C., Wei, Y., and Wu, S. Strategies and current trends on etiologic prevention of liver cancer. *In;* C. C. Harris (ed.), Biochemical and Molecular Epidemiology of Human Cancer, UCLA Symposium Series, Academic Press, New York (in press).
17. Wu, S., Wu, Y., and Sun, T. High affinity monoclonal antibody against aflatoxin M1: practical and theoretical applications (to be published).
18. Muller, R. Calculation of average antibody affinity in anti-hapten sera from data obtained by competitive radio-immunoassay. J. Immunol. Methods, *34*: 345–352, 1980.
19. Chen, G. Epidemiological trend of liver cancer in Qidong County during 1972–1981. Qidong Liver Cancer Research, pp. 8–14, Qidong Cancer Institute, Jiangsu, 1983 (in Chinese).
20. Wu, S., Liu, G., and Sun, T. Extended observation on urinary excretion of aflatoxin M1 in people at high risk of liver cancer, in preparation (in Chinese).
21. Qian, K., Xu, G., and Liu, Y. Cancer induction in experimental animals by moulded corns and insecticides. Qidong Liver Cancer Research, pp. 68–76, Qidong Liver Cancer Institute, Jiangsu, 1976 (in Chinese).
22. Sun, T., Chu, Y., Wang, L., Xia, Q., Wang, N., and Zhang, Y. Immunological approach to natural history, early diagnosis and etiology of human hepatocellular carcinoma. Cold Spring Harbor Conf. Cell Prolif., *7*: 471–480, 1980.
23. Chu, Y. AFP sero-survey and early diagnosis of liver cell cancer in the Qidong field. Chin. J. Oncol., *3*: 35–38, 1981 (in Chinese).

24. Sieber, S. M. and Adamson, R. H. Chemical carcinogenesis in non-human primates and attempt at prevention. *In;* P. Chandra (ed.), Anti-viral Mechanism in the Control of Neoplasm, pp. 455–480, Plenum, New York, 1978.
25. Xia, Q., Li, T., Li, Y., Wang, N., Cao, L., Zhang, S., and Sun, T. Liver cell cancer in the young and adolescent and its relation to HBV. Chin. J. Oncol., *6*: 413–416, 1984 (in Chinese).
26. Wang, N., Sun, Z(T)., Pan, Q., and Hsia, C. Liver cancer, liver disease background and virus-like particles in serum among ducks from high incidence area of human hepatocellular carcinoma. Chin. J. Oncol., *2*: 174–176, 1980 (in Chinese).

DIET, NUTRITION AND CANCER, Y. HAYASHI ET AL. (EDS.),
JAPAN SCI. SOC. PRESS, TOKYO/VNU SCI. PRESS, UTRECHT, PP. 237–245, 1986

Vitamin A and Selenium Intake in Relation to Human Cancer Risk

Walter WILLETT

Department of Epidemiology, Harvard School of Public Health, and the Channing Laboratory, Department of Medicine, Harvard Medical School and Brigham and Women's Hospital, Boston, Massachusetts 02115, U.S.A.

Abstract: The possibility that dietary intake of certain vitamins and minerals may influence the occurrence of human cancer is receiving considerable scientific attention. One prominent hypothesis, that increased dietary intake of vitamin A reduces the occurrence of cancer, has received support from a large number of epidemiologic studies in which an inverse association was observed. The largest body of evidence relates to lung cancer. However, when examined in further detail, this apparent protective effect appears primarily attributable to higher intakes of green and yellow vegetables, which contain the carotenoid precursors of vitamin A. In contrast, there is little evidence to support an association between preformed vitamin A intake and cancer risk. In several studies based on prospectively collected sera, retinol levels were inversely related to subsequent cancer risk. However, these have not been supported by further investigations and appear to be the result of methodologic artifact. Available evidence thus suggests that factors associated with green and yellow vegetables provides modest protection against certain forms of cancer; beta-carotene is a likely candidate and is the focus of considerable research activity.

Stimulated by the results of many animal studies and ecologic comparisons, we and other investigators have examined the association of serum selenium levels with subsequent risk of cancer. In the three published prospective studies an inverse association was observed, with a 2-to 5-fold increase in overall cancer risk among those with lowest selenium levels. As would be predicted by animal studies, the combination of low selenium and low vitamin E appears to be particularly deleterious. Selenium is known to have other important nutritional interactions, which are certain to complicate future research in this area.

Preformed vitamin A and Carotene

Since vitamin A plays a major physiologic role in the regulation of cell differentiation, and loss of differentiation is a basic feature of cancer, it is understandable that this nutrient is presently a topic of substantial research. A large body of animal data indicate that retinol and synthetic retinoids can decrease the incidence of cancers that are induced by a variety of agents and that occur at different sites (*1*, *2*).

Epidemiologic interest in vitamin A and cancer was stimulated by a prospective study of Bjelke *et al.*, conducted among approximately 14,000 Norwegians. They observed that those with an index of vitamin A intake below average experienced approximately half the rate of lung cancer compared with those consuming above average amounts (*3*). In a very large, prospective study among 265,118 men and women in Japan, Hirayama observed that individuals consuming green and yellow vegetables on a daily basis had lower rates of cancer at several sites, including lung, prostate, and stomach (*4*).

Similar findings were observed in studies from Singapore (*5*), further follow-up of the Norweigan cohort (*6*), and in investigations from the U.S. which examined cancers of the bladder (*7*), esophagus (*8*), breast (*9*), lung (*10*), and cervix (*11*). Although higher intake of green and yellow plants was associated with a decreased risk of prostatic cancer in data from Japan (*4*) and Minnesotá (*12*), the converse was seen among a subgroup of older men living in Hawaii (*13*). We recently reported on the follow-up of an elderly population who had, for other reasons, completed a simple dietary questionnaire several years previously. As others have observed in younger adults, we found an inverse relationship between intake of green and yellow fruits and vegetables and the overall risk of cancer after accounting for cigarette smoking and other risk factors (*14*).

While these studies have been construed by some as evidence that vitamin A reduces cancer risk, this interpretation is complicated by the diversity of vitamin A sources. Plants, principally green and yellow fruits and vegetables, contain no preformed vitamin A, but rather a series of carotenoid compounds. Only a minority of these can be metabolized to form retinol, the physiologically active form of vitamin A. Beta-carotene, the most plentiful carotenoid with potential vitamin A activity, is a dimer that is potentially cleaved after absorption to form two molecules of retinol. Since the questionnaires employed in most studies of vitamin A and cancer have inquired primarily about plant, rather than animal, sources of vitamin A, these investigations provide stronger support for a protective effect of carotenoids than for an effect of preformed vitamin A itself.

More recently, several epidemiologists have attempted to separate the effects of preformed vitamin A intake from that of provitamin A, or carotenoid, intake. In a prospective study among approximately 2,000 men, Shekelle and others found that the overall protective effect of vitamin A was limited to lung cancer and that this protective effect was entirely attributable to carotene intake rather than pre-formed vitamin A intake (*15*). Gregor *et al.* observed a protective effect of preformed vitamin A intake among men, but not women, in a small case-control study of lung cancer (*16*). However, in two much larger case-control studies of lung cancer by Samet *et al.* (*17*) and by Zeigler *et al.* (*18*) a protective association with carotene intake was observed, but no effect of preformed vitamin A intake was seen. Although information was not presented separately for preformed vitamin A in another large case-control study reported by Hinds *et al.*, a similar conclusion can be inferred from their data (*19*). Neither preformed vitamin A nor carotene intake were associated with risk of gastrointestinal cancer in a study from Israel, despite a protective association with fruit and vegetable consumption (*20*). Thus, available data provide little evidence that preformed vitamin A intake is related to cancer incidence among populations that are basically well nourished. It must be noted, however, that many cancer sites have not been carefully studied with respect to preformed vitamin A intake. In contrast, a remarkably consistent association has been observed between low intake of green and yellow fruits and vegetables and the risk of lung cancer. Indeed, of all the diet and cancer relationships that have been investigated to date, this relationship is the only one that has been consistently supported in a substantial number of investigations.

Beta-carotene has unique antioxidant properties (*21*) and, along with preformed vitamin A, has reduced micronuclei in the buccal smears of betel nut chewers (*22*). It has thus been hypothesized that the inverse associations with cancer risk discussed above are due to an effect of beta-carotene (*23*). While it is a strong possibility that beta-carotene does have an anti-cancer effect, it must be kept in mind that the plant sources of carotenoids also contain a vast number of other chemicals. A number of these substances exhibit anti-cancer effects and thus provide alternative explanations for the beneficial effects of fruit and vegetable consumption (*24*, *25*).

If certain foods are consistently found to be associated with lower cancer incidence, it would be possible to make recommendations regarding increased consumption, even though the specific cancer inhibitary substance was yet unknown. Unfortunately, many studies that have examined the relationships of vitamin A and cancer have not reported complete data relating specific foods to cancer risk. Thus, it is presently not possible to identify with confidence specific foods that predictably reduce the risk of cancer.

Biochemical Measurements of Retinol and Carotenoids in Relation to Cancer Risk

An alternate epidemiologic approach to the study of vitamin A and cancer is to measure levels of retinol, retinol building protein (RBP), or carotenoids in the blood of individuals who subsequently develop cancer, and compare these levels with those of individuals who remain cancer-free. The prospective aspect of these studies is important, since it is unlikely that biochemical measurements in blood collected before the diagnosis of cancer will have been affected by the disease or its treatment.

In the two initial studies of serum retinol and subsequent risk of cancer (*26*, *27*) a strong inverse relationship observed; individuals with the lowest serum retinol levels experienced the highest risk of cancer at all sites combined. The strongest protective effect in these studies was for lung cancer. These findings stimulated great excitement since it appeared that a single measurement of serum retinol might be predictive of cancer in the same way that serum cholesterol is predictive of coronary heart disease. However, subsequent investigations have not supported the findings. Among a group of over 4,000 men and women participating in a national trial of blood pressure treatment (the Hypertension Detection and Follow-up Program, HDFP), 111 cases of cancer occurred over a five-year period (*28*). Among these individuals, we found no relation between serum retinol or RBP level and risk of cancer. The lack of association has now been confirmed in a number of investigations (*29–31*), including the continued follow-up of the cohort initially reported by Kark *et al.* (*32*). It is not completely clear why inverse associations were found in the two original studies, but in the Kark study, it may have been the result of differences in processing the sera of cancer cases and non-cases. In the study reported by Wald *et al.* (*27*), pre-clinical disease may have caused lower serum levels since the duration of follow-up was two years at the most. It must be noted that lack of association between serum retinol level and cancer incidence does not constitute strong evidence against a relationship between preformed vitamin A intake and cancer risk. Intake of this vitamin and serum retinol are almost completely unrelated in well nourished populations due to tight homeostatic regulation by the liver; we observed only minimal changes in plasma retinol when large supplements of retinyl esters were consumed daily for 16 weeks (*33*).

Fewer data are available from prospectively collected serum analyses relevant to carotenoid levels and cancer risk. This is partly due to the relative instability of carotenoids in frozen sera; specimens must be preserved at ultra-low temperatures (*e.g.*, −70 °C) to avoid rapid loss (*34*). In the HDFP study we observed no association between total carotenoid level and subsequent risk of cancer at all sites combined (*28*). However, an effect of beta-carotene, if present, might well have not been detected because beta-carotene typically represents only 10–30 percent of the total carotenoid. Stahelin *et al.* have reported an inverse association between a specific measure of serum beta-carotene and risk of lung cancer among a cohort of Swiss residents (*29*). A similar association was recently reported among a cohort of Japanese-Hawaiian men (*31*).

Epidemiologic data based on biochemical measurements, like that derived from dietary intake measures, should be interpreted cautiously since such findings are susceptible to confounding. For example, even if serum beta-carotene levels are consistently associated with low cancer incidence, it is still conceivable that some other factors in the fruits and vegetables containing the carotenoids is actually the substance that reduced cancer risk. For this reason, definitive conclusions regarding an anti-cancer effect of beta-carotene can probably only be obtained from randomized trials. Several trials are now ongoing in the U.S. and elsewhere. The largest of these is a study in which over 21,000 male physicians have been allocated to alternate-day beta-carotene supplements or placebo (*35*).

Developments in Dietary Assessment

The capacity to measure the dietary intakes of large numbers of individuals is fundamental to progress in elucidating nutritional relationships with human cancer. Most investigators involved in this type of research have converged to the use of some form of a food frequency questionnaire. An example of this approach was that employed in the pioneering prospective cohort study conducted by Hirayama, which was appended to a Japanese census (*4*). In that study subjects were asked how often they consumed each of a list of specific foods. Subsequent rates of cancer were then related to previous dietary intake.

Many have questioned the validity of information collected in this manner. However, recent studies indicate that these simple questionnaires can provide data that are as reproducible as traditional epidemiologic measures such as blood pressure or serum cholesterol, and that correlate reasonably with carefully weighed meal-by-meal recordings of dietary intake. To evaluate the validity of a semiquantitative food frequency questionnaire employed in a prospective study of approximately 100,000 women, we compared responses to this questionnaire with four one-week diet weighed diet records collected over the preceeding one-year period. After adjustment for caloric intake, correlation coefficients between the two methods ranged from 0.47 to 0.75 for the nutrients examined (*36*). In addition, the questionnaire measurement of beta-carotene was able to predict plasma carotenoid level (*37*). Others have demonstrated that responses to simple questionnaires are reproducible over periods of up to 24 years (*38*). These findings indicate that the use of such methods in study designs that minimize bias, such as Hirayama's prospective cohort, can provide important and valid information on many, but not all, dietary factors associated with cancer incidence.

Extreme caution is required in the conduct and interpretation of case-control studies of diet and cancer. Typically, even important and statistically significant differences between cases and controls are small, representing differences in mean values on the order of 5%. Except in unusual circumstances, it will be difficult to exclude the possibility that differences of this magnitude are not the result of differential dietary recall due to cancer or its treatment. This is increasingly problematic in an environment where the relation of diet with cancer is receiving wide-spread publicity.

Selenium and Cancer Incidence

For approximately four decades animal experimentalists have known that high selenium intake can reduce tumorigenesis in a number of animal models. Ip (*39*) has recently reviewed 35 such studies; in 31 selenium decreased tumorigenesis, in three no effect was seen, and in one selenium increased tumor incidence (pancreatic ductal cancer in male hamsters). Internationally, rates of cancers at many sites, particularly breast and colon, are inversely related to selenium levels measured in pooled blood samples (*40*).

Stimulated by these findings, we and other epidemiologists have examined the association of serum selenium levels with subsequent risk of cancer; three such studies have been published to date. In analyses based on the HDFP study, those with serum selenium levels in the lowest 20% experienced a two-fold higher risk of cancer compared with those having higher levels (*41*). In two separate studies from Finland, low selenium level was associated with a 3-to 5-fold higher risk of cancer (*42, 43*). In our cohort as well as the one Finish study that examined the issue, the combination of low vitamin E and low selenium levels was even more strongly associated with cancer risk.

Clark *et al.* conducted a case-control study of serum selenium level in relation to risk of basal and squamous cell skin cancer (*44*). These small, localized tumors were selected for study specifically to avoid any plausible effect of cancer on the biochemical assessment of selenium. In this study serum selenium levels in the lowest 10% were associated with a five-fold increase in risk of skin cancer.

Collectively, these studies suggest that selenium intake may play a role in human cancer under some circumstances. However, it remains possible that some other constituent of selenium-containing foods is actually contributing to lower cancer risk. The experience of animal nutritionists indicates that selenium interacts with a wide variety of other dietary and non-dietary factors. If also true in humans, it is possible that the associations of selenium level with cancer may vary depending on a variety of other factors, such as vitamin E status or heavy metal intake.

Assessment of Selenium Intake in Epidemiologic Studies

Calculation of nutrient intakes from food consumption information is based on the assumption that the nutrient composition of a food is approximately constant. In the case of selenium, this assumption appears to be violated. Depending on the soil from which the food was produced, the concentration can range from extremely low to very high, even to the point of being toxic. Thus some form of tissue biochemical measurement will usually be needed to measure the previous selenium intake.

Because blood is usually the most available biological specimen, epidemiologic studies have typically utilized serum measurements. Based on a number of supplementation studies, it is clear that serum levels do reflect dietary intake of selenium, with a half-life of several weeks. Although serum selenium levels are affected by other factors, such as albumin level, they do provide a useful epidemiologic tool. Hair reflects dietary selenium and has been successfully employed in China to assess intake of this element (*45*). Unfortunately, the wide-spread use of anti-dandruff shampoos based on selenium hinders their use in the U.S.

Since blood collection is sometimes logistically difficult in large epidemiological studies, we have explored the use of nail clippings to measure selenium intake. These have the conceptual advantage of being environmentally sheltered and reflecting selenium intake during formation of the nail matrix many months before. Indeed, if nails are clipped from all toes simultaneously, these specimens provide a time-integrated measure of intake over much of the preceeding year.

We have evaluted the capacity of nails to reflect selenium in several ways. First, we collected specimens from geographic areas known to differ in their selenium intake (*46*). These did accurately reflect the known geographical distribution of selenium; for example, specimens from New Zealand, an area known to have low selenium intake, were uniformly lower than any in the U.S. In addition, we have measured selenium levels in the toenails of women taking various levels of selenium supplements; a linear dose-response relationship was observed (unpublished data). Nail selenium measurements thus provide an alternative method for assessing selenium status. We have, requested participants in the Nurses' Health Study to mail their toenails clippings to us; more than 68,000 women have done so. These have been catalogued and will provide the basis for future analyses relating selenium intake to cancer risk.

REFERENCES

1. Sporn, M. B. and Roberts, A. B. Role of retinoids in differentiation and carcinogenesis. J. Natl. Cancer Inst., *73*: 1381–1386, 1984.
2. Sporn, M. B., Dunlop, N. M., Newton, D. L., and Smith, J. M. Prevention of chemical carcinogenesis by vitamin A and its synthetic analogs (retinoids). Fed. Proc., *35*: 1332–1338, 1976.
3. Bjelke, E. A. Dietary vitamin A and human lung cancer. Int. J. Cancer, *15*: 561–565, 1975.
4. Hirayama T. Diet and cancer. Nutr. Cancer, *1*(3): 67–81, 1979.
5. MacLennan, R., Da Costa, J., Day, N. E., Law, C. H., Ng, Y. K., and Shanmugaratnam, K. Risk factors for lung cancer in Singapore Chinese, a population with high female incidence rates. Int. J. Cancer, *20*: 854–860, 1977.
6. Kvale, G., Bjelke, E., and Gart, J. J. Dietary habits and lung cancer risk. Int. J. Cancer, *31*: 397–405, 1983.
7. Mettlin, C. and Graham, S. Dietary risk factors in human bladder cancer. Am. J. Epidemiol., *110*: 255–263, 1979.
8. Mettlin, C., Graham, S., Priore, R., Marshall, J., and Swanson, M. Diet and cancer of the esophagus. Nutr. Cancer, *2*: 143–147, 1980.
9. Graham, S., Marshall, J., Mettlin, C., Rzepka, T., Nemoto, T., and Byers, T. Diet in the epidemiology of breast cancer. Am. J. Epidemiol., *116*: 68–75, 1982.
10. Mettlin, C., Graham, S., and Swanson, M. Vitamin A and lung cancer. J. Natl. Cancer. Inst., *62*: 1435–1438, 1979.
11. Marshall, J. R., Graham, S., Byers, T., Swanson, M., and Brasure, J. Diet and smoking in the epidemiology of cancer of the cervix. J. Natl. Cancer. Inst., *70*: 847–851, 1983.
12. Schuman, L. M., Mandell, J. S., Radke, A., Seal, U., and Halberg, F. Some selected features of the epidemiology of prostatic cancer: Minneapolis-St. Paul, Minnesota case-control study, 1976–1979. *In;* K. Magnus (ed.), Trends in Cancer Incidence: Causes and Practical Implications, pp. 345–354, Hemisphere Publishing, Washington, D.C., 1982.
13. Kolonel, L., Hankin, J., and Lee, J. Diet and prostate cancer (abstr.). Am. J. Epidemiol., *118*: 454, 1983.
14. Colditz, G. A., Branch, L. G., Lipnick, R. J., Willett, W. C., Rosner, B., Posner, B. M., and Hennekens, C. H. Increased green and yellow vegetable intake and lowered cancer deaths in an elderly population. Am. J. Clin. Nutr., *41*: 32–36, 1985.
15. Shekelle, R. B., Liu, S., Raynor, W. R., Jr., Lepper, M., Maliza, C., Rosof, A. H., Paul, O., Shyrock, A. M., and Stamler, J. Dietary vitamin A and risk of cancer in the Western Electric Study. Lancet, *2*: 1185–1189, 1981.
16. Gregor, A., Lee, P. N., Roe, F.J.C., Wilson, M. J., and Melton, A. Comparison of dietary histories in lung cancer cases and controls with special reference to vitamin A. Nutr. Cancer, *2*: 93–97, 1980.
17. Samet, J. L., Skipper, B. J., Humble, C. G., and Pathak, D. R. Lung cancer risk and vitamin A consumption in New Mexico. Am. Rev. Respir. Dis., *131*: 198–202, 1985.
18. Ziegler, R. G., Mason, T. J., Stemhagen, A., Hoover, R., Schoenberg, J. B., Gridley, G., Virgo, P. W., Altman, R. and Fraumeni, J. F., Jr. Dietary carotene and vitamin A and risk of lung cancer among white men in New Jersey. J. Natl. Cancer Inst., *73*: 1429–1435, 1984.
19. Hinds, M. W., Kolonel, L. N., Hankin, J. E., and Lee, J. Dietary vitamin A, carotene, vitamin C and risk of lung cancer in Hawaii. Am. J. Epidemiol., *119*: 227–237, 1984.

20. Modan, B., Cuckle, H., and Lubin, F. A note on the role of dietary retinol and carotene in human gastro-intestinal cancer. Int. J. Cancer, *28*: 241–244, 1981.
21. Burton, G. W. and Ingold, K. U. Beta-carotene: an unusual lipid antioxidant. Science, 11: 569–573, 1984.
22. Stich, H. F., Stich, W., Rosin, M. P., and Vallejera, M. O. Use of the micronucleus test to monitor the effect of vitamin A, beta-carotene and canthaxanthin on the buccal mucosa of betel nut/tobacco chewers. Int. J. Cancer, *34*: 745–750, 1984.
23. Peto, R., Doll, R., Buckley, J. D., and Sporn, M. D. Can dietary beta-carotene materially reduce human cancer rates? Nature, *290*: 201–208, 1981.
24. Newmark, H. L. A hypothesis for dietary components as blocking agents of chemical carcinogenesis: plant phenols and pyrrole pigments. Nutr. Cancer, *6*: 58–70, 1984.
25. Wattenberg, L. W. and Loub, W. D. Inhibition of polycyclic aromatic hydrocarbon-induced neoplasia by naturally occurring indoles. Cancer Res., *38*: 1410–1413, 1978.
26. Kark, J. D., Smith, A. H., Switzer, B. R., and Hames, C. G. Serum vitamin A (retinol) and cancer incidence in Evans County, Georgia. J. Natl. Cancer Inst., *66*: 7–16, 1981.
27. Wald, N., Idle, M., Boreham, J., and Bailey, A. Low serum-vitamin-A and subsequent risk of cancer: preliminary results of a prospective study. Lancet, *2*: 81–85, 1980.
28. Willett, W. C., Polk, B. F., Underwood, B. A., Stampfer, M. J., Pressel, S., Rosner, B., Taylor, J. O., Schneider, K., and Hames, C. G. Relation of serum vitamins A and E and carotenoids to the risk of cancer. N. Engl. J. Med., *310*: 430–434, 1984.
29. Stahelin, H. B., Rosel, R., Buess, E., and Brubacker, G. Cancer, vitamins, and plasma lipids: Prospective basel study. J. Natl. Cancer Inst. *73*: 1463–1468, 1984.
30. Wald, N. J., Boreham, J., Hayward, J. L., and Bulbrook, R. D. Plasma retinol, β-carotene and vitamin E levels in relation to the future risk of breast cancer. Br. J. Cancer, *49*: 321–324, 1984.
31. Nomura, A.M.Y., Stemmermann, G. N., Heilbrun, L. K., Salkeld, R. M., and Vuilleumier, J. P. Serum vitamin levels and the risk of cancer of specific sites in men of Japanese ancestry in Hawaii. Cancer Res., *45*: 2369–2372, 1985.
32. Peleg, I., Heyden, S., Knowles, M., and Hames, C. G. Serum retinol and risk of subsequent cancer: extension of the Evans County, Georgia, Study. J. Natl. Cancer Inst., *73*: 1455–1458, 1984.
33. Willett, W. C., Stampfer, M. J., Underwood, B. A., Taylor, J. O., and Hennekens, C. H. Vitamins A, E, and carotene: effects of supplementation in their plasma levels. Am. J. Clin. Nutr., *38*: 559–566, 1983.
34. Roth, M. M. and Stampfer, M. J. Some factors affecting determination of carotenoids in serum. Clin. Chem., *30*: 459–461, 1984.
35. Stampfer, M. J., Buring, J., Willett, W. C., Rosner, B., Eberlein, K., and Hennekens, C. H. The 2×2 factorial design: its application to a randomized trial of aspirin and carotene among U. S. physicians. Stat. Med., *4*: 111–116, 1985.
36. Willett, W. C., Sampson, L., Stampfer, M. J., Rosner, B., Bain, C., Witschi, J. Hennekens, C. H., and Speizer, F. E. Reproducibility and validity of a semiquantitative food frequency questionnaire. Am. J. Epidemiol., *122*: 51–65, 1985.
37. Willett, W. C., Stampfer, M. J., Underwood, B. A., Speizer, F. E., Rosner, B., and Hennekens, C. H. Validation of a dietary questionnaire with plasma carotenoid and alpha-tocoperol levels. Am. J. Clin. Nutr., *38*: 631–639, 1983.
38. Byers, T. E., Rosenthal, R. I., Marshal, J. R., Rzepka, K., Cummings, M., and Graham, S. Dietary history from the distant past: a methodologic study. Nutr.

Cancer., *5*: 59–77, 1983.

39. Ip, C. The chemopreventive role of selenium in carcinogenesis. Presented at the NCI workshop on "Strategies Needed to Develop Selenium Compounds for Cancer Preventive Agents," Bethesda, MD, February 8, 1985.
40. Schrauzer, G. N., White, D. A., and Schneider, C. J. Cancer mortality correlation studies-III: statistical associations with dietary selenium intakes. Bioinorg. Chem., *7*: 23–34, 1977.
41. Willett, W. C., Polk, B. F., Morris, J. S., Stampfer, M. J., Pressel, S., Rosner, B., Taylor, J. O., Schenider, K., and Hames, C. G. Prediagnostic serum selenium and risk of cancer. Lancet, *2*: 130–134, 1983.
42. Salonen, J. T., Alfthan, G., Huttunen, J. K., and Puska, P. Association between serum selenium and the risk of cancer. Am. J. Epidemiol., *120*: 342–349, 1984.
43. Salonen, J. T., Salonen, R., Lappetlainen, R., Maenpaa, P. H., Alfthan, G., and Puska, P. Risk of cancer in relation to serum concentrations of selenium and vitamins A and E: matched case-control analysis of prospective data. Br. Med. J., *290*: 417–420, 1985.
44. Clark, L. C., Graham, G., and Bray, J. Nonmelanoma skin cancer and plasma selenium: a prospective cohort study. Am. J. Epidemiol., *122*: 528 (abstr.), 1985.
45. Keshan Disease Research Group of the Chinese Academy of Medical Sciences. Observations of effect of selenium and Keshan disease. Chin. Med. J., *92*: 477–487, 1979.
46. Morris, J. S., Stampfer, M. J., and Willett, W. Toenails as an indicator of dietary selenium. Biol. Trace Elem. Res., *5*: 529–537, 1983.

DIET, NUTRITION AND CANCER, Y. HAYASHI ET AL. (EDS.),
JAPAN SCI. SOC. PRESS, TOKYO/VNU SCI. PRESS, UTRECHT, PP. 247–253, 1986

Dietary Fibre in the Japanese Diet

Masanori KURATSUNE,*[1] Teruko HONDA,*[2] H. N. ENGLYST,*[3] and J. H. CUMMINGS*[3]

Department of Food and Nutrition, Nakamura Gakuen College, Fukuoka 814, Japan,[1] Department of Food and Nutrition, Yamaguchi Women's University, Yamaguchi 753, Japan,*[2] and Dunn Clinical Nutrition Centre, Cambridge CB2 1QL, England*[3]*

Abstract: The low risk of colon cancer among the Japanese suggests their high intake of dietary fibre. Composite diet mixtures for 1959, 1970, and 1979 were prepared utilising the food consumption data from the National Nutrition Survey in Japan and analysed for non-starch polysaccharides (NSP) at the Dunn Clinical Nutrition Centre in Cambridge. The results showed that average intake of NSP by a Japanese in the above years did not exceed 13 g per day, which is as low as the corresponding intake by the Scandinavians and the British whose risk of colon cancer is known to be high. Since all the materials were analysed by the same methods and at the same laboratory, the results are well comparable and indicate that the low risk of colon cancer among the Japanese cannot readily be explained by their intake of NSP alone.

The incidence of many diseases is different in Japan when compared with that in Europe and North America. Particular attention has been focussed recently on the low incidence and mortality from large bowel cancer in Japan compared with other industrialised countries (*1–3*). In quite a few epidemiological studies it has been suggested that dietary fibre protects against the development of bowel cancer (*4–6*) from which it might be predicted that fibre intake by the Japanese would be high.

Preparation of Composite Diets and Analytical Methods of Dietary Fibre

Proper comparison of the fibre content of the diet of different populations has been impeded by the lack of an agreed definition of fibre and a suitable method for its analysis. Recently, however, accurate chemical techniques have become available for the measurement of the non-starch polysaccharides (NSP) in food (*7*, *8*). NSP are the polysaccharides of the plant cell wall and comprise such substances as cellulose, pectin, hemicellulose, *etc.* They are the prime, and in the case of most human foods the only, components of dietary fibre in the diet. We have used this technique to measure the dietary fibre content of the Japanese diet and compared it with that in other populations determined using the same method.

Food consumption data for the years 1959, 1970, and 1979 in Japan have been obtained from the National Nutrition Survey (*9*) and composite mixtures for analysis were made up of the vegetables and vegetable-containing foods ingested per head

TABLE 1. Foods Used for Preparation of the Composite Diets for 1959, 1970, and 1979

Food	Amount (g)		
	1959	1970	1979
Rice, polished	362.0	306.1	222.9
Barley, pressed	34.0	3.3	1.6
Wheat flour	9.0	6.3	6.0
White bread	22.9	28.3	49.3
"Udon" noodle,			
boiled	20.5	14.8	33.1
dried	10.9		
Spaghetti			5.0
Chinese noodle,			
precooked		13.7	2.8
"Yaki-fu,"			
baked gluten		0.9	
Bread crumbs		0.9	
Sesame	0.6	1.0	0.7
Peanut	0.7	1.0	0.7
Sweet potato	40.7	2.6	10.0
White potato	18.7	22.3	24.0
Taro	26.4	3.0	15.9
"Konnyaku,"			
devil's tongue	13.3	9.8	14.0
White sugar	11.3	19.0	13.0
Strawberry jam	0.4	0.7	0.5
Soybean	1.5		
"Miso," fermented			
soybean,			
"Rice-kohji"	23.4	19.3	15.0
"Barley-kohji"	5.9	4.8	3.8
"Tohfu," soybean			
curd	33.1	38.9	34.4
Fried "Tohfu"			
"Abura-age"			2.0
"Nama-age"			3.4
"Ganmodoki"			2.8
"Natto," fermented			
soybean with			
B. natto	0.7	4.0	5.8
Kidney bean	5.2	4.3	2.4
Satsuma mandarin	15.3	22.1	87.1
Japanese			
persimmon	30.9	17.2	38.7
Apple	32.0	24.2	28.0
Banana		13.9	6.5
Strawberry			0.4
Satsuma mandarin,			
canned		3.6	
juice	0.6	29.9	5.8
Carrot	13.0	14.5	20.2
Spinach	23.8	26.4	25.2
Pumpkin	6.1	6.8	
Sweet pepper		2.5	5.6
Japanese radish	40.5	10.2	34.8
Cabbage	9.3	24.7	24.8
Chinese cabbage	27.3	3.7	21.4
Welsh onion	16.7	21.0	11.6
Lettuce		4.0	4.6
Edible burdock			
root	4.9	6.6	4.2
Turnip		9.8	3.8
Onion	18.4	18.4	21.5
Lotus root		2.9	2.9
Cucumber	1.7	10.2	10.1
Bamboo shoot		1.7	1.8
Eggplant	0.2	1.0	1.7
Cauliflower		3.5	2.0
Tomato	0.2	25.3	10.6
Bean sprout		10.2	8.2
Garden pea,			
immature pod		2.9	1.6
"Shiitake," fungi,			
raw		2.4	8.8
dried	1.5	4.3	0.9
Algae,			
"Nori," dried			
and seasoned	0.5	0.5	0.5
"Wakame," dried	4.1	6.4	4.8
Pickle,			
"Takuanzuke,"			
pickled radish	23.8	18.2	15.6
Salted Chinese			
cabbage	23.8	18.2	13.6

per day. Most of the consumption data was for single food items but some were available only for groups of food items. In the case of the latter, one or two of the most popular items of each food group were chosen as being representative of the group and used for the preparation of the diet composites. Among green-yellow vegetables, only carrot, spinach, and pumpkin had individual consumption figures, the others being shown only as a group. Those designated as "others" were assumed to consist of carrot, spinach, and pumpkin in proportion to their respective consumption figures. Individual consumption figures were available for a number of the most popular vegetables such as Japanese radish, cabbage, Chinese cabbage, tomato, and some others, but the rest were grouped as "other vegetables." This group was assumed to consist of about 10 popular vegetables on the basis of the national average purchase figures of vegetables by households which were available from the Annual Report on the Family Income and Expenditure Survey conducted by the Statistics Bureau, Prime Minister's Office in 1959, 1970, and 1979 (*10*). The food items and their amounts used for preparation of the composite diets are listed in Table 1.

All food items used were purchased raw, cooked, or prepared at food stores in Fukuoka in November and December in 1982. Except for fruits and certain vegetables that are usually eaten raw, all the vegetables used were boiled together for about 30 min. Rice was cooked separately with water in the usual way. All the foods and the water used for boiling were then homogenised with a Waring blender, freeze-dried and representative portions shipped to the MRC Dunn Clinical Nutrition Centre for analysis. The method for measuring NSP has been described in detail previously (*7*, *8*).

After identical enzymic treatment to remove starch, three samples of food were

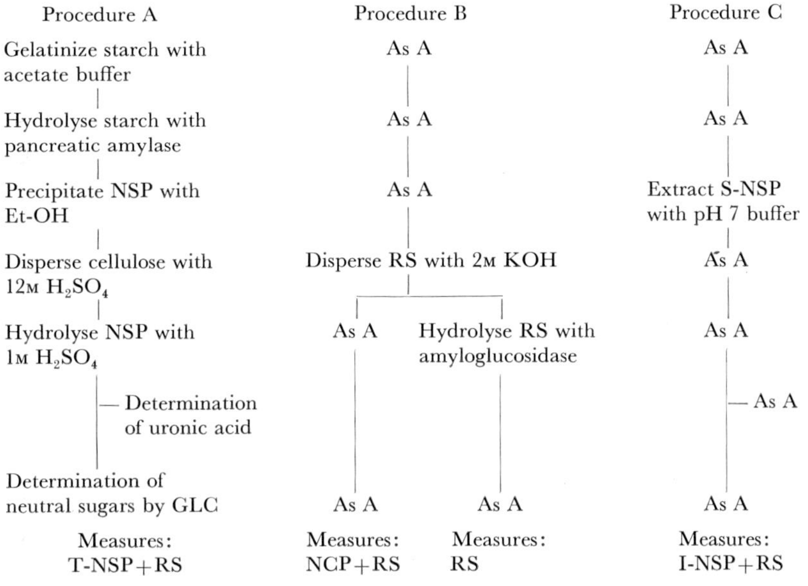

FIG. 1. Measurement of NSP and resistant starch. T-NSP, total NSP; NCP, non-cellulosic polysaccharides; S-NSP, soluble NSP; I-NSP, insoluble NSP; RS, resistant starch.

TABLE 2. Energy and Nutrient Intake per Head per Day by the Japanese in 1959, 1970, and 1979

Year	Calorie (kcal)	Protein			Fat (g)	Carbohydrate (g)
		Total (g)	Animal (g)	Vegetable (g)		
1959	2,148	70.1	24.3	45.8	24.0	413
1970	2,210	77.6	34.2	43.4	46.5	368
1979	2,113	78.4	39.4	39.0	54.8	315

TABLE 3. Results of Analysis of the Composite Diets for Dietary Fibre

Year	Dietary fibre	Constituent sugars													
		Total	Ara	Xyl	Man (g/day)	Gal	Glu	U.Ac.	Total	Ara	Xyl	Man (g/1,000 cal of diet)	Gal	Glu	U.Ac.
1959	Cellulose	3.14					3.14		1.46					1.46	
	Soluble NCP	5.56	0.74	0.48	0.04	0.65	1.52	2.13	2.59	0.34	0.22	0.02	0.30	0.71	0.99
	Insoluble NCP	3.65	0.65	0.96	0.52	0.26	0.87	0.39	1.70	0.30	0.45	0.24	0.12	0.41	0.18
	Total NSP	12.35	1.39	1.44	0.56	0.91	5.53	2.52	5.76	0.65	0.67	0.26	0.42	2.58	1.17
1970	Cellulose	3.58					3.58		1.62					1.62	
	Soluble NCP	5.16	0.79	0.59	0.15	0.64	0.69	2.30	2.33	0.36	0.27	0.07	0.29	0.31	1.04
	Insoluble NCP	4.12	0.54	0.88	0.54	0.29	1.28	0.59	1.86	0.24	0.40	0.24	0.13	0.58	0.27
	Total NSP	12.86	1.33	1.47	0.69	0.93	5.55	2.89	5.81	0.60	0.67	0.31	0.42	2.51	1.31
1979	Cellulose	3.02					3.02		1.43					1.43	
	Soluble NCP	4.07	0.50	0.16	—	0.74	—	2.67	1.93	0.24	0.08	—	0.35	—	1.26
	Insoluble NCP	3.81	0.43	0.74	0.54	0.31	1.32	0.47	1.80	0.20	0.35	0.26	0.15	0.62	0.22
	Total NSP	10.90	0.93	0.90	0.54	1.05	4.34	3.14	5.16	0.44	0.43	0.26	0.50	2.05	1.48

NSP, non-starch polysaccharide=Dietary fibre; NCP, non-cellulosic polysaccharides; Ara, arabinose; Xyl, xylose; Man, mannose; Gal, galactose; Glu, glucose; U.Ac., uronic acids.

subject to the different but complementary procedures, A, B and C. Figure 1 summarises the sequence of steps in the three procedures.

Energy and Nutrient Intake by the Japanese

Table 2 shows that the present day Japanese diet contains about 2,100 kcal of which 80% comes from carbohydrate, 15% from protein, and 23% from fat. Since 1959 the relative amounts of the diet have changed significantly with fat intake rising from 24 to 55 g/day and from 10 to 23% of total calorie intake. Conversely, carbohydrate intake has fallen whilst protein intake remains approximately the same, although slightly more protein now comes from animal rather than vegetable sources.

In regard to carbohydrate intake, Table 1 also shows a similar fact that the Japanese used to eat a very large amount of rice but during the fast 20 years their rice consumption together with the consumption of barley and sweet potato has declined markedly, while the consumption of white bread has considerably increased. This indicates the rather rapid westernisation tendency of the Japanese diet. It is also notable that the Japanese eat various soybean products and some peculiar traditional foods such as algae and pickles. It is interesting to note that the intake of these specific foods has been fairly stable in spite of the overall westernisation of their diet.

TABLE 4. NSP consumption in 9 regions of Britain

Sample	Dietary fibre	Constituent sugars (g/day)							
		Total	Rha	Ara	Xyl	Man	Gal	Glu	U.Ac.
Scotland	Cellulose	2.45						2.45	
	Soluble NCP	6.40	0.12	1.49	1.11	0.20	1.29	0.85	1.34
	Insoluble NCP	3.00	—	0.76	1.05	0.41	0.32	0.26	0.20
	Total NSP	11.85	0.12	2.25	2.16	0.61	1.61	3.56	1.54
North	Cellulose	2.37						2.37	
	Soluble NCP	5.75	0.12	1.48	1.07	0.21	1.45	0.53	1.42
	Insoluble NCP	4.17	—	0.86	1.13	0.42	0.36	0.60	0.27
	Total NSP	12.29	0.12	2.34	2.20	0.63	1.81	3.50	1.69
York and Humberside	Cellulose	2.71						2.71	
	Soluble NCP	6.53	0.11	1.60	1.17	0.23	1.43	0.54	1.45
	Insoluble NCP	3.39	—	0.76	1.06	0.40	0.34	0.57	0.26
	Total NSP	12.63	0.11	2.36	2.23	0.63	1.77	3.82	1.71
Northwest	Cellulose	2.46						2.46	
	Soluble NCP	6.33	0.16	1.53	1.06	0.22	1.42	0.49	1.45
	Insoluble NCP	3.36	—	0.76	1.07	0.41	0.30	0.60	0.22
	Total NSP	12.15	0.16	2.29	2.13	0.63	1.72	3.55	1.67
East Midlands	Cellulose	2.84						2.84	
	Soluble NCP	6.72	0.17	1.67	1.12	0.20	1.47	0.45	1.64
	Insoluble NCP	3.61	—	0.80	1.09	0.43	0.37	0.66	0.26
	Total NSP	13.17	0.17	2.47	2.21	0.63	1.84	3.93	1.90
West Midlands	Cellulose	2.65						2.65	
	Soluble NCP	6.40	0.16	1.48	1.10	0.21	1.36	0.56	1.53
	Insoluble NCP	3.07	—	0.75	0.99	0.37	0.32	0.40	0.24
	Total NSP	12.12	0.16	2.23	2.09	0.58	1.68	3.61	1.77
Southwest	Cellulose	2.52						2.52	
	Soluble NCP	6.07	0.11	1.40	1.05	0.22	1.32	0.38	1.59
	Insoluble NCP	3.41	—	0.78	1.02	0.40	0.35	0.62	0.24
	Total NSP	12.00	0.11	2.18	2.07	0.62	1.67	3.52	1.83
Southeast	Cellulose	2.95						2.95	
	Soluble NCP	6.70	0.16	1.45	1.05	0.27	1.39	0.75	1.63
	Insoluble NCP	3.11	—	0.78	1.02	0.40	0.35	0.27	0.29
	Total NSP	12.76	0.16	2.23	2.07	0.67	1.74	3.97	1.92
Wales	Cellulose	2.84						2.84	
	Soluble NCP	6.87	0.11	1.63	1.18	0.28	1.30	0.76	1.52
	Insoluble NCP	3.07	—	0.79	1.07	0.37	0.34	0.25	0.25
	Total NSP	12.69	0.11	2.42	2.25	0.65	1.64	3.85	1.77

Rha, rhamnose; other abbreviations are same as in Table 3.

Intake of Dietary Fibre in Japan

Table 3 shows that none of the three composite diets tested contained more than 13 g of NSP. The intake of NSP was 12.4 g/day in 1959 and had fallen to 10.9 g/day in 1979, probably largely as a result of the decline in the intake of cereals. Decreased content of xylose and arabinose, the principal NSP in cereals was also found in 1979, whilst the uronic acid content was high indicating an increased consumption of vegetables during the observation period. A major fraction of the NSP intake was from glucose-containing polymers, particularly cellulose. Practically the same facts were seen when the data were presented as per 1,000 kcal of diet.

Intake of Dietary Fibre and Risk of Colon Cancer

Minowa *et al.* (*11*) calculated the Japanese daily dietary fibre intake per capita as 20.2 and 19.4 g in 1970 and 1979, respectively, using the British food tables to estimate the dietary fibre content of Japanese foods. These figures are considerably larger than ours. Since, we used the same food consumption figures as theirs, both being derived from the National Nutrition Survey and no technical failure allowing loss of any NSP components before analysis is conceivable, the discrepancy noted must be due to the different approaches for estimation of the fibre content, probably mainly due to the recently improved procedures for removal of starch from the samples.

The latest figures for intake of NSP by the Scandinavian populations range from 13 to 18 g/day (*12*) and the mean NSP intake for the British is 12.5 g/day (range 11.8–13.2) (Table 4) (*13*). The results of this study show that NSP intake by the Japanese is as low as it is in western countries where large bowel cancer rates are much higher. The intake of pentose (arabinose and xylose) which was reported to be associated with the risk of colon cancer (*14*) was from 1.83 to 2.83 g in the present study, considerably less than those for the Scandinavians (4.5–7.4 g) and the British (4.25–4.68 g) (*13*). The lower risk of colon cancer among the Japanese can therefore not be explained by a higher intake of NSP pentose.

When analysed by the present method, the total intake of NSP is very similar for Britain and for Japan. However, the Japanese diet, when compared with the British (*13*), contains only about half the amount of pentose, but twice as much uronic acid (measuring as soluble non-cellulosic polysaccharides (NCP)). Such difference in the monosaccharide pattern of NSP suggests a lower intake of wheat products and a higher intake of vegetables and rice in Japan. Diets with a high content of vegetables have previously been associated with low risk of colon cancer (*15*).

Finally, it should be noted that colon cancer has rapidly increased in Japan recently, although it is still less common than in most western countries. Such increase might be connected in some way with the present findings, that is, the rather smaller intake of dietary fibre by the Japanese.

ACKNOWLEDGMENTS

This investigation was supported in part by a grant-in-aid to an international cooperative study on dietary fibre and colon neoplasm and other diseases (organizer: Professor Shun-ichi Yamamoto) from the Japan Society for the Promotion of Science, and by a grant-in-aid from the Princess Takamatsu Cancer Research Fund.

REFERENCES

1. Doll, R. The geographical distribution of cancer. Br. J. Cancer, *23*: 1–8, 1969.
2. Segi, M. Age-adjusted Death Rates for Cancer for Selected Sites in 48 Countries in 1975, Segi Institute of Cancer Epidemiology, Nagoya, 1980.
3. Waterhouse, J., Muir, C., Shanmugaratnam, K., and Powell, J. Cancer Incidence in Five Continents, vol. IV, Int. Agency Res. Cancer, Lyon, 1982.
4. Burkitt, D. P. Epidemiology of cancer of the colon and rectum. Cancer, *28*: 3–13, 1971.
5. Graham, S., Dayal, H., Swanson, M., Mittelman, A., and Wilkinson, G. Diet in the epidemiology of cancer of the colon and rectum. J. Natl. Cancer Inst., *61*: 709–714, 1978.
6. Jensen, O. M., MacLennan, R., and Wahrendorf, J. Diet, bowel function, fecal characteristics and large bowel cancer in Denmark and Finland. Nutr. Cancer, *4*: 5–19, 1982.
7. Englyst, H., Wiggins, H. S., and Cummings, J. H. Determination of the non-starch polysaccharides in plant foods by gas-liquid chromatography of constituent sugars as alditol acetates. Analyst, *107*: 307–318, 1982.
8. Englyst, H. N. and Cummings, J. H. Simplified method for measurement of total non-starch polysaccharides by gas-liquid chromatography of constituent sugars as alditol acetates. Analyst, *109*: 937–942, 1984.
9. Ministry of Health and Welfare, Current Nutritional Status of the Nation. Results of National Nutrition Survey, 1959, 1970, and 1979, Daiichi Publishing Co., Tokyo, 1961, 1972, and 1981.
10. Statistics Bureau, Prime Minister's Office. Annual Report on the Family Income and Expenditure Survey, 1959, 1970, and 1979, Japan Statistics Association, Tokyo, 1961, 1971, and 1980.
11. Minowa, M., Bingham, S., and Cummings, J. H. Dietary fibre intake in Japan. Hum. Nutr. Appl. Nutr., *37A*: 113–119, 1983.
12. Englyst, H. N., Bingham, S. A., Wiggins, H. S., Southgate, D.A.T., Seppänen, R., Helms, P., Anderson, V., Day, K. C., Choolum, R., Collinson, E., and Cummings, J. H. Nonstarch polysaccharide consumption in four Scandinavian populations. Nutr. Cancer, *4*: 50–60, 1982.
13. Englyst, H. N. Intake of non-starch polysaccharides in Britain. *In;* Dietary polysaccharide breakdown in the gut of man. Ph.D. Thesis, University of Cambridge, pp. 177–182, 1985.
14. Bingham, S., Williams, D.R.R., Cole, T. J., and James, W.P.T. Dietary fibre and regional large bowel cancer mortality in Britain. Br. J. Cancer, *40*: 456–463, 1979.
15. Bjelke, E. Case control study of cancer of the stomach, colon and rectum. *In;* R. L. Clark, R. W. Cumley, J. E. McCay, and M. M. Copeland (eds.). Oncology 1970, Proc. 10th Int. Cancer Congr. vol. V, 320 pp. Year Book Medical Publishers Inc. Chicago, Ill, 1971.

DIET, NUTRITION AND CANCER, Y. HAYASHI ET AL. (EDS.),
JAPAN SCI. SOC. PRESS, TOKYO/VNU SCI. PRESS, UTRECHT, PP. 255–263, 1986

Dietary Fat in Relation to Mammary Carcinogenesis

Kenneth K. CARROLL
Department of Biochemistry, University of Western Ontario, London, Ontario N6A 5C1, Canada

Abstract: The first evidence that dietary fat influences mammary carcinogenesis was provided by Tannenbaum, who showed that mice fed a high-fat diet developed spontaneous tumors more readily than those fed a low-fat diet. Similar observations have been made with various other animal models. Polyunsaturated vegetable oils enhance carcinogenesis more effectively than saturated fats, because of their higher linoleate content. Diets containing high levels of polyunsaturated fish oils do not stimulate carcinogenesis, however, perhaps because their polyunsaturated fatty acids belong mainly to the linolenate family. Dietary fat acts primarily as a promoting agent, but the exact mechanism is still unclear. The requirement for linoleate and the fact that the fat effect can be blocked by prostaglandin biosynthesis inhibitors suggests that it may be mediated by biologically-active compounds derived from linoleate. Other possibilities include changes in hormonal balance, alterations in the fatty acids of membrane lipids, effects on the immune system, modulation of intercellular communications, and metabolic alterations related to differences in fat and caloric intake. Interest in the role of dietary fat in mammary carcinogenesis has been greatly stimulated by epidemiological evidence of a strong, positive correlation between breast cancer and dietary fat. In these epidemiological data, total dietary fat shows a better correlation than fat from either plant or animal sources individually, and there is no apparent correlation with the polyunsaturated fatty acid content of the diet. Further studies are needed to investigate more thoroughly this apparent difference between experimental and epidemiological data. A reduction in dietary fat intake has been recommended in recent dietary guidelines designed to reduce cancer risk, and feasibility trials are in progress to examine the possible use of low-fat diets for treatment of breast cancer.

The studies described in this article were initiated in the early 1960's in collaboration with Drs. E. R. Plunkett and E. B. Gammal. They were studying effects of hormones on mammary tumors induced in rats by 7, 12-dimethylbenz(α)anthracene (DMBA) as described by Huggins *et al.* (*1*), and were interested in moni-

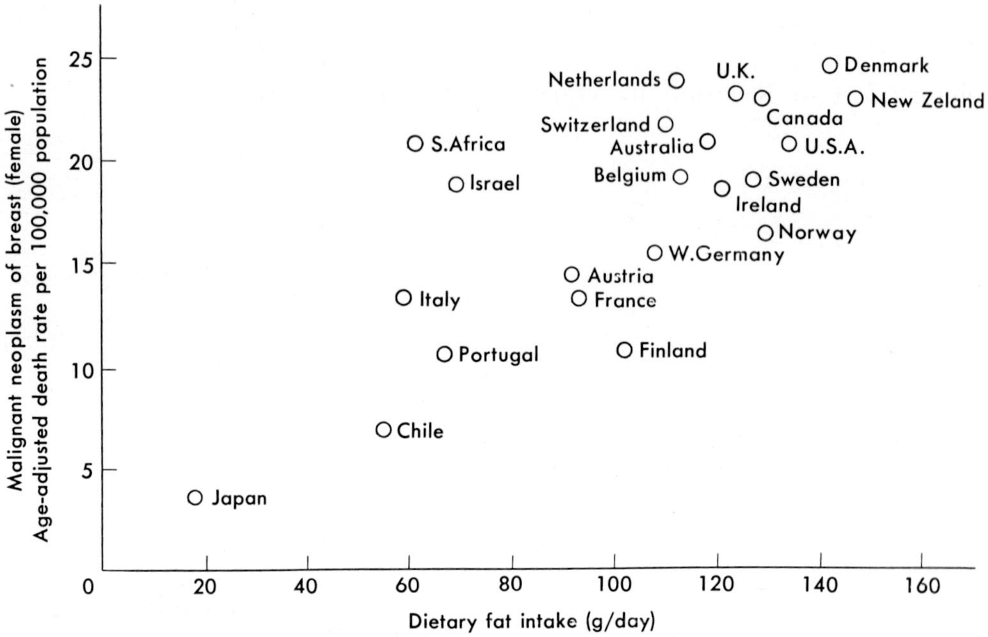

FIG. 1. A plot of age-adjusted mortality from breast cancer in females against dietary fat in different countries. Reproduced from Carroll *et al.* (*7*) with permission of the publisher.

toring levels of DMBA in the mammary gland. We therefore collaborated in developing an assay for this purpose (*2*).

DMBA, like other polycyclic hydrocarbons, is fat-soluble, and it had been suggested that this might be a factor in their ability to function as mammary carcinogens, since the adipose tissue of the mammary gland could serve as a storage depot for such compounds, increasing exposure to the susceptible epithelial tissue (*3*). With this in mind, it seemed to us that it might be possible to influence the ability of DMBA to produce mammary tumors by feeding diets containing different amounts and types of dietary fat. An experiment was therefore designed in which weanling, female, Sprague-Dawley rats were fed semipurified diets containing 0.5% corn oil, 20% coconut oil, or 20% corn oil by weight and given 10 mg of DMBA intragastrically at about 50 days of age. The diets were continued for another 4 months during which time the rats were palpated regularly for mammary tumors. At autopsy, the results showed that rats fed the 20% corn oil diet had developed about twice as many tumors as those fed either of the other two diets (*4*).

A search of the literature showed that this was not a new finding. Experiments in a number of different laboratories, beginning with those of Tannenbaum, had shown that mice and rats fed high-fat diets develop mammary tumors more readily than those fed low-fat diets (*5, 6*). The consistency of these results led us to consider whether breast cancer in humans might be influenced by dietary fat, and examination of available epidemiological data showed a strong, positive correlation between age-adjusted mortality from breast cancer and the amount of fat available for consumption in different countries (Fig. 1). This correlation was noted at about the same time by Lea (*8*) and by Wynder (*9*).

Our original idea that dietary fat might influence mammary carcinogenesis by altering the distribution and metabolism of DMBA appeared on further examination to be incorrect. Measurement of DMBA in mammary glands of rats fed the 3 different diets used in our first experiment failed to show differences that could explain the higher tumor yield in rats fed the 20% corn oil diet (*10*). Furthermore, it was found in subsequent experiments that this diet increased the tumor yield even when it was first fed one or two weeks after the DMBA was given, by which time most of the carcinogen had been cleared from the mammary gland. This led us to think that dietary fat was affecting the promotional stage of carcinogenesis rather than initiation (*11*, *12*).

Effects of Different Types of Dietary Fat

Our first experiment indicated that the type of dietary fat had an important bearing on its ability to influence mammary carcinogenesis. Later studies provided further evidence that polyunsaturated fats increased tumor yields, while saturated fats had little or no effect (*13*). Of the polyunsaturated fats tested, only rapeseed oil failed to produce a marked increase in tumor yield. This lack of effect may have been due to the high erucic acid content of rapeseed oil, since a subsequent experiment showed that rapeseed oil with a low erucic acid content promoted carcinogenesis about as effectively as other polyunsaturated oils (*14*).

The differing effects of saturated and polyunsaturated fats on mammary carcinogenesis were found to be related to a requirment for n-6 essential fatty acids. When a small amount of polyunsaturated fat (*15*) or purified ethyl linoleate (*16*) was fed with a saturated fat, the mixture promoted mammary carcinogenesis as effectively as the polyunsaturated fat itself when they were fed as 20% by weight of the diet.

In one experiment, a mixture of a small amount of polyunsaturated fish oil and a saturated fat gave a relatively high tumor yield (*16*), suggesting that the requirement for essential fatty acids could also be satisfied by n-3 polyunsaturated fatty acids derived from linolenate. More recent studies, however, have shown that higher levels of fish oil do not promote the development of mammary tumors induced by carcinogens (*17*, *18*).

Studies on Mammary Tumors Induced by Various Carcinogenic Agents

The experiments on mammary carcinogenesis in our laboratory have all involved the use of mammary tumors induced in rats by DMBA. Other workers, however, have observed the promoting effect of dietary fat in mice and rats bearing spontaneous mammary tumors or tumors induced by a synthetic estrogen, by other chemical carcinogens or by exposure to X-rays (*19*). Recently, a sabbatical leave provided an opportunity to collaborate with Dr. R. L. Noble in testing the effect of dietary fat on mammary tumors induced in rats by subcutaneous implantation of estrone, a natural estrogen (*20*). In this experiment, tumors appeared somewhat earlier in rats fed a diet containing 20% corn oil compared to one containing only 3% corn oil, but the tumor incidence was similar in both groups at the end of the experiment (*21*). The reason for the relative lack of effect of dietary fat is not clear and may warrant further investigation.

Mechanism of Action of Dietary Fat

Although there is good evidence that dietary fat acts mainly at the promotional stage of mammary carcinogenesis, the exact mechanism is not clear. Our own studies (*16*) failed to provide support for a hormonal mechanism, although hormones are undoubtedly needed for development of mammary tumors as well as for normal development of the mammary gland (*22*).

There is increasing evidence that icosanoids derived from polyunsaturated fatty acids may be involved in the promotion of mammary tumors by dietary fat. Inhibitors of prostaglandin biosynthesis have been shown to counteract this effect in both carcinogen-induced tumors (*23*) and transplantable tumors (*24*). The recent studies with dietary fish oils provide additional support for this suggestion (*17*, *18*, *25*). The polyunsaturated fatty acids in fish oils consist mainly of n-3 fatty acids, in contrast to the linoleic acid of vegetable oils, in which the double bond nearest the methyl end of the chain is in the n-6 position. Linoleic acid can be converted to arachidonic acid, which gives rise to the 2-series of icosanoids. The corresponding n-3 fatty acid, icosapentaenoic acid, which occurs in fish oils, can be converted into a 3-series of icosanoids, but probably exerts greater effects by competing with arachidonic acid for the enzymes involved in this conversion, thus decreasing the formation of icosanoids of the 2-series (*26*).

This could explain why dietary fish oils fail to promote carcinogenesis in the same way as vegetable oils. However, fish oils are more susceptible to lipoperoxidation than vegetable oils and it is also possible that non-specific peroxidation products of fish oils exert toxic effects that tend to counteract any promotional effects. This might explain why low levels of fish oil in the diet appear to promote while higher levels are ineffective (*17*, *18*).

Our experiments on promotional effects of dietary fat indicated that mammary carcinogenesis was influenced by the amount as well as the type of fat in the diet. Thus, in addition to specific effects of polyunsaturated fatty acids or compounds derived from them, there may be other, less specific effects related to the level of fat in the diet.

The relative importance of caloric intake and dietary fat was considered by early workers in the field and is still under investigation (*27*). Since high-fat diets have a higher caloric density than low-fat diets, experimental animals and humans on high-fat diets tend to consume more calories and it is thus difficult to dissociate effects of caloric intake from those of dietary fat. This is particularly true in epidemiological studies which show a strong, positive correlation between breast cancer mortality and caloric intake as well as dietary fat (*12*).

In experiments with animals, it can be clearly demonstrated that dietary fat has an effect that is independent of caloric intake, but this does not mean that caloric intake has no influence on carcinogenesis. It is, in fact, well-known that carcinogenesis can be greatly inhibited by caloric restriction in experimental animals (*6*, *28*). In most such experiments, however, caloric intake has been restricted to between one-half and three-quarters of normal intake, and it is less clear whether variations due to feeding high- or low-fat diets have a significant effect. More experimental work on this aspect of the problem is clearly indicated.

Since dietary fat appears to act mainly at the promotional stage of carcinogenesis, it appears to be providing a more favourable environment for tumor growth. This may occur by prolonging the life of individual cells as well as by increasing the number of new cells, since tumor growth depends on the balance between the rates at which cells accumulate and are lost (*29*).

Relevance of Studies in Animal Models to Breast Cancer in Humans

A major goal of our experimental studies on mammary carcinogenesis has been to provide information that may be useful in the prevention or treatment of breast cancer in humans. It has therefore seemed to us important to try to relate our experimental findings to epidemiological data on cancer in human populations.

Since the studies on DMBA-induced carcinogenesis showed that polyunsaturated vegetable oils promote mammary carcinogenesis more effectively than saturated fats, it was disturbing to find in analyzing epidemiological data that breast cancer mortality correlated best with the total amount of fat available for consumption, and showed little or no correlation with dietary vegetal fat which tends to be more unsaturated than fats from animal sources (*28*).

We attempted to explain this apparent discrepancy between experimental and epidemiological data in the following way (*30*): the diets of most countries contain levels of polyunsaturated fatty acids that are comparable to the relatively small amount required for maximum promotion of DMBA-induced mammary tumors in rats. If this requirement is provided by most national diets, it would not be expected to show a correlation with breast cancer mortality. The observed positive correlation could thus be due to a separate, less specific requirement for dietary fat, as discussed in the section on mechanism of action. A recent report by Ip *et al.* (*31*) indicates, however, that the threshold for polyunsaturated fatty acids may be higher than originally thought, and this is supported by the results of further studies in our own laboratory (*32*). There is thus a need for further investigation of the relative effects of total dietary fat and degree of unsaturation on tumor promotion in rats.

In other studies, we have investigated the effects on subsequent tumor yield of reducing the level of dietary fat after a period of promoting with high-fat diet in rats exposed to DMBA. Decreasing the level of dietary fat from 20% to 10% and changing to more saturated types of fat significantly decreased the number of new tumors that developed (*33*). In a subsequent experiment, rats given DMBA were fed a diet containing 20% by weight of a fat blend having a fatty acid composition similar to that of the American diet. This blend contained 18% linoleic acid and significantly increased the mammary tumor yield relative to a low-fat diet. After 2 months on the diet containing 20% of this fat blend, the level was reduced to 15, 10, or 5% without changing the composition of the blend. This had relatively little effect on subsequent tumor yields (Carroll, K.K., Kalamegham, R., Braden, L. M. and Bell, J. A., unpublished data).

Feasibility trials are currently underway in the United States to study the possibility of using low-fat diets as a means of reducing the risk of breast cancer in human populations. Comparison of data on the recurrence as well as the occurrence of breast cancer in Japan relative to the United States offers hope that low-fat diets may offer some protection (*34*). The primary aim of the feasibility trials is to reduce total fat intake, but in view of the results of animal experiments described above, some reduction in degree of unsaturation of the dietary fat may help to maximize the possibility of observing any beneficial effects in such trials.

Development of Dietary Guidelines

The evidence relating diet to carcinogenesis has also led in recent years to the development of dietary guidelines designed to reduce the risk of cancer (*35*). Because of the strong and consistent association between excessive dietary fat and various types of cancer, these dietary guidelines have typically included a recommendation to reduce dietary fat intake.

Perhaps the most influencial dietary guidelines have been those proposed by the Committee on Diet, Nutrition, and Cancer of the U.S. National Research Council (*36*). After a thorough review of the literature, this committee recommended a reduction in fat intake from approximately 40% to 30% of total calories as a moderate and practical target that might have a beneficial effect. The committee recognized, however, that the evidence could justify an even greater reduction.

A somewhat similar recommendation was included in the Consensus Statement on Provisional Dietary Guidelines developed at a workshop on Diet and Human Carcinogenesis held in Aarhus, Denmark in June 1985, under the sponsorship of the European Organization for Cooperation in Cancer Prevention Studies (ECP) and the International Union of Nutritional Sciences (IUNS). This reads as follows:

"Decrease the intake of saturated and unsaturated fat in countries where, on average, fat constitutes more than 30% of total food energy (calories). In other countries, people should maintain their lower fat intake" (*37*).

The wording of this and other dietary recommendations in the consensus statement was designed to provide guidelines on an international rather than a national basis.

In developing dietary guidelines, it is important to keep a broad perspective so that recommendations with regard to a particular disease are consistent with general good health. The above recommendations regarding dietary fat were developed specifically for cancer, but there seems to be little risk associated with a reduced fat intake, and there could be other beneficial effects as well. Excessive dietary fat may contribute to various chronic diseases of Western civilization, including cardiovascular disease, diabetes and gall bladder disease as well as cancer. The high caloric density of high-fat diets also encourages overeating and probably contributes to obesity. For these various reasons, recommendations to avoid high-fat diets appear to be well-justified.

REFERENCES

1. Huggins, C., Grand, L. C., and Brillantes, F. P. Mammary cancer induced by a single feeding of polynuclear hydrocarbon, and its suppression. Nature, *189*: 204–207, 1961.
2. Gammal, E. B., Carroll, K. K., Muhlstock, B. H., and Plunkett, E. R. Quantitative estimation of 7,12-dimethylbenz(α)anthracene in rat mammary tissue by gas liquid chromatography. Proc. Soc. Exp. Biol. Med., *119*: 1086–1089, 1965.
3. Dao, T. L., Bock, F. G., and Crouch, S. Level of 3-methylcholanthrene in mammary glands of rats after intragastric instillation of carcinogen. Proc. Soc. Exp. Biol. Med., *102*: 635–638, 1959.
4. Gammal, E. B., Carroll, K. K., and Plunkett, E. R. Effects of dietary fat on mammary carcinogenesis by 7,12-dimethylbenz(α)anthracene in rats. Cancer Res., *27*: 1737–1742, 1967.
5. Tannenbaum, A. The genesis and growth of tumors III. Effects of a high-fat diet. Cancer Res., *2*: 468–475, 1942.
6. Tannenbaum, A. Nutrition and cancer *In;* F. Homburger (ed.), The Physiopathology of Cancer, 2nd ed., pp. 517–562, Hoeber-Harper, New York, 1959.
7. Carroll, K. K., Gammal, E. B., and Plunkett, E. R. Dietary fat and mammary cancer. Can. Med. Assoc. J., *98*: 590–594, 1968.
8. Lea, A. J. Dietary factors associated with death-rates from certain neoplasms in man. Lancet, *2*: 332–333, 1966.
9. Wynder, E. L. Identification of women at high risk for breast cancer. Cancer, *24*: 1235–1240, 1969.
10. Gammal, E. B., Carroll, K. K., and Plunkett, E. R. Effects of dietary fat on the uptake and clearance of 7,12-dimethylbenz(α)anthracene by rat mammary tissue. Cancer Res., *28*: 384–385, 1968.
11. Carroll, K. K. and Khor, H. T. Effects of dietary fat and dose level of 7,12-dimethylbenz(α)anthracene on mammary tumor incidence in rats. Cancer Res., *30*: 2260–2264, 1970.
12. Carroll, K. K. and Khor, H. T. Dietary fat in relation to tumorigenesis. Progr. Biochem. Pharmacol., *10*: 308–353, 1975.
13. Carroll, K. K. and Khor, H. T. Effects of level and type of dietary fat on incidence of mammary tumors induced in female Sprague-Dawley rats by 7,12-dimethylbenz(α)-anthracene. Lipids, *6*: 415–420, 1971.
14. Carroll, K. K. Dietary factors in hormone-dependent cancers. *In;* M. Winick (ed.), Current Concepts in Nutrition, vol. 6, Nutrition and Cancer, pp. 25–40, John Wiley & Sons, New York, 1977.
15. Hopkins, G. J. and Carroll, K. K. Relationship between amount and type of dietary fat in promotion of mammary carcinogenesis induced by 7,12-dimethylbenz(a)-anthracene. J. Natl. Cancer Inst., *62*: 1009–1012, 1979.
16. Hopkins, G. J., Kennedy, T. G., and Carroll, K. K. Polyunsaturated fatty acids as promoters of mammary carcinogenesis induced in Sprague-Dawley rats by 7,12-dimethylbenz(a)anthracene. J. Natl. Cancer Inst., *66*: 517–522, 1981.
17. Carroll, K. K. and Braden, L. M. Dietary fat and mammary carcinogenesis. Nutr. Cancer, *6*: 254–259, 1985.

18. Jurkowski, J. J. and Cave, W. T., Jr. Dietary effects of menhaden oil on the growth and membrane lipid composition of rat mammary tumors. J. Natl. Cancer Inst., *74*: 1145–1150, 1985.
19. Carroll, K. K. The role of dietary fat in carcinogenesis. *In;* E. G. Perkins and W. J. Visek (eds.), Dietary Fats and Health, pp. 710–720, American Oil Chemists' Society, Champaign, IL, 1983.
20. Noble, R. L., Hochachka, B. C., and King, D. Spontaneous and estrogen-produced tumors in Nb rats and their behaviour after transplantation. Cancer Res., *35*: 766–780, 1975.
21. Carroll, K. K. and Noble, R. L. Effects of dietary fat on induction of mammary and prostatic carcinoma in Nb rats by implanted hormone pellets. Abstrs. Fourth Int. Conf. on Environmental Mutagens, Stockholm, June 24–28, p. 75, 1985.
22. Welsch, C. W. and Aylsworth, C. F. Enhancement of murine mammary tumorigenesis by feeding high levels of dietary fat: a hormonal mechanism? J. Natl. Cancer Inst., *70*: 215–221, 1983.
23. Carter, C. A., Milholland, R. J., Shea, W., and Ip, M. M. Effect of the prostaglandin synthetase inhibitor indomethacin on 7,12–dimethylbenz(a)anthracene-induced mammary tumorigenesis in rats fed different levels of fat. Cancer Res., *43*: 3559–3562, 1983.
24. Hillyard, L. A. and Abraham, S. Effect of dietary polyunsaturated fatty acids on growth of mammary adenocarcinomas in mice and rats. Cancer Res., *39*: 4430–4437, 1979.
25. Karmali, R. A., Marsh, J., and Fuchs, C. Effect of omega-3 fatty acids on growth of a rat mammary tumor. J. Natl. Cancer Inst., *73*: 457–461, 1984.
26. Oliw, E., Granström, E., and Änggård, E. The prostaglandins and essential fatty acids. *In;* C. Pace-Asciak and E. Granström (eds.), New Comprehensive Biochemistry, vol. 5, pp. 1–44, Elsevier, Amsterdam, 1983.
27. Kritchevsky, D., Weber, M. M., and Klurfeld, D. M. Dietary fat versus caloric content in initiation and promotion of 7,12-dimethylbenz(a)anthracene-induced mammary tumorigenesis in rats. Cancer Res., *44*: 3174–3177, 1984.
28. Carroll, K. K. Experimental evidence of dietary factors and hormone-dependent cancers. Cancer Res., *35*: 3374–3383, 1975.
29. Gabor, H., Hillyard, L. A., and Abraham, S. Effect of dietary fat on growth kinetics of transplantable mammary adenocarcinoma in BALB/c mice. J. Natl. Cancer Inst., *74*: 1299–1305, 1985.
30. Carroll, K. K., Hopkins, G. J., Kennedy, T. G., and Davidson, M. B. Essential fatty acids in relation to mammary carcinogenesis. Prog. Lipid Res., *20*: 685–690, 1981.
31. Ip, C., Carter, C. A., and Ip, M. M. Requirement of essential fatty acid for mammary tumorigenesis in rats. Cancer Res., *45*: 1997–2001, 1985.
32. Braden, L. M. and Carroll, K. K. Dietary polyunsaturated fat in relation to mammary carcinogenesis in rats. Lipids, *21*: 285–288, 1986.
33. Kalamegham, R. and Carroll, K. K. Reversal of the promotional effect of high-fat diet on mammary tumorigenesis by subsequent lowering of dietary fat. Nutr. Cancer, *6*: 22–31, 1984.
34. Wynder, E. L. and Cohen, L. A. A rationale for dietary intervention in the treatment of postmenopausal breast cancer patients. Nutr. Cancer, *3*: 195–199, 1982.
35. Disogra, C. A. and Disogra, L. K. Nutrition and cancer prevention. A perspective on dietary recommendations. *In;* J. Weininger and G. M. Briggs (eds.), Nutrition Update, Vol. 2, pp. 3–27, John Wiley & Sons, New York, 1985.

36. Committee on Diet, Nutrition, and Cancer. Assembly of Life Sciences, National Research Council. Diet, Nutrition, and Cancer, pp. 1–14 to 1–16, National Academy Press, Washington, D.C., 1982.
37. Proceedings of ECP-IUNS Workshop on Diet and Human Carcinogenesis. Nutr. Cancer, *8*: 1–40, 1986.

DIET, NUTRITION AND CANCER, Y. HAYASHI ET AL. (EDS.),
JAPAN SCI. SOC. PRESS, TOKYO/VNU SCI. PRESS, UTRECHT, PP. 265–274, 1986

Cancer Risk in Relation to Fat and Energy Intake among Hawaii Japanese: A Prospective Study

Grant N. STEMMERMANN, Abraham M. Y. NOMURA, and Lance K. HEILBRUN

Japan-Hawaii Cancer Study, Kuakini Medical Center, Honolulu, Hawaii 96817, U.S.A.

Abstract: This study assesses the impact of fat and energy consumption upon cancer risk in a prospective study of 8,006 Japanese men, who have developed 885 incidence cancers since initial examinations were completed between 1965 and 1968. Energy intake was not related to any incidence cancer. The mean total fat intake was unrelated to the risk of developing cancers in the stomach ($n=130$), lung ($n=145$), urinary bladder ($n=51$), pancreas ($n=25$), prostate ($n=141$), liver ($n=22$). There was a weak inverse association between mean fat intake and colon cancer. There was a statistically significant inverse relation between mean daily fat intake and all other cancers ($n=118$). There was a weakly positive association between fat intake and rectal cancer ($n=71$). When assessed on the basis of quartiles of fat intake, there was a statistically significant negative association with colon cancer risk ($p=0.03$); and weaker negative trends for lung cancer ($p=0.076$) and all other cancers (0.076). These findings are in essential agreement with the results of a 10-year mortality study of the cohort. The fat intake of men who have developed cancer is substantially lower than that of men who have developed coronary heart disease. These findings cast doubt upon the importance of fat intake as a risk factor for cancer at sites other than the rectum.

In a monograph devoted to assessing the impact of diet on cancer risk (*1*), a committee of the National Research Council stated that the dietary change most likely to be protective against cancer would entail a reduction of fat intake by the entire population. This recommendation was made in the face of the committee's observation that only two sites (breast and colon) showed any putative positive association between fat intake and cancer risk. We have tested this hypothesis with a cohort of Japanese men living in Hawaii. The people of Japanese ancestry in Japan and the United States consume comparable levels of dietary calories, but the Japanese Americans consume more sucrose and less complex carbohydrate; more animal fat and more animal protein than do the Japanese in Japan (*2*). The observed differences in the patterns of disease among native and migrant Japanese have been attributed in part to increased fat consumption by the migrants (*3*). This hypothesis would appear to be confirmed in respect to the increased frequency of coronary heart disease (CHD) (*4*) among Hawaii Japanese, although the impact of dietary fat upon this disease is weaker than that of other variables (*e.g.*, hypertension, smoking, obesity, serum cholesterol). The influence of fat intake on cancer risk in this migrant population is less clear. A retrospective case-control study of large bowel cancer in Hawaii Japanese of both sexes indicated that this tumor was positively associated with meat consumption (*5*), a finding that implied increased fat consumption. Our prospective study of Hawaii Japanese men showed that men with the lowest fat intake experienced the highest risk of developing colon cancer, while rectal cancer showed a weak positive association with fat intake (*6*). A 10-year mortality study of these Hawaii Japanese men indicated that those with the lowest fat intake experienced higher mortality risk than those with higher fat intake, and that this increase in risk could be attributed to death from cancer and stroke (*7*). The inverse relation between cancer mortality and fat intake was actually strengthened when deaths occurring within 5 years were eliminated. Thus, it is unlikely that undiagnosed cancer influenced fat intake. We then undertook to test the validity of the mortality study, using cancer incidence data accumulated over a longer time period. This permitted us to assess a much larger number of cancers and to weigh the relative impact of fat and energy intake upon cancer risk at different primary sites.

A Description of the Study Population

This cohort of Hawaii Japanese consists of 8,006 Japanese men who lived in Honolulu County in 1965, who were born in the years 1900–19, and who registered

TABLE 1. Nutrient Intake Based on the 7-Day and 24-Hour Diet Records (*9*)

	Men supplying a 7-day record (*n*=329)		First examination diet interview (*n*=7,677)	
	Average of 7-day records		24-hour recall	
	Mean	S.D.	Mean	S.D.
Calories	2,300	511	2,273	742
Protein (g)	94.7	21.4	93.9	35.7
Carbohydrate (g)	265.3	75.1	260.1	97.1
Fat (g)	84.6	23.0	85.1	39.2
Saturated fatty acids (g)	30.7	9.0	31.3	15.6
Monounsaturated fatty acids (g)	32.7	9.3	32.5	15.6
Polyunsaturated fatty acids (g)	15.1	5.1	15.2	9.6
Cholesterol (mg)	529.0	182.9	545.5	317.1
Alcohol (g)	14.3	23.3	13.2	30.3
Starch (g)	160.3	57.1	163.2	72.6
Protein (%)	16.7	2.6	16.7	4.0
Carbohydrates (%)	46.4	7.3	46.4	11.0
Fat (%)	33.0	6.2	33.3	9.4
Saturated fatty acids (%)	11.9	2.5	12.2	4.1
Alcohol (%)	3.9	6.1	3.7	7.9

for military service in the years 1940–42 (*8*). After initial examinations and interviews from 1965–68, they have been followed for the development of incident events of CHD and cancer, as well as mortality. Follow-up has been through surveillance of hospital discharge summaries, state obituary records, and by matching with entries into the Hawaii Tumor Registry. We have excluded 81 prevalence cancer cases and 90 suspected incidence cases that were not confirmed histologically.

The dietary information was obtained by a 24-hour recall interview administered to each study subject during his regular clinic visit. Food models and serving utensils were used to illustrate portion sizes. Food composition values were compiled from the best available sources to calculate calories and nutrients (*9*). We also excluded 490 men whose 24-hour dietary intake was considered "atypical" for them. This left 7,345 study eligible men for analysis, including 885 tissue-confirmed cancer incidence cases as of April, 1985. A 7-day diet record was also administered to a subsample of 329 men to assess the validity of the 24-hour dietary questionnaire (*10*). Dietitians secured the cooperation of subjects and instructed them and their wives in standard methods for keeping written records of all foods eaten for 7 consecutive days. The two methods yielded similar levels of nutrient intake (Table 1).

Cancer Risk and Fat Intake at Different Primary Sites

Tumor sites with 20 or more accessioned cases were selected for individual analysis. Tumors of the oral cavity, larynx, and esophagus were combined since they share the common feature of high risk from the use of alcohol and cigarettes. Sites with fewer than 20 incidence events were also combined. These included tumors of biliary tree, kidney, endocrine glands, central nervous system, soft tissue, reticulo-endothelial system, and leukemia. Men who developed cancer subsequent to their examination in the years 1965–68 did not differ significantly from controls in respect

TABLE 2. Age-adjusted Means of Dietary and Energy Variables by Cancer Site

Cancer site		Calories/day	S.E.	Total Fat (g/day)	S.E.	Sat. Fat (g/day)	S.E.
Stomach	(130)[a]	2,350	60.4	89.3	3.2	62.9	2.8
Colon	(138)	2,224	58.6	80.0*	3.1	54.8*	2.7
Rectum	(71)	2,396	81.7	89.1	4.3	64.5	3.7
Pancreas	(25)	2,235	137.6	85.9	7.4	58.5	6.3
Liver	(22)	2,416	146.6	82.4	7.9	59.5	6.7
Lung	(145)	2,290	57.1	81.1	3.1	57.7	2.6
Mouth, larynx, esophagus	(44)	2,397	103.7	82.1	5.6	57.6	4.7
Prostate	(141)	2,297	58.1	83.6	3.1	54.4	2.7
Urinary bladder	(51)	2,320	96.3	86.6	5.2	59.5	4.4
All other sites	(118)	2,238	63.3	79.3**	3.4	56.0	2.9
All cancers	(885)	2,300	23.3	83.5*	1.3	58.8	1.1
Controls	(6,460)	2,283	8.5	86.0	0.5	59.6	0.4

Age-adjusted means were computed and compared to the mean of controls using unbalanced, one-factor analysis of covariance models. [a] Numbers in parentheses indicate number of men in each group. * $p<0.1$; ** $p\leq0.05$.

to mean caloric intake (Table 2). This applied to cancer cases considered altogether or separately by site. The age-adjusted mean total fat intake was significantly lower for the entire group of men who developed cancer when compared to the controls. The lowest total fat intakes were observed in men who subsequently developed cancer of the colon, lung and at miscellaneous sites not listed separately. The differences from the controls were statistically significant ($p<0.05$) only for cancers at miscellaneous sites. Men who developed gastric and rectal cancer consumed more fat than controls, but the differences were not impressive. The intake of saturated fat was similar to that of total fat for all tumor sites, although none of the differences between cancers (except for colon) and controls were noteworthy. This is understandable in view of the high correlation ($r=0.84$) between total fat and saturated fat intake in our study subjects. Accordingly, we have not analyzed saturated fat in great detail here.

The "dose-response" relationship of total fat intake in gm/day to the risk of site-specific cancer was explored using estimated odds ratios (OR) for varying levels of fat intake. OR's of site-specific cancer were derived from logistic models (*11*), while adjusting for covariates (age at examination, and for certain cancers, either alcohol intake or smoking). A separate model was fit for each cancer site. Men in the lowest fat quartile were chosen as the referent group. Quartiles were determined using the least common cancer cases only (liver, pancreas), in order to conserve the number of such cases per quartile. Indicator variables were used in the logistic models to denote membership of subjects in their respective quartiles of fat intake. The OR derived from the logistic model coefficient for a given indicator variable estimates the risk of cancer for men in that (fat intake) quartile, as compared to men in the referent group.

The covariate-adjusted odds ratios of cancer for each site and for each quartile of total fat intake are indicated in Table 3. The test for linear trend in the logit of

TABLE 3. Age-adjusted OR of Cancer by Site and by Quartile or Total Fat Intake

Cancer site	n	Fat intake (g/day)				Test for trend	
		<50	50–74.9	75–99.9	100+	$\hat{\beta}$[d]	p-value
Colon	138	1.0	0.87	0.53[a]	0.64	−0.182	0.029
Rectum[b]	71	1.0	1.07	1.17	1.63	+0.168	0.153
Stomach	130	1.0	1.17	1.23	1.16	+0.041	0.639
Lung[c]	145	1.0	1.14	0.66	0.76	−0.145	0.076
Prostate	141	1.0	1.15	1.36	0.89	−0.017	0.835
Bladder[c]	51	1.0	1.73	0.85	1.37	−0.004	0.976
Pancreas	25	1.0	0.81	0.93	0.68	−0.093	0.634
Liver	22	1.0	0.92	0.61	1.0	−0.012	0.955
Mouth, larynx, esophagus[b]	44	1.0	0.95	0.76	0.73	−0.118	0.414
All other cancers	118	1.0	0.67	0.68	0.58[a]	−0.160	0.076
All cancers[c]	885	1.0	1.03	0.85	0.87	−0.066	0.062
Controls	6,460						

[a] OR is significantly different from unity ($p<0.05$). [b] OR adjusted for age at examination and ethanol intake (oz/month). [c] OR adjusted for age at examination and smoking (current cigarettes/day). [d] Logistic regression coefficient.

TABLE 4. Age-adjusted OR of Cancer by Site, by Time Interval from Examination to Diagnosis, and by Quartile of Total Fat Intake

Cancer site and time interval		n	Total fat intake (g/day)				Test for trend	
			<50	50–74.9	75–99.9	100+	$\hat{\beta}$[d]	p-value
Colon	<10 years	47	1.00	0.92	0.58	0.44	−0.294	0.041
	≥10 years	91	1.00	0.85	0.50[a]	0.76	−0.126	0.218
Rectum[b]	<10 years	45	1.00	0.93	0.91	0.91	−0.030	0.835
	≥10 years	26	1.00	1.77	2.45	5.06[a]	+0.539	0.011
Lung[c]	<10 years	51	1.00	0.89	0.66	0.66	−0.157	0.249
	≥10 years	94	1.00	1.32	0.66	0.84	−0.138	0.170
Other (miscellaneous)								
Cancers	<10 years	59	1.00	0.62	0.32[a]	0.61	−0.212	0.097
	≥10 years	59	1.00	0.73	1.19	0.55	−0.110	0.387
All cancers[c]	<10 years	369	1.00	0.82	0.70[a]	0.80	−0.079	0.131
	≥10 years	516	1.00	1.21	0.99	0.92	−0.058	0.192
Controls		6,460						

a)–d) are same as in Table 3.

cancer risk was obtained from a separate model involving the quartile-coded (0, 1, 2, 3) fat intake and the appropriate variable (*12*). There was a positive, monotonic trend of increasing rectal cancer risk with total fat intake, but the association was not statistically significant. Negative trends were observed for colon cancer ($p=0.029$), lung cancer ($p=0.076$), and for all other cancer sites ($p=0.076$).

We then calculated the cancer risk for each quartile of fat intake using patients diagnosed less than 10 years from examination and those diagnosed 10 or more years after examination (Table 4). There was no difference in the dose response curves for these time periods for all cancers, for cancers at miscellaneous sites or for lung cancer. Rectal cancer showed no relation to fat intake in the patients who acquired the disease within 10 years of examination, but there was a significant positive trend ($p<=0.01$) in men whose tumor appeared after 10 years. In the case of colon cancer, the inverse trend for fat intake remained in men diagnosed more than 10 years after examination, but was no longer statistically significant.

The Evidence That Different Tumor Sites Have Different Risk Associations with Fat Intake

Evidence that post migratory changes in fat consumption by Hawaii Japanese account for the observed changes in cancer incidence or mortality are scanty in this longitudinal study. Since this is a male cohort, the relation of fat intake to breast cancer cannot be weighed. There is no evidence that excess fat consumption increases the risk for cancer in general. More specifically, the findings do not support the findings of the case-control studies of colon cancer (*13, 14*) upon which the National Research Council based its conclusions that high fat intake favors the development of colon cancer. Fat intake showed an inverse association with colon cancer risk, although this inverse relation is no longer statistically significant in men who developed their tumor more than 10 years after examination. Only rectal and gastric cancers have shown a direct relation to fat intake. In the case of rectal cancer, this

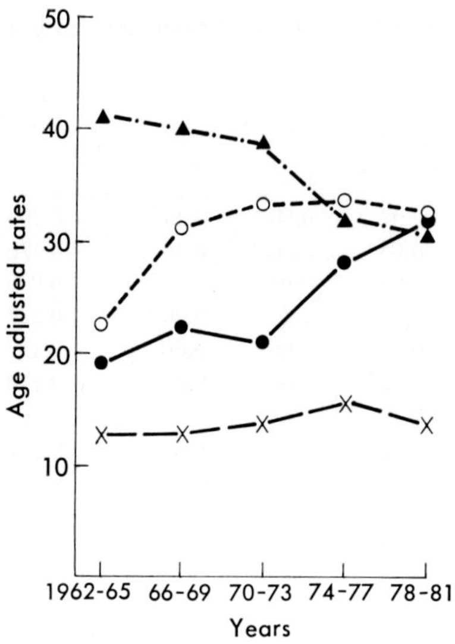

FIG. 1. Recent trends, gastrointestinal cancer, Hawaii Japanese males. Data are from Hawaii tumor registry and rates are adjusted for Hawaii population. ●, colon; ▲, stomach; ○, lung; ×, rectum.

is a statistically significant ($p=0.01$) in men who have developed cancer more than 10 years after examination.

It would seem reasonable to expect cancer trends to follow fat intake if this nutrient had an important biologic role in cancer induction. During the period of observation covered by this study, the tumors with a direct relation to fat intake have decreased in frequency (stomach) or shown a stable secular trend (rectum); while colon cancer, despite its inverse association with fat intake, has shown a steady increase (Fig. 1). Lung cancer has also steadily increased in frequency in Japanese males, and this tumor shows an inverse relation to fat intake that approaches statistical significance.

The Basis for Inconsistent Findings in Respect to Fat Intake and Cancer Risk

It is important that we seek reasons for the conflicting observations between fat intake and cancer risk in various studies. Cancer requires many years of incubation. In retrospective case-control studies, a diet interview may reflect food habits distorted by existing disease. Prospective dietary studies are not faced with the same limitation, but a single interview may not reflect food use over an extended period of time. However, studies have shown 24-hour diet recalls are representative of usual dietary pattern when large numbers of subjects are compared, as was done in this study. It is also possible that small differences in consumption might be biologically significant over the long term. For example, a small daily excess of energy consumption over expenditure might result in substantial obesity if the excess were consistently maintained for 30 or more years. Lastly, the association between cancer and fat intake could be obscured by limitations of our analytic methods. For example, an attempt to weigh the independent impact of potassium, calcium, protein and milk upon blood pressure in this cohort has not been possible because of the high degree of intercorrelation (multicollinearity) among these dietary factors (*15*). The same constraints apply to fat and protein intake in respect to CHD risk as well (*16*) since they are highly correlated ($r=0.77$). It is unlikely that the relation of fat to cancer risk differs in this respect.

Gordon and co-workers (*17*) in a paper dealing with three geographically separated longitudinal studies found that total fat and caloric intake were related significantly and inversely to total mortality. In our cohort the incidence of cancer in general has shown a similar inverse relation to fat intake, but has demonstrated no significant differences in caloric intake between controls and subjects with cancer at any primary site. Colon cancer was positively associated with weight gain from 25 years of age among cohort men who were over 55 years of age at the time of examination (*18*). It is reasonable to assume that these men have consumed more energy than they spent. Since their energy intake was similar to that of controls, their increased body mass could be explained by a positive energy balance. This is consistent with the observation that colon cancer is associated with sedentary occupation (*19*, *20*) and high family income (*21*). It is therefore reasonable to question whether any of the known functions of body fat, such as estrogen production (*22*) or vitamin transport and storage, might affect the risk of colon cancer.

Fat Intake and Serum Cholesterol Levels as Cancer Risk Factors

The association of a low fat intake with increased cancer mortality (*7*) and increased cancer incidence in these Hawaii Japanese men accompanies other unexpected findings in this population. For example, low serum cholesterol levels are associated with increased cancer mortality in this cohort (*23*); and adherence to Japanese diet patterns and traditional Japanese cultural practices are associated with increased cancer incidence (*24*). Low serum cholesterol levels have also been tied to an increased risk of colon cancer in cohort men (*25*). Men who had had a subtotal gastrectomy at the time of examination had very low serum cholesterol levels and are now at especially high risk of developing lung cancer, even after adjusting for smoking (*26*). Keys *et al.* (*27*) observed an increased risk of cancer in national populations with low mean cholesterol levels (Japan and Southern Europe). He proposed that a level of 170 mg/dl might constitute a threshold below which cancer risk increased. Considered as a whole, the U-shaped curves of mortality rates for fat intake (*7*) and serum cholesterol (*23*) are remarkably similar in this population, with high levels of both variables associated with CHD mortality and low levels with death from cancer, and in neither case can the result be attributed to disease at the time of examination.

In spite of the similarity between the disease patterns associated with different fat intake and serum cholesterol levels, it should not be assumed that the high fat content of the western diet accounts for the post imgrational increase in serum cholesterol levels of Hawaii Japanese. Yano *et al.* (*28*) have shown that dietary fat has no significant association with the serum cholesterol level in this cohort, whether measured in grams per day or calculated as the percent of calories. These results are generally similar to those of other studies (*29–32*). Although methods for dietary assessment may be imprecise, the role of diet in determining the levels of serum lipids should be greater than such studies indicate. Carbohydrates are the only major nutrients that are significantly related to serum cholesterol. Total carbohydrate and starch are inversely related to serum cholesterol, and sucrose is directly related to serum cholesterol. This said, the internal and external consistency of prospective studies relating fat and cholesterol to cancer risk indicates that it is time to reconsider the role played by fat in the induction and promotion of cancer. It may have no role at all, but may reflect the effect of an excess or deficiency of a nutrient with which it is highly correlated (*e.g.*, protein); or it may mask the effect of another nutrient. For example, Hawaii Japanese consume more fat and sucrose than do indigenous Japanese, but sucrose is the only nutrient that is directly associated with serum cholesterol (*28*).

This long-term study indicates that, if fat intake does influence cancer risk, the nature and degree of the influence will probably differ from one primary site to another. It suggests that we should also determine whether excessively low levels of dietary fat or serum cholesterol may actually increase the risk of certain cancers; or whether a confounding factor, highly correlated with fat intake (*e.g.*, the absorption of fat soluble vitamins), might account for the results obtained in this and other prospective studies.

In summary, it is premature to make recommendations aimed at lowering cancer risk by dietary manipulations; or to make drastic dietary alterations in order to lower CHD risk, since it is possible that the pursuit of dietary extremes may exchange one risk for the other.

ACKNOWLEDGMENT

The manuscript was written while the authors were supported by a grant from the National Cancer Institute, R01-CA-33644.

REFERENCES

1. Diet, Nutrition and Cancer. Committee on Diet, Nutrition and Cancer. Assembly of Life Sciences, National Research Council, 439 pp., National Academy Press, Washington, D.C., 1982.
2. Kagan, A., Harris, B. R., Winkelstein, W., Johnson, K. G., Kato, H., Syme, S. L., Rhoads, G. G., Gay, M. L., Nichaman, M. Z., Hamilton, H. B., and Tillotson, J. Epidemiologic studies of coronary heart disease and stroke in Japanese men living in Japan, Hawaii, and California: Demographic, physical, dietary and biochemical characteristics. J. Chron. Dis., *27*: 345–364, 1974.
3. Keys, A. Coronary heart disease in seven countries. Circulation, *41* (Suppl. I): I 1–221, 1970.
4. McGee, D. L., Reed, D. M., Yano, K., Kagan, A., and Tillotson, J. Ten year incidence of coronary heart disease in the Honolulu Heart Program. Am. J. Epidemiol., *119*: 667–676, 1984.
5. Haenszel, W., Berg, J. W., Segi, M., Kurihara, M., and Locke, F. B. Large bowel

cancer in Hawaiian Japanese. J. Natl. Cancer Inst., *51*: 1765–1779, 1973.

6. Stemmermann, G. N., Nomura, A.M.Y., and Heilbrun, L. K. Dietary fat and the risk of colorectal cancer. Cancer Res., *44*: 4633–4637, 1984.
7. McGee, D., Reed, D., Stemmermann, G., Rhoads, G., Yano, K., and Feinlieb, M. The relationship of dietary fat and cholesterol to mortality in 10 years: The Honolulu Heart Program. Int. J. Epidemiol., *14*: 97–105, 1985.
8. Worth, R. and Kagan, A. Ascertainment of men of Japanese ancestry in Hawaii through World War II Selective Service Registration. J. Chron. Dis., *23*: 389–397, 1970.
9. Tillotson, J. L., Kato, H., Nichaman, M. Z., Miller, D. C., Gay, M. L., Johnson, K. G., and Rhoads, G. G. Epidemiology of coronary heart disease and stroke in Japanese men living in Japan, Hawaii, and California: Methodology for comparison of diet. Am. J. Clin. Nutr., *26*: 177–184, 1973.
10. McGee, D., Rhoads, G., Hankin, J., Yano, K., and Tillotson, J. Within-person variability of nutrient intake in a group of Hawaiian men of Japanese ancestry. Am. J. Clin. Nutr., *36*: 657–663, 1982.
11. Breslow, N. E. and Day, N. E. Statistical Methods in Cancer Research, Vol. 1: The Analysis of Case-Control Studies, International Agency for Research on Cancer, Lyon, France, 1980.
12. Schlesselman, J. J. Case-Control Studies: Design, Conduct, Analysis. Oxford University Press, New York, 1982.
13. Reddy, B. S. Dietary fat and its relation to large bowel cancer. Cancer Res., *41*: 3700–3705, 1981.
14. Jain, M., Cook, G. M., Davis, F. G., Grace, M. G., Howe, G. R., and Miller, A. B. A case control study of diet and colorectal cancer. Int. J. Cancer, *76*: 757–768, 1980.
15. Reed, D., McGee, D., Yano, K., and Hankin, J. Diet, blood pressure and multicollinearity. Hypertension, *7*: 405–410, 1985.
16. McGee, D., Reed, D., and Yano, K. The results of logistic analysis when the variables are highly correlated: An empirical example using diet and CHD incidence. J. Chron. Dis., *37*: 713–719, 1984.
17. Gordon, T., Kagan, A., Garcia-Palmieri, M., Kannel, W. B., Zenkel, W. J., Tillotson, J., Sorlis, P., and Hjortland, M. Diet and its relation to coronary heart disease and death in three populations. Circulation, *63*: 500–515, 1981.
18. Nomura, A., Heilbrun, L., and Stemmermann, G. Body mass index as a predictor of cancer in men. J. Natl. Cancer Inst., *74*: 319–323, 1985.
19. Lynch, H. T., Guigis, H., Lynch, J., Brodkey, F., and Magee, H. Cancer of the colon: Socioeconomic variables in a community. Am. J. Epidemiol., *102*: 119–127, 1975.
20. Garabrant, D. H., Peters, J. M., Mack, T. M., and Bernstein, L. Job activity and colon cancer risk. Am. J. Epidemiol., *119*: 1005–1013, 1984.
21. Haenszel, W., Correa, P., and Cuello, C. Social class differences among patients with large bowel cancer in Colombia. J. Natl. Cancer Inst., *54*: 1031–1035, 1975.
22. Siiteri, P. K. Extraglandular oestrogen formation and serum binding of oestradiol: Relationship to cancer. J. Endocrin., *89*: 119–129, 1981.
23. Kagan, A., McGee, D. L., Yano, K., Rhoads, G., and Nomura, A. Serum cholesterol and mortality in a Japanese-American population. Am. J. Epidemiol., *114*: 11–20, 1981.
24. Joffres, M., Reed, D. M., and Nomura, A.M.Y. Psychosocial processes and cancer incidence among Japanese men in Hawaii. Am. J. Epidemiol., *121*: 488–500, 1985.

25. Stemmermann, G. N., Nomura, A., Heilbrun, L. K., Pollack, E. S., and Kagan, A. Serum cholesterol and colon cancer in Hawaiian Japanese men. J. Natl. Cancer Inst., *67*: 1179–1182, 1981.
26. Stemmermann, G. N., Heilbrun, L., Nomura, A.M.Y., Rhoads, G. G., and Glober, G. A. Late mortality after partial gastrectomy. Int. J. Epidemiol., *13*: 299–303, 1984.
27. Keys, A., Aravanis C., Blackburn, H., Bergina, R., Dontas, A. S., Fidanza, F., Karvonen, M. J., Menotti, A., Nedeljkovic, S., Punsar, S., and Toshima, H. Serum cholesterol and cancer mortality in the seven countries study. Am. J. Epidemiol., *12*: 870–883, 1985.
28. Yano, K., Reed, D. M., Curb, J. D., Hankin, J., and Albers, J. J. Biological and dietary correlates of plasma lipids and lipoproteins in elderly Hawaiian Japanese men. (submitted for publication).
29. Kahn, H. A., Medalie, J. H., Neufeld, H. N., Riss, E., Balagh, M., and Groen, J. J. Serum cholesterol: its distribution and association with dietary and other variables in a survey of 10,000 men. Isr. J. Med. Sci., *5*: 1117–1127, 1969.
30. Strelb, S. C., McDonough, J. R., Greenberg, B. G., and Hames, C. G. The relationship of nutrient intake and exercise to serum cholesterol levels in white males in Evans County, Georgia. Am. J. Clin. Nutr., *16*: 238–242, 1965.
31. Kannel, W. B. and Gordon, T. (eds.) The Framingham Study: an epidemiologic investigation of cardiovascular disease. Section 24. Diet and Regulation of Serum Cholesterol, Government Printing Office, Washington, 1970.
32. Nichols, A. B., Ravenscroft, C., Lamphiear, D. E. and Oshander, L. D. Independence of serum lipid levels and dietary habits. The Tecumseh Study. J. Am. Med. Assoc., *236*: 1948–1953, 1976.

DIET, NUTRITION AND CANCER, Y. HAYASHI ET AL. (EDS.),
JAPAN SCI. SOC. PRESS, TOKYO/VNU SCI. PRESS, UTRECHT, PP. 275–290, 1986

Dietary Influences upon Colon Carcinogenesis

A. J. McMichael and J. D. Potter

Division of Human Nutrition, Commonwealth Scientific and Industrial Research Organization (Australia), Adelaide, South Australia 5000, Australia

Abstract: A succession of case-control studies of diet and colon cancer, predominantly in developed countries, has produced varied and generally inconsistent findings.

The somatic mutation theory of carcinogenesis has dominated much of cancer research for the past 30 years, encouraging emphasis on exogenous genotoxic agents capable of inducing malignant transformation *via* heritable damage to DNA. Increased risks of human cancers due to various potent chemical carcinogens (found in certain occupations), ionizing radiation, and sunlight have corroborated this "toxicological" view of cancer. Recently, however, greater emphasis has been paid to cancer as a disorder of growth control. The stimulation or derepression of cell growth, *via* hormones or proto-oncogene activation respectively, is likely to reflect "metabolic" disturbances—such as can be caused by diet.

If diet influences large bowel carcinogenesis *via* mediating metabolic or biochemical factors such as intracolonic pH, production of bile acid metabolites, and fermentative production of volatile fatty acids (which appear to influence mucosal cell stability), then a variety of configurations of diet may have an equivalent net effect upon bowel carcinogenesis. Further, non-specific aspects of diet (such as total energy intake and frequency of eating) may be important; indeed, those two factors were found to be positively and independently associated with large bowel cancer (LBC) risk in our Adelaide case-control study.

The accumulating evidence that other factors that alter sex hormonal status and/or hepatobiliary metabolism, and physical aspects of bowel function, are also associated with altered risk of LBC adds further credence to this metabolic model. Such factors are: gender, reproductive history and oral contraceptive usage in women, cholecystectomy, and physical activity.

Throughout the 1970's the main stimulus to research into the causes of colon cancer came from the two (related) hypotheses about the cancer-enhancing effects of diets low in dietary fibre or high in total fat (*1*). Epidemiologists around the world compared the diets of persons with and without colon cancer and those of populations at high and low risk of colon cancer; meanwhile, laboratory scientists examined the influence of dietary fat and fibre upon colon tumourigenesis, and also the postulated biochemical mediation of their effects *via* bile acid secretion and degradation. Results from the earliest of these experiments stimulated a rapid growth in "metabolic epidemiology," seeking to go beyond identifying dietary differences between groups of humans (in relation to colon cancer) by studying etiologically relevant "metabolic" differences between them.

This research effort appears recently to have lost momentum in the laboratory, and to have lost a sense of direction in the epidemiologist's domain where the findings from case-control studies around the world have been hugely varied, often inconsistent, and sometimes paradoxical (*e.g.*, the handful of studies that have found *positive* associations of dietary fibre or high-fibre foods with colon cancer risk).

A number of etiological hares have thus been set running, and now are being pursued—fibre, fat, protein, cruciferous vegetables, vitamin A, vitamin C, selenium, pyrolysate mutagens, endogenously-formed nitrosamines, and more. Epidemiological research into colon cancer appears to have been diverging rather than converging, and there is now a perceptible mood of frustration. Are our dietary instruments deficient? Is colon carcinogenesis extravagantly multi-factorial? Or does diet really have a lot less to do with colon cancer than we assume?

Perhaps we are being guided by the wrong star. The star of experimental chemical carcinogenesis has shone strongly from the laboratories for several decades, imparting an emphasis to the role of "initiation" in carcinogenesis. A twin star has shone from the early successes of cancer epidemiology in relation to specific occupational exposures (aromatic amines, *etc.*), ionising radiation, cigarette smoking, and sunlight exposure. Clearly, potent single-factor exposures of an "initiating" kind are important in some human (and many artifactual animal) cancers; but perhaps they are in the minority—despite their ready visibility.

What we wish to propose here is a different formulation of the etiology of colon cancer. The ingredients are not all new, but the assemblage and the emphases are. It also incorporates a dynamic element, wherein a factor that might have a potent promoting (accelerating) effect will be strongly associated with cancers occurring at a young age—but may, paradoxically, therefore be inversely associated with cancers occurring at older ages (such a phenomenon has been identified in several recent studies of oral contraceptives in the etiology of gallstone disease. It has been described as a rapid depletion of the metabolically susceptible from the exposed population, and is manifested as a reversal of risk with increasing time since commencement of exposure).

Before embarking on this search for a new star to guide our etiological thinking, let us consider the evidence accrued so far.

Empirical Facts, and Their Likely Interpretations

1. Descriptive epidemiology

1\. 1. Colon cancer (CC) incidence varies at least 10-fold around the world. It can also vary up to 5-fold *within* countries (*1*, *2*).

1. 2. Variations by time, place, and category of person (religion, vegetarianism, *etc.*) are associated with a range of dietary variations (*3–5*).

1. 3. Changes in CC rates typically occur abruptly—and soon after population dietary change. Two examples are:

   i. Migrants from low-risk populations to U.S. and Australia (in the latter case, the risk of CC in the migrating generation increased markedly towards that of the host population with increased duration of residence (*6*)).

   ii. Post-World War II reduction in CC rates in U.K. (*7*).

1. 4. *Interpretation of 1. 1.–1. 3.:* There is a strong environmental influence on CC

TABLE 1. Dietary Factors and Their Relation to Cancer of Colon and Rectum—Main Findings of Case-control Studies

Factor	Relationship to risk	No. of studies showing relationship	No. of studies done
Vegetables			
Vegetables	−	8	11
String beans	+	1	2
Fibre			
Fibre	−	3	7
Fibre	+	2	7
Cereals			
Rice	−	1	3
Rice	+	2	3
Pasta and Rice	+	1	1
Cereals	+	2	6
Fruit			
Fruit	−	2	9
Fruit	+	1	9
Vitamin C	−	3	3
Vitamin A	−	1	3
Meat, fat, protein, energy			
Meat	+	6	11
Saturated fat / total fat	+	5	12
Protein	+	2	3
Energy	+	2	3
Non-meat protein foods			
Non-meat protein	−	1	1
Fish	−	1	6
Fish	+	2	6
Milk	−	1	5
Milk	+	1	5
Other dairy products	+	2	6
Eggs	+	2	3
Other			
Japanese meals	−	1	1
Coffee	−	1	2
Coffee	+	1	2

Source: from refs. *11–29*.

etiology, most probably dietary factors. This influence predominantly acts at later stages of carcinogenesis.

1. 5. The most consistent dietary correlate in descriptive studies of CC is high fat (*8*) and/or low fibre intake (*9*). Rates of cancers of breast, endometrium, and prostate are also correlated with CC (*2*).

1. 6. CC rates are higher in women than men prior to sixth decade of age. The female excess is most marked, and persistent across age, for right-sided CC (*2, 10*).

2. Analytic epidemiology

2. 1. Case-control studies of CC and diet have mostly produced weak and inconsistent results (Table 1). Stronger (but varied) associations are seen in studies conducted within populations undergoing dietary transition. The few cohort studies reported have also produced varied results, and have not corroborated the case-control studies (Table 2).

2. 2. *Possible interpretations of 2. 1.:*

   i. Dietary etiology of CC varies between populations, for genetic reasons (*e.g.*, lactase deficiency (*33*)).

   ii. Some particular dietary factors have a dominant influence upon CC etiology (*e.g.*, total energy intake; frequency of eating; antioxidant consumption), but in different populations different aspects of diet are the most visible "marker" of this dominant factor (*34*).

   iii. Since most studies have been done in populations whose individual members have a generally high intake of an "affluent" diet, there may be little residual inter-individual variation in risk attributable to diet. This could reflect two things:

      a. Lack of dietary heterogeneity between individuals; or

      b. Location of study population above risk-saturation point on exposure-response graph (*34*).

2. 3. There is an increased risk of CC after cholecystectomy (especially of right colon in women) (*35*).

TABLE 2. Dietary Factors and Their Relation to Cancer of Colon and Rectum—Main Findings of Cohort Studies

Factor	Relationship to risk	No. of studies	Authors
Vegetables, cereals, fruit			
Rice and wheat	–	1	Hirayama (*30*)
Meat, fat, protein			
Meat	–	1	Hirayama (*30*)
Saturated fat	– (colon) + (rectum)	1	Stemmermann *et al.* (*31*)
Non-meat protein foods			
Eggs	+	1	Phillips and Snowdon (*32*)
Other			
Coffee	+	1	Phillips and Snowdon (*32*)

2. 4. In women, decreased risk of CC (especially right-sided) is associated with being parous, with early age at first completed pregnancy, and with use of oral contraception (*10*, *36*).

2. 5. Additional observations are:
   i. Men with prostate cancer treated with estrogens have an altered CC subsite distribution, with shift from left to right (*37*).
   ii. Men with "naturally" low blood cholesterol level have increased risk of CC (especially right colon) (*38*, *39*).

2. 6. *Interpretation of 2. 3.–2. 5.:* Host factors which affect production, secretion, and re-circulation of bile acids (BA) alter the risk of CC. (Alternatively, estrogens may (also) affect bowel carcinogenesis directly, *via* mucosal receptors (*40*).)

2. 7. Diet-associated increases in CC risk are maximal in *young* women and in *older* men (*29*, *34*).

3. Metabolic epidemiology of CC

3. 1. Fecal concentrations of BA and, especially, secondary BA, are:
   i. Higher in cases than controls (some inconsistency) (see, *e.g.*, refs. *41* and *42*)
   ii. Higher in high-risk population samples (see, *e.g.*, refs. *43* and *44*)

3. 2. Fecal concentrations of neutral sterols are:
   i. Higher in cases than controls (see, *e.g.*, ref. *45*)
   ii. Higher in high-risk population samples (see, *e.g.*, ref. *44*)

3. 3. Fecal concentrations of anaerobic colonic bacteria and BA-degrading bacterial enzymes are:
   i. Higher in cases than controls (see, *e.g.*, ref. *46*)
   ii. Higher in high-risk population samples (see, *e.g.*, ref. *47*).

3. 4. There is not a consistent association of bowel transit time with CC risk. However, fecal bulk is inversely associated with CC risk in ecological studies (*48*).

3. 5. Fecal mutagenicity is higher in high-risk populations (*49*, *50*), and is higher in non-vegetarian than vegetarian individuals (*51*). (Recently, Gupta *et al.* (*52*) have identified a mutagenic lipid metabolite, fecapentaene, the production of which is stimulated by a high concentration of BA.)

3. 6. Experimental modulation of diet in humans affects these above-mentioned parameters. For example:
   i. Increased animal fat causes increased fecal total BA (see, *e.g.*, ref. *53*).
   ii. Increased dietary fibre causes increased fecal total BA, but also increased fecal bulk and therefore decreased fecal BA concentration (*54*). It also causes a reduction of the ratio of secondary to primary BA (*55*) possibly *via* decreased activity of 7α-dehydroxylase. (*However*, there is some evidence that males and females respond differently to dietary fibre (*56*).)

3.7. *Interpretation of 3. 1.–3. 6.:* Factors affecting the fecal concentration of secondary BA (*i.e.*, fecal bulk, total BA, degradative bacteria and their specific enzymes) are consistently associated with the occurrence of, or a high risk of, CC.

4. Relevant data from clinical and experimental studies

4. 1. Compared to men, women have:

   i. Slower bowel transit (*57*)

   ii. Lesser fecal bulk (*57*)

   iii. Higher fecal pH (*57*)

   iv. A lower rate of BA synthesis (?) (*58*)

   v. A smaller pool of circulating BA (?) (*58*)

   vi. Higher ratio of secondary to primary BA in bile (*59*)

   Because of lessening of male-female differences in sex hormonal profile after the age of female menopause, these biological differences are anticipated to be greater at younger ages. This would be in accord with various studies which have shown the composition of human bile to be influenced by exogenous sex hormones and, in women, by pregnancy and the menstrual cycle (*36*) (*Q*. Do women have higher concentrations of fecal BA than men? Does any such difference vary with age?).

4. 2. Many animal experiments, using varied techniques to increase BA concentrations in the large bowel, have shown an increased yield of carcinogen-induced bowel tumours (see, *e.g.*, refs. *60* and *61*). This effect appears strongest when BA exposure *follows* (rather than precedes or concurs with) administration of the chemical carcinogen.

4. 3. In experimental animals, the production of volatile fatty acids (VFA) by fermentation of fibre (and other diet components, including starch) within the cecum is sensitive to dietary manipulation. Dietary restriction (in rats) causes reduced VFA production, and, therefore, reduced absorption into the hepatic portal vein. The reduced absorption is maximal for butyrate (compared to acetate and propionate). There is a concomitant increase in intracecal pH (Topping, D., personal communication).

4. 4. *Interpretation of 4. 3.:* There is preferential use of butyrate as a metabolic fuel by colonic cells (*62*). Since other data indicate an "antineoplastic" influence of butyrate (*63*), a critical reduction in the availability of butyrate should enhance colon carcinogenesis. Further, the correlation between fibre fermentation (*i.e.*, VFA production) and lowered intra-colonic pH could help account for an increased bacterial production of secondary BA in individuals consuming low fibre diets (*64*) (see also 4. 6. ii. below).

4. 5. Studies of human colostomy subjects indicate that the concentration of VFA's decreases between the proximal and distal colon. These concentrations also vary in response to altered amounts and types of dietary fibre intake (*65*).

4. 6. *Interpretation of 4.5.:* The gradient in intra-colonic VFA concentration (especially butyrate) may contribute to the gradient in CC subsite incidence rates. Also:

   i. Since women appear to have lower capacity for fermentation of dietary fibre (and therefore have lower fecal-bacterial bulk, since fermentation "fuels" bacterial proliferation) their production of VFA may be less (which would accord with their known higher faecal pH). This could help explain their higher CC risk at younger ages, particularly in the right colon.

ii. The VFA concentration gradient might also affect carcinogenesis *via* altered pH, by:
   a. A direct influence on cell metabolism.
   b. The enhanced conversion of primary BA to secondary BA that accompanies an increase in pH (*64*). Secondary BA are colon tumour promoters in experimental animal models.

5. Oncogenes and colon cancer

5. 1. Within the past five years, the role of oncogenes in human carcinogenesis has been extensively defined and clarified. The sequence of this evidence, in summary, has been:
   i. Neoplastic transformation of normal mouse cells by insertion of human tumour cell DNA.
   ii. Elucidation of mechanisms of activation (derepression) of proto-oncogenes —point mutation, proximity to chromosomal breakage points.
   iii. Identification of oncogenes in various human tumours.
   iv. Characterisation of oncogene products—growth factors, cell membrane receptors.

5. 2. Immunohistochemical studies with antibodies to the H-*ras* (oncogene) product have clearly distinguished between many types of benign and malignant lesions in the colon (*66*).

5. 3. *Interpretation:* The likely involvement of oncogenes in all human cancers, and the widely demonstrated mutagenicity of fecal samples and fecal constituents, suggest that proto-oncogene activation within colon mucosa cells proceeds readily. (Indeed, it is also possible that the rich and varied microbial population within the colon would facilitate, *via* plasmid-mediated transfer, DNA recombination within mucosal cells. This, too, could activate proto-oncogenes (Mathews, J. D., personal communication).)
   Early-stage events ("initiation") may therefore be frequent within the colon mucosa, in most if not all populations; this step would therefore not be rate-limiting.

An Overall Interpretation

Now, from all of this evidence it seems plausible that there may be *three* major front-line influences upon colon carcinogenesis:

1. There are early-stage ("initiating") events that entail either mutation or chromosomal breakages, and that switch on the necessary oncogenes. The destiny of the cell is thus re-directed towards immortalisation, with its associated aberrant growth and proliferation. The gene-altering factors responsible are probably widespread within the human diet, and may therefore not be a rate-limiting element. (The role of DNA repair mechanisms, and their relationship to diet, remains unexplored.)
2. There are dietary (and other) factors that result in an enhancement of the later

stages of carcinogenesis ("promotion"). These factors increase the secretion and degradation of BA, and increase the colonic mucosal contact with those metabolites.

3. There are stabilising, or antineoplastic, influences upon the colon mucosal cells. Butyric acid, a major product of the fermentation of fibre and starches within the cecum, is a prime candidate. (Vitamin A may also be involved—although this would be mediated *via* blood retinol concentrations, not enterally.)

Such a formulation is consistent with the available empirical evidence, and with the various specific interpretations offered above. The formulation suggests a final common "metabolic" pathway for the influence of diet upon colon carcinogenesis; indeed, it takes the emphasis off the widely-prevalent "toxicological" model, which seeks specific exogenous carcinogenic factors and which anticipates an approximately linear dose-response model, and offers instead a "metabolic" model. A variety of dietary configurations can influence both the fermentation processes in the large bowel and the metabolic fate of the BA. These two processes are in turn influenced by a range of host characteristics (*35*, *37*).

Additional Reconciliations of Existing Data

This metabolic model suggests the following additional reconciliations of various unexplained empirical observations:

1. The higher rates of CC in young (*i.e.*, pre-menopausal) women than in young men, especially of the right-sided colon (*57*), reflect their lesser capacity for fermentation of fibre and starches—thereby raising pH and increasing the production of secondary BA, while lowering the production of butyrate. A diet high in fat (or indeed in energy density, since the provocation of cholecystokinin release is not exclusive to fat) will increase the amount of BA entering the bowel, and a diet low in fibre (*e.g.*, low in cereal, vegetables, or fruit) will minimise fermentation by denial of substrate to cecal bacteria. Thus, compared to men, young women will be more susceptible to (or will experience at an accelerated rate) the cancer-enhancing effects of such diets.
2. Diets high in *cereal* fibre do not appear to be protective against CC (*13*, *14*, *22*, *25*, *26*, *29*). This may reflect the fact that cereal fibre undergoes less fermentation than many other types of fibre. Although the fibre itself adds physical bulk, thus diluting BA, the production of secondary BA will not be impeded since fermentation-induced lowering of pH does not occur. Neither will much VFA production occur. A high intake of cereal fibre may therefore *not* significantly protect against colon cancer—although the dilution of BA may retard tumor promotion, thus deferring clinical presentation until older ages and thereby producing a paradoxical positive association between high fibre intake and CC risk at older age (*29*).
3. More generally, diets causing minimal intra-cecal fermentation (and therefore minimal increase in the fecal bacterial mass) will result in raised fecal concentrations of not only BA but of all other ingredients, including mutagens. Thus, the raised concentrations of fecal mutagens observed in relation of CC may simply

be a physical index of reduced fecal bulk, and may have no special etiological significance (This argument is advanced on the assumption that mutagens are ever-present within the colon, and therefore not likely to be rate-limiting in colon carcinogenesis).

4. The fact that, at older age, CC rates in men exceed those of women, may represent a selective deferral (*i.e.*, slowing down of carcinogenesis) of male cases to older ages because of a different equilibrium of promoting and retarding influences upon carcinogenesis. Although males appear to produce more BA, their lesser degradation to secondary BA and their dilution by greater fecal bacterial mass may slow down the later stages of colon carcinogenesis in men. That is, it is *not* that males at older ages are exposed to more of the "carcinogen" than are females—rather the time-course of exposure differs. Males thus may manifest diet-induced increases in risk of CC at older ages than do women (*29*). This accords with the descriptive epidemiological observation of an increasing male: female ratio of CC with increasing age (*57*).
5. The rapid change in population risk of CC following a change in dietary "environment," and the results of experimental studies of the effects of varying the timing of administered dietary change or BA, indicate that dietary influences upon fermentation and BA metabolism affect the later stages of colon carcinogenesis. Metabolic phenomena are more likely to be involved in late-stage (probably non-mutagenic) effects.
6. The varied and inconsistent results from case-control studies may reflect the surrogate nature of itemised dietary intake data as a measure of etiological influence. A wide range of dietary profiles could have the same net metabolic effect, because of their equal net influence upon the intra-colonic carcinogenesis promoting and retarding processes.
7. Relatedly, the intake of total dietary energy may be of importance *per se*, independent of its contributory nutrients. Likewise, so might the temporal pattern of freeding. (Our case-control study showed independent effects of these two dietary parameters upon risk of CC (*29*). There are also animal experimental data to support such effects upon colon tumorigenesis—see below.) Within the gastrointestinal tract, the total dietary intake will have both functional and biochemical influences by, for example, effects upon the gastrocolic reflex and the composite secretion of enterohormones and digestive juices. The frequency of feeding determines the frequency of enterohepatic recirculation of BA, which in turn is positively related to the proportion of secondary BA within the total BA pool.

There may also be a systemic (*i.e.*, not intra-colonic) influence of total energy intake upon colon carcinogenesis since there appears to be such an effect on carcinogenesis in general. The evidence for a generalised effect of dietary energy upon carcinogenesis is reviewed below.

Irrespective of whether any effect of dietary energy upon CC risk is local or systemic, it is (again) plausible that the variety of findings from case-control studies reflects population differences in which particular foods or nutrients are the best local index of a high-energy diet.

Dietary Energy and Carcinogenesis

There is epidemiological evidence suggesting that the growth and proliferation of cancer cells, like normal cells, are influenced by hormonal and metabolic processes. Examples include the stimulatory effect of endogenous estrogens upon some breast cancers (and its vitiation by anti-estrogen therapy) (*67*), the increased incidence of thyroid carcinoma associated with increased levels of thyroid stimulating hormone (TSH) (*67*), the generalised increase in cancer risk (including cancers of the breast and endometrium in women) in obese persons (*68*), and, perhaps, the increased risk of colon cancer in persons (men) with naturally low concentrations of blood cholesterol (*39*). The total intake of dietary energy, and the time pattern of its intake, appears to directly influence hormonal, immunological, and general metabolic status. However, relatively little is known of these phenomena.

Experimentally, the pioneering observations upon dietary energy and carcinogenesis were made by Rous in 1914 (*69*). He reported that mice fed a diet restricted in calories developed fewer spontaneous tumours than did mice fed an unrestricted diet. He also found that the nutrient composition of the diet influences tumour yield. Further experiments in the 1930's confirmed the importance of caloric intake upon tumour yield. Subsequently, nutritional influence on spontaneous and induced carcinogenesis became the object of intense study in the 1940's (see, *e.g.*, refs. *70* and *71*). (Interestingly, the role of nutrition in carcinogenesis was eclipsed by other theories of carcinogenesis after 1950, and has only re-emerged strongly during the 1970's.)

From the work of Tannenbaum and others in the 1940's it became apparent that:

1. Underfeeding, directed at achieving a pre-selected weight reduction, reduced tumour incidence. This reduction was commensurate with the weight deficit, and was evident in animals of varied age.
2. Caloric restriction *per se* (irrespective of weight deficit) reduced tumour yield.
3. Caloric restriction during tumour "promotion" caused greater reduction in tumour yield than did restriction during "initiation."
4. Nutrient composition of diet had an independent effect upon tumour yield. One specific finding was that, in several factorial experiments, an increase in dietary energy had approximately twice the tumour-enhancing effect of an increase in dietary fat.
5. These several effects of diet varied considerably with tumour site/type.
6. Studies of induced weight loss by *non*-dietary means indicated that weight loss *per se* reduced tumour yield. (However, a more recent study (*72*) has shown that intestinal tumourigenesis in methylazoxy methanol (MAM)-treated rats is more responsive to the pattern of under-feeding than to restriction of either energy or weight *per se*. While this applies to tumours induced by MAM (which requires metabolic activation) it does not apply to direct-acting methylnitroso urea (MNU) (73).)

In general, experimental research over the past decade (Kritchevsky, D., personal communication) has supported these early findings. Further, some recent epidemiological research (see ref. *29*) has suggested a primary role for dietary energy intake in colon carcinogenesis.

In the meantime, however, more information about the biological effects of dietary energy restriction is needed, particularly in relation to the hormonal or immune status of the organism.

A Broader View: Digesta as "Culture Medium"

As stated above, epidemiological and, particularly, experimental research into colon cancer etiology indicates that dietary factors influence carcinogenesis primarily at stages other than "initiation." The evidence that occurrence of colonic cancer or polyps and the consumption of diets associated with increased risk of colon cancer are accompanied by generalised morphological and histochemical changes in the bowel mucosa suggests that colon cancer may not be simply a focal aberration, but instead may be the "tip of the iceberg" of an altered pattern of cell growth, proliferation, and metabolism throughout the colonic mucosa.

It has long been known that, experimentally, chyme, bile, pancreatic secretions, and secretions of the stomach cause enlargement of the villi in the rat intestine (*74*). More recently, Tasman-Jones (*75*) has shown that components of dietary fibre have specific effects upon the morphology of the rat small bowel: the fingerlike arrangement of the villi in rats fed fibre-free diets approximates the jejunal mucosa observed in adults in Western countries, whereas the more convoluted ridging occurring in rats fed fibre-supplemented diets approximates the mucosa seen in adults in non-Western tropical countries—and in healthy vegetarians in Western society (*76*).

In a study comparing colon mucosal samples from 15 patients with colon cancer and 15 patients without cancer, Shamsuddin *et al.* (*77*) observed that all the cancer patients, but none of the controls, showed definite abnormalities in the mucosa remote from the tumour. The abnormalities included dilatation and distortion of crypts, cellular overcrowding, and increased cellular production of sialomucin (*vs.* the usual sulfomucin).

Bile appears to have a short-term trophic effect upon the bowel mucosa (*78*). Reasoning teleologically, this effect is presumably beneficial in certain circumstances, as in settings where food supply is sporadic or seasonal. Survival would be enhanced by maximising the absorption of nutrients; this would be achieved by a synchronous intermittent mucosal hyperplasia induced by bile. However, in circumstances of a continuous supply of energy-dense food causing, in turn, a continuous stimulus from bile and related enteric secretions, as in Western society, this mucosal hyperplasia may become chronic. This could increase the likelihood of occurrence of premalignant cells, or might increase the rate of progression and proliferation of established neoplastic lesions.

This invites the view that dietary fibre, fat, other dietary factors, and total food intake—and the resultant biochemical characteristics of the intracolonic contents may collectively influence colon cancer risk by acting as a long-term "culture medium" in which the probability of disordered and malignant mucosal growth is altered. This metabolic view of colon carcinogenesis differs qualitatively from earlier prevailing views of human carcinogenesis as being primarily the result of single or multiple "hits" by exogenous, discrete, factors.

REFERENCES

1. Zaridze, D. G. Environmental etiology of large bowel cancer. J. Natl. Cancer Inst., *70*: 389–400, 1983.
2. Boyle, P., Zaridze, D. G., and Smans, M. Descriptive epidemiology of colorectal cancer. Int. J. Cancer, *36*: 9–18, 1985.
3. IARC Intestinal Microecology Group. Dietary fibre, transit time, faecal bacteria, steroids, and colon cancer in two Scandinavian populations. Lancet, *2*: 207–211, 1977.
4. McMichael, A. J., Potter, J. D., and Hetzel, B. S. Time trends in colorectal cancer mortality in relation to food and alcohol consumption: United States, United Kingdom, Australia and New Zealand. Int. J. Epidemiol., *8*: 295–303, 1979.
5. Phillips, R. L., Garfinkel, L., Kuzma, J. W., Beeson, W. L., Lotz, T., and Brin, B. Cancer mortality among California Seventh-day Adventists for selected cancer sites. J. Natl. Cancer Inst., *65*: 1087–1107, 1980.
6. McMichael, A. J., McCall, M. G., Hartshorne, J. M., and Woodings, T. L. Patterns of gastrointestinal cancer in European migrants to Australia: the role of dietary change. Int. J. Cancer, *25*: 431–437, 1980.
7. Powles, J. W. and Williams, D.R.R. Trends in bowel cancer in selected countries in relation to wartime changes in flour milling. Nutr. Cancer, *6*: 40–48, 1984.
8. Armstrong, B. and Doll, R. Environmental factors and cancer incidence and mortality in different countries, with special reference to dietary practices. Int. J. Cancer, *15*: 617–631, 1975.
9. McKeown-Eyssen, G. and Bright-See, E. Relationship between colon cancer mortality and fibre consumption: an international study (abstract). *In;* G. Wallace and L. Bell (eds.), Fibre in Human and Animal Nutrition, Proceedings of a symposium, New Zealand, May 1982, p. 35, Royal Society of New Zealand, Wellington, 1983.
10. McMichael, A. J. and Potter, J. D. Reproduction, endogenous and exogenous sex hormones, and colon cancer: a review and hypothesis. J. Natl. Cancer Inst., *65*: 1201–1207, 1980.
11. Stocks, P. Cancer incidence in North Wales and Liverpool region in relation to habits and environment. Brit. Emp. Cancer Campaign 35th Annual Report Suppl. to part 2, pp. 1–127, London, 1957.
12. Pernu, J. An epidemiological study on cancer of the digestive organs and respiratory system. A study based on 7078 cases. Ann. Med. Intern. Fenn., *49* (Suppl. 3): 1–117, 1960.
13. Higginson, J. Etiological factors in gastrointestinal cancer in man. J. Natl. Cancer Inst., *37*: 527–545, 1966.
14. Wynder, E. L. and Shigematsu, T. Environmental factors of cancer of the colon and rectum. Cancer, *20*: 1520–1561, 1967.
15. Wynder, E. L., Kajitani, T., Ishikana, S., Dodo, H., and Takano, A. Environmental factors of cancer of the colon and rectum II. Japanese epidemiological data. Cancer, *23*: 1210–1220, 1969.
16. Haenszel, W., Berg, J. W., Segi, M., Kurihara, M., and Locke, F. B. Large bowel cancer in Hawaiian Japanese. J. Natl. Cancer Inst., *51*: 1765–1779, 1973.

17. Bjelke, E. Epidemiological studies of cancer of the stomach, colon and rectum. vol. III, Case-control study of gastrointestinal cancer in Norway. vol. IV, Case-control study of digestive tract cancers in Minnesota. Ann Arbor Univ. Microfilms, 1973.
18. Phillips, R. Role of lifestyle and dietary habits in risk of cancer among Seventh-day Adventists. Cancer Res., *35*: 3513–3522, 1975.
19. Modan, B., Barell, V., Lubin, F., Modan, M., Greenberg, R. A., and Graham, S. Low-fiber intake as an etiologic factor in cancer of the colon. J. Natl. Cancer Inst., *55*: 15–18, 1975.
20. Graham, S., Dayal, H., Swanson, M., Mittelman, A., and Wilkinson, G. Diet in the epidemiology of cancer of the colon and rectum. J. Natl. Cancer Inst., *61*: 709–714, 1978.
21. Dales, L. G., Friedman, G. D., Ury, H. K., Grossman, S., and Williams, S. R. A case-control study of relationships of diets and other traits to colorectal cancer in American blacks. Am. J. Epidemiol., *109*: 132–144, 1979.
22. Martinez, I., Torres, R., Frias, Z., Colon, J. R., and Fernandez, N. Factors associated with adenocarcinomas of the large bowel in Puerto Rico. *In;* Advances in Medical Oncology, Research and Education, vol. III, pp. 45–52, Pergamon, Oxford, 1979.
23. Haenszel, W., Locke, F. B., and Segi, M. A case-control study of large bowel cancer in Japan. J. Natl. Cancer Inst., *64*: 17–22, 1980.
24. Jain, M., Howe, G. R., Johnson, K. C., and Miller, A. B. Evaluation of a diet history questionnaire for epidemiologic studies. Am. J. Epidemiol., *111*: 212–219, 1980.
25. Miller, A. B., Howe, G. R., Jain, M., Craib, K.J.P., and Harrison, L. Food items and food groups as risk factors in a case-control study of diet and colorectal cancer. Int. J. Cancer, *32*: 155–161, 1983.
26. Manousos, O., Day, N. E., Trichopoulos, D., Gerovassilis, F., Tzonou, A., and Polychronopoulou, A. Diet and colorectal cancer: a case-control study in Greece. Int. J. Cancer, *32*: 1–5, 1983.
27. Berta, J.-L., Coste, T., Rautureau, J., Guilloud-Bataille, M., and Péquignot, G. Alimentation et cancers rectocoliques. Résultats d'une étude "cas-témoin." Gastroenterol. Clin. Biol., *9*: 348–353, 1985.
28. Macquart-Moulin, G., Riboli, E., Cornée, J., Charnay, B., Berthezène, P., and Day, N. Case-control study on colorectal cancer and diet in Marseilles. Int. J. Cancer (in press.)
29. Potter, J. D. and McMichael, A. J. Diet and cancer of the colon and rectum. A case-control study. J. Natl. Cancer Inst. *76*: 557–569, 1986.
30. Hirayama, T. A large-scale cohort study on the relationship between diet and selected cancers of the digestive organs. *In;* W. R. Bruce, et al. (eds.), Gastrointestinal Cancer: Endogenous Factors. Banbury Report No. 7, pp. 409–429, Cold Spring Harbor Laboratory, New York, 1981.
31. Stemmermann, G. N., Nomura, A.M.Y., and Heilbrun, L. K. Dietary fat and the risk of colorectal cancer. Cancer Res., *44*: 4633–4637, 1984.
32. Phillips, R. L. and Snowdon, D. A. Dietary relationships with fatal colorectal cancer among Seventh-day Adventists. J. Natl. Cancer Inst., *74*: 307–317, 1985.
33. Smith, A. H., Pearce, N. E., and Joseph, J. G. Major colorectal cancer aetiological hypotheses do not explain mortality trends among Maori and non-Maori New Zealanders. Int. J. Epidemiol., *14*: 79–85, 1985.
34. McMichael, A. J. and Potter, J. D. Diet and colon cancer: integration of the descriptive, analytic and metabolic epidemiology. NCI Monogr., *69*: 223–228, 1985.
35. McMichael, A. J. and Potter, J. D. Host factors in carcinogenesis: certain bile-acid

metabolic profiles that selectively increase the risk of proximal colon cancer. J. Natl. Cancer Inst., *75*: 185–191, 1985.

36. Potter, J. D. and McMichael, A. J. Large bowel cancer in women in relation to reproductive and hormonal factors: a case-control study. J. Natl. Cancer Inst., *71*: 703–709, 1983.
37. Davidson, M., Yoshizawa, C. N., and Kolonel, L. N. Do sex hormones affect colorectal cancer? Br. Med. J., *290*: 1868, 1985.
38. Stemmermann, G. N., Nomura, A.M.Y., Heilbrun, L. K., Pollack, E. S., and Kagan, A. Serum cholesterol and colon cancer incidence in Hawaiian Japanese men. J. Natl. Cancer Inst., *67*: 1179–1182, 1981.
39. McMichael, A. J., Jensen, O. M., Parkin, D. M., and Zaridze, D. G. Dietary and endogenous cholesterol and human cancer. Epidemiol. Rev., *6*: 192–216, 1984.
40. Tutton, P.J.M. and Barkla, D. H. Differential effects of oestrogenic hormones on cell proliferation in the colonic epithelium and in colonic carcinomata of rats. Anticancer Res., *2*: 199–202, 1982.
41. Hill, M. J., Drasar, B. S., Williams, R.E.O., Meade, T. W., Cox, A. G., Simpson, J.E.P., and Morson, B. C. Faecal bile-acids and clostridia in patients with cancer of the large bowel. Lancet, *1*: 535–539, 1975.
42. Murray, W. R., Blackwood, A., Trotter, J. M., Calman, K. C., and MacKay, C. Faecal bile acids and clostridia in the aetiology of colorectal cancer. Br. J. Cancer, *41*: 923–928, 1980.
43. Reddy, B. S. and Wynder, E. L. Large bowel carcinogenesis—fecal constituents of populations with diverse incidence rates of colon cancer. J. Natl. Cancer Inst., *50*: 1437–1442, 1973.
44. Hill, M. J. and Aries, V. C. Faecal steroid composition and its relationship to cancer of the large bowel. J. Pathol., *104*: 129–139, 1971.
45. Reddy, B. S., Mastromarino, A., and Wynder, E. L. Further leads on metabolic epidemiology of large bowel cancer. Cancer Res., *35*: 3403–3406, 1975.
46. Mastromarino, A., Reddy, B. S., and Wynder, E. L. Fecal profiles of anaerobic microflora of large bowel cancer patients and patients with nonhereditary large bowel polyps. Cancer Res., *38*: 4458–4462, 1978.
47. Goldin, B. R., Swenson, L., Dwyer, J., Sexton, M., and Gorbach, S. L. Effect of diet and lactobacillus acidophilus supplements on human fecal bacterial enzymes. J. Natl. Cancer Inst., *64*: 255–261, 1980.
48. Jensen, O. M., MacLennan, R., and Wahrendorf, J. Diet, bowel function, faecal characteristics and large bowel cancer in Denmark and Finland. Nutr. Cancer, *4*: 5–19, 1982.
49. Erich, M., Ashell, J. E., Van Tassell, R. L., Wilkins, T. D., Walker, A.R.P., and Richardson, N. J. Mutagens in the feces of three South African populations at different levels of risk for colon cancer. Mutat. Res., *64*: 231–240, 1979.
50. Reddy, B. S., Sharma, C., Darby, L., Laakso, K., and Wynder, E. L. Metabolic epidemiology of large bowel cancer. Fecal mutagens in high- and low-risk population for colon cancer. A preliminary report. Mutat. Res., *72*: 511–522, 1980.
51. Nader, C. J., Potter, J. D., and Weller, R. Diet and DNA-modifying activity in human fecal extracts. Nutr. Rep. Int., *23*: 113–117, 1981.
52. Gupta, I., Suzuki, K., Bruce, W. R., Krepinski, J. J., and Yates, P. A model study of faecapentaenes: mutagens of bacterial origin with alkylating properties. Science, *225*: 521–523, 1984.
53. Cummings, J. H., Wiggins, H. S., Jenkins, D.J.A., Houston, H., Jivraj, T., Drasar,

B. S., and Hill, M. S. Influence of diets high and low in animal fat on bowel habit, gastrointestinal transit time, fecal microflora, bile acid and fat excretion. J. Clin. Invest., *61*: 953–962, 1978.
54. Walters, R. L., McLean Baird, I., Davies, P. S., Hill, M. J., Drasar, B. S., Southgate, D.A.T., Green, J., and Morgan, B. Effects of two types of dietary fibre on faecal steroid and lipid excretion. Br. Med. J., *2*: 536–538, 1975.
55. Ullrich, I. H., Lai, H.-Y., Vona, L., Reid, R. L., and Albrink, M. J. Alterations of fecal steroid composition induced by changes in dietary fiber consumption. Am. J. Clin. Nutr. *34*: 2054–2060, 1981.
56. Stasse-Walthuis, M., Albers, H.F.F., Van Jeveren, J.G.C., De Jong, J. W., Hautvast, J.G.A.J., Hermus, R.J.J., Katan, M. B., Brydon, W. G., and Eastwood, M. A. Influence of dietary fiber from vegetables and fruits, bran or citrus pectin on serum lipids, fecal lipids, and colonic function. Am. J. Clin. Nutr. *33*: 1745–1756, 1980.
57. McMichael, A. J., and Potter, J. D. Do intrinsic sex differences in lower alimentary tract physiology influence the sex-specific risks of bowel cancer and other biliary and intestinal diseases? Am. J. Epidemiol., *118*: 620–627, 1983.
58. Einarsson, K., Nilsell, K., Leijd, B., and Angelin, B. Influence of age on secretion of cholesterol and synthesis of bile acids by the liver. N. Engl. J. Med., *313*: 277–282, 1985.
59. Fisher, M. M. and Yousef, I. M. Sex differences in the bile acid composition of human bile: studies in patients with and without gallstones. Can. Med. Assoc. J., *109*: 190–193, 1973.
60. Reddy, B. S., Watanabe, K., Weisburger, J. H., and Wynder, E. L. Promoting effect of bile acids in colon carcinogenesis in germ-free and conventional F344 rats. Cancer Res., *37*: 3238–3242, 1977.
61. Williamson, R.C.N. and Rainey, J. B. The relationship between intestinal hyperplasia and carcinogenesis. Scand. J. Gastroenterol., *19* (suppl. 104): 57–76, 1984.
62. Roediger, W.E.W. The effect of bacterial metabolites on nutrition and function of the colonic mucosa: symbiosis between man and bacteria. *In;* H. Kasper and H. Goebell (eds.), Colon and Nutrition, pp. 11–24, MTP Press Ltd., Lancaster, 1982.
63. Sakata, T. and Yajima, T. Influence of short chain fatty acids on the epithelial cell division of digestive tract. Quart. J. Exp. Physiol., *69*: 639–648, 1984.
64. Thornton, J. R. High colonic pH promotes colorectal cancer. Lancet, *1*: 1081–1083, 1981.
65. Mitchell, B. L., Lawson, M. J., Davies, M., Kerr Grant, A., Roediger, W.E.W., Illman, R. J., and Topping, D. L. Volatile fatty acids in the human intestine: studies in surgical patients. Nutr. Res., *5*: 1089–1092, 1985.
66. Thor, A., Hand, P. H., Wunderlich, D., Caruso, A., Muraro, R., and Schlom, J. Monoclonal antibodies define differential ras gene expression in malignant and benign colonic diseases. Nature, *311*: 562–565, 1984.
67. Armstrong, B. K. Endocrine factors in human carcinogenesis. *In;* H. Bartsch and B. Armstrong (eds.), Host Factors in Human Carcinogenesis, IARC Scientific Publ. No. 39, pp. 193–221, IARC, Lyon, 1982.
68. Lew, E. A. and Garfinkel, L. Variations in mortality by weight among 750,000 men and women. J. Chron. Dis., *32*: 563–576, 1979.
69. Rous, P. The influence of diet on transplanted and spontaneous tumors. J. Exp. Med., *20*: 433–451, 1914.
70. Tannenbaum, A. The initiation and growth of tumours. Introduction I. Effects of underfeeding. Am. J. Cancer, *38*: 335–350, 1940.

71. Tannenbaum, A. The dependence of tumor formation on the degree of caloric restriction. Cancer Res., *5*: 609–615, 1945.
72. Pollard, M., Luckert, P. H., and Pan, G.-Y. Inhibition of intestinal tumorigenesis in MAM-treated rats by dietary restriction. Cancer Treat. Rep., *68*: 405–408, 1984.
73. Pollard, M. and Luckert, P. H. Tumorigenic effects of direct- and indirect-acting chemical carcinogens in rats on a restricted diet. J. Natl. Cancer Inst., *74*: 1347–1349, 1985.
74. Altmann, G. G. Influence of bile and pancreatic secretions on the size of intestinal villi in the rat. Am. J. Anat., *132*: 167–178, 1971.
75. Tasman-Jones, C. Effects of dietary fibre in the structure and function of the small intestine. *In;* G. A. Spiller and R. M. Kay (eds.), Medical Aspects of Dietary Fiber. pp. 67–74, Plenum Press, New York, 1980.
76. Owen, R. L. and Brandborg, L. L. Jejunal morphologic consequences of vegetarian diet in humans. Gastroenterology, *72*: A88–A111, 1977.
77. Shamsuddin, A.K.L., Weiss, L., Phelps, P. C., and Trump, F. Colon epithelium IV. Human colon carcinogenesis. Changes in human colon mucosa adjacent to and remote from carcinomas of the colon. J. Natl. Cancer Inst., *66*: 413–419, 1981.
78. Williamson, R.C.N. Intestinal adaptation. Mechanisms of control. N. Engl. J. Med., *298*: 1444–1450, 1978.

DIET, NUTRITION AND CANCER, Y. HAYASHI ET AL. (EDS.),
JAPAN SCI. SOC. PRESS, TOKYO/VNU SCI. PRESS, UTRECHT, PP. 291–294, 1986

The Effect of Calcium on the Pathogenicity of High Fat Diets to the Colon

W. R. BRUCE, R. P. BIRD, and J. J. RAFTER

Ludwig Institute for Cancer Research, Toronto Branch, Departments of Medical Biophysics and Nutrition, University of Toronto, Toronto, Ontario M4Y IM4, Canada

Abstract: The pathogenicity of lipid components and dietary fats on the colonic epithelium have been studied with five model systems in experimental animals, with rectal perfusion of bile acids, colonic perfusion of bile acid solutions, dietary supplementation with cholic acid, oral boluses of fat, and diets with various levels of fats. The lipid or fat led to colonic epithelial cell cytotoxicity and/or an increase in cell proliferation which was inhibited by supplementary calcium. These results could mean that calcium may reduce the toxicity of high fat diets to the colon and reduce the colonic cancer risk associated with high fat diets.

It has become evident that dietary fat can be toxic to the colonic epithelium and that dietary calcium can significantly reduce this pathogenicity. Here we will review our studies of the interaction of fat and calcium on the colonic epithelium and suggest that the interaction may explain the correlations of diets high in fat with colonic cancer risk.

Intrarectal Instillation of Bile Acids

In our first study of the effects of lipids on the colonic epithelium, sodium deoxycholic acid was administered intrarectally at a concentration approximating the concentration of bile acid in the fecal stream. The degree of epithelial damage was intense and resembled that seen with EDTA (*1*). It was subsequently demonstrated that calcium, as the lactate or carbonate salt, was able to inhibit the damage when it was given orally shortly before the bile salt (*2*). Similar results were observed with the salts of fatty acids (*3*). These results prompted us to define a hypothesis for colon cancer causation which proposed that the bile and fatty acids in the "ionized" form were responsible for the disease and that the oral calcium produced its effect by leading to the formation of insoluble calcium soaps which reduced the concentration of the toxic soluble salts (*4*).

Colonic Perfusion Studies

The effect of the bile acids can be examined in a more defined way with the use of a colon perfusion model (*5*). Groups of rats were perfused with graded doses of the bile acid, deoxycholic acid, for 80 min and the degree of damage was quantitated by cell morphometry and by measuring the degree of loss of DNA in the perfusion stream. A marked effect of 2 mM of the bile acid was evident. When a 4 mM concentration was used the damage was extensive. This damage was however inhibited if the bile acid was mixed with a calcium salt solution prior to the perfusion (*6*). Under this condition the concentration of the bile acid in solution was greatly reduced and the damage reflected the concentration of the solution and not the total mass of the deoxycholate in the perfusion fluids.

Dietary Cholic Acid

While the perfusion studies allow a detailed analysis of the effects of the bile acids, they are quite removed from the physiological exposure. As a step closer to the reality of dietary regimens we utilised the oral cholic acid model (*3*). In this system cholic acid added at a low concentration to the diet, results in an increase in the concentration of deoxycholic acid in the fecal stream, an increase in the proliferation rate in colonic epithelial cells and an increase in the yield of colon tumors in animals given colon carcinogens (*7*, *8*). We determined the effect of dietary calcium on this model (*9*). Laboratory animal diets are usually formulated to contain in excess of 0.5% calcium by weight. We examined the effects of synthetic diets with from 0.1 to 1.0% calcium, that is the levels more closely approximating the levels in human diets to the levels seen commonly in animal chows were examined. At the lower dietary calcium levels the colonic epithelium was extremely sensitive to dietary cholic acid (*1*) and proliferation was markedly increased at cholic acid levels as low as 0.1%. This degree of proliferation was reduced to the normal rate when higher levels of calcium were incorporated in the diet mix. These results showed that dietary calcium could have a significant effect on the effects of bile acids in feeding studies.

Oral Boluses of Fat

Oral boluses of fat provided another method for examining the action of fat and calcium on the colon. In the first experiments mice were given oral boluses of 0.4 ml of beef tallow (*10*). Two hours later there was a substantial loss of cells from the surface of the colonic epithelium and 12 hr later there was a compensatory wave of mitotic activity at the base of the crypts. By 24 hr after the bolus the damage was repaired and there was no evidence of the exposure to fat. Similar though less pronounced effects were observed following 0.1 and 0.2 ml of beef tallow after corn oil. These first experiments were carried out with mice consuming a diet with about 1% calcium. Subsequent experiments repeated on animals fed lower levels of dietary calcium showed that low dietary calcium could augment the effects of the oral fat (manuscript in preparation). The damage evident in animals receiving 0.1% calcium was substantially greater than that seen in animals receiving 1% calcium.

Dietary Fat

Our dietary fat intake is in the form of meals that may resemble boluses, but fat is more usually found with other components in the diet and not just by itself. Consequently we began to examine the effect of defined diets that differed in their contents of fat and calcium. Although these studies have not been carried out over long periods of time or over many different dietary conditions, it is already clear that animals on high levels of dietary fat have a higher proliferation rate than those on a low level. Presumably the fecal stream for animals on high fat diets contains lipid components such as bile acids in solution that are cytotoxic to the colonic epithelium and result in a compensatory hyperplasia.

The Human Fecal Stream

The effect of dietary fat and calcium on the concentration of bile acids in solution in the human fecal stream can be examined in humans. A preliminary study has been carried out with twenty volunteers. They were randomized into two groups. One group was given a low fat, high calcium diet, while the other a high fat, low calcium Western diet. Prior to the study and at the end of a four day diet, the volunteers collected fecal samples that were analysed for their levels of bile acids. The level of bile acids in the low fat group dropped to an average of 0.2 mM, while that in the high fat group increased to 0.5 mM, with a range from 0.15 to 0.8 (*11*). Studies are now being carried out over longer periods of time and with independent control of fat, calcium, and fiber levels.

DISCUSSION

These observations taken together suggest that many potentially toxic compounds are formed within the intestinal lumen during the digestion of dietary fat. These compounds could include fatty acids, mono and diglycerides, lysophospholipids as well as the deconjugated bile acids such as deoxycholic acid. Most of the lipids involved in digestion are absorbed in the small bowel, some remain in the feces in the solid phase, but small but measurable amounts appear in the aqueous phase in the colon. These soluble lipids pose a risk to the colonic epithelium. Their detergent activity reduces the protection of the mucin layer and disorganizes the membranes of the surface cells. The loss of these surface cells results in a compensatory proliferation of the cells in the crypts as the damage produced by the lipids is repaired.

How do these results bear on the origin of colon cancer and on the relation between dietary fat and colon cancer incidence? High fat, low calcium diets likely lead to an increase in the proliferation of cells in the crypts. Many studies have suggested that such an increase in proliferation leads to an increased risk for carcinogenesis.

High fat, low calcium diets also likely lead to an increased permeability of the epithelium and a loss in the integrity of the mucosal barrier. A reduction of this protection could lead to an increase in the exposure of the stem cells of the crypt to toxic chemicals in the fecal stream. Thus we might expect that high fat, low calcium diets would increase the sensitivity of the colon to chemical carcinogens although the importance of these mechanisms to colon carcinogenesis remains to be determined. If they are correct, an important biochemical risk factor for colon cancer is likely to be the concentration of bile acids in solution in the fecal stream. The important dietary factors are likely to be high fat, low calcium and perhaps also high phosphate and low fibre diets. The importance of the mechanisms can thus be tested with ecological and case control studies. If these add support to the hypothesis, intervention studies of calcium supplementation on colon cancer incidence might be initiated.

REFERENCES

1. Cassidy, M. M. and Tidball, C. S. Cellular mechanism of intestinal permeability alterations produced by chelation depletion. J. Cell Biol., *32*: 685–698, 1970.
2. Wargovich, M. J., Eng, V.W.S., Newmark, H. L., and Bruce, W. R. Calcium ameliorates the toxic effect of deoxycholic acid on colonic epithelium. Carcinogenesis, *4*: 1205–1207, 1983.
3. Wargovich, M. J., Eng., V.W.S., and Newmark, H. L. Calcium inhibits the damaging and compensatory effects of fatty acids on mouse colon epithelium. Cancer Lett., *23*: 253–258, 1984.
4. Newmark, H. L., Wargovich, M. J., and Bruce, W. R. Colon cancer and dietary fat, phosphate and calcium: a hypothesis. J. Natl. Cancer Inst., *72*: 1323–1325, 1984.
5. Mekhjian, H. S. and Phillips, S. F. Perfusion of the canine colon with unconjugated bile acids. Effect on water and electrolyte transport, morphology and bile acid absorption. Gastroenterology, *69*: 380–386, 1975.
6. Rafter, J., Eng, V.W.S., Furrer, R., Medline, A., and Bruce, W. R. The effect of calcium and pH on the mucosal damage produced by deoxycholic acid in the rat colon. Gut (in press).
7. Deschner, E. E., Cohen, B. I., and Raicht, R. F. Acute and chronic effect of dietary cholic acid on colonic epithelial cell proliferation. Digestion, *21*: 290–296, 1981.
8. Cohen, B. I., Raicht, R. F., Deschner, E. E., Takahashi, M., Sarwal, A. N., and Fazzini, E. Effect of cholic acid feeding on N-methyl-N-nitrosourea induced colon tumors and cell kinetics in rats. J. Natl. Cancer Inst., *64*: 573–578, 1980.
9. Bird, R. P. Effect of dietary components on the pathobiology of colonic epithelium: possible relationship with colon tumorigenesis. Lipids, *21*: 289–291, 1986.
10. Bird, R. P., Medline, A., Furrer, R., and Bruce, W. R. Toxicity of orally administered fat to the colonic epithelium of mice. Carcinogenesis, *6*: 1063–1066, 1985.
11. Rafter, J. J., Child, P., Anderson, A. M., Alder, R., Env, V., and Bruce, W. R. Cellular toxicity of fecal water depends on diet (submitted for publication).

DIET, NUTRITION AND CANCER, Y. HAYASHI ET AL. (EDS.),
JAPAN SCI. SOC. PRESS, TOKYO/VNU SCI. PRESS, UTRECHT, PP. 295-303, 1986

Risk Evaluation of Tumor-Inducing Substances in Foods

Yuzo HAYASHI, Yuji KUROKAWA, and Akihiko MAEKAWA

Department of Pathology, National Institute of Hygienic Sciences, Tokyo 158, Japan

Abstract: Carcinogenic risk assessment of chemicals consists of four phases, namely, 1) hazard identification, 2) exposure assessment, 3) hazard assessment or dose-response assessment, and 4) risk characterization. The third phase of risk assessment is the evaluation of both hazard and exposure information to estimate the mathematical probability that the carcinogenic potential associated with an agent will be realized in the human population under defined conditions of exposure. The estimation of virtually safe dose (VSD) is regarded as a component of the third phase. Carcinogenic risk assessment is still at an embryonal stage of development, and there remains a number of problems to be clarified. With regard to the estimation of VSD, it is an important task to establish the principle for selection of appropriate mathematical models. The intention of this paper is to illustrate the estimation of VSD of two tumor-inducing chemicals, N-ethyl-N-nitrosourea and potassium bromate based on the dose-response data in animals and to discuss the biological implication of the estimated VSD in relation to the risk assessment of the chemicals in humans.

Motivated by the epidemiological evidence suggesting that the majority of human cancers result from exposure to environmental carcinogens, an international cooperative project aiming at the primary prevention of cancers has been in progress. A number of chemicals have already been tested for potential carcinogenicity in animals. With accumulation of the test data, it has become apparent that for assessment of carcinogenic risk in humans one sometimes needs additional information concerning the mechanism of action and potency of the test chemicals. With regard to the mechanistic issues, extensive studies are now being conducted in various countries to establish a scientific basis for classifying chemical carcinogens into 3 categories: genotoxic (primary) carcinogens, epigenetic (secondary) carcinogens, and promoters. As an approach to quantitative risk assessment of carcinogens, attempts have been made to estimate the virtually safe dose (VSD) or doses of extremely low risk by the downward extrapolation of animal dose-response data (*1*). The intention of the present paper is to illustrate the estimation of VSDs of some carcinogenic chemicals based on the dose-response data in animals and to discuss the biological implication of the estimated VSDs in relation to the risk assessment of the chemicals in humans.

Dose-response Study of N-Ethyl-N-nitrosourea in F344 Rats

Since the discovery of the carcinogenic action of dimethylnitrosamine (DMNA) (*2*), over a hundred nitrosamines and other N-nitroso compounds have been found to be carcinogenic in experimental animals. Epidemiological evidence suggests that nitrosamines or other nitroso compounds may be involved in the causation of human bilharzial bladder cancer (*3*), nasopharyngeal cancer in Hong Kong (*4*), and esophageal cancer in Northern China (*5*). Carcinogenic nitroso compounds are also known to be strongly mutagenic in various *in vitro* or *in vivo* short-term test systems, thus demonstrating the compounds to be typical of genotoxic carcinogens.

Several investigators reported the widespread occurrence of low levels (ppb) of nitrosamines in a variety of human contact sources (*6*) and in human blood (*7*). The evidence raises the question of whether long-term exposure to such small amounts of nitroso compounds can contribute to cancer risk. In this context, we selected N-ethyl-N-nitrosourea (ENU) as a test compound, and a dose-response carcinogenicity study was conducted to estimate its VSD (*8*).

Eight week-old F344/DuCrj rats were divided into 5 groups of 52 males and 52 females each. ENU was dissolved in distilled water at concentrations of 0, 0.3, 1, 3, and 10 ppm, and rats had access to 20 ml of these solutions/rat/day from a light-proof plastic bottle in place of drinking water for 2 years. The first rat with tumors was autopsied in week 41. Therefore, all rats that survived beyond this week were included in effective numbers. From the daily water consumption, the mean daily intakes of ENU/rat were calculated to be as follows: group 1 (0.3 ppm)-males 0.006 mg, females 0.0045 mg; group 2 (1 ppm)-males 0.02 mg, females 0.015 mg; group 3 (3 ppm)-males 0.06 mg, females 0.045 mg; group 4 (10 ppm)-males 0.2 mg, females 0.15 mg. Table 1 shows the incidences of tumors and mean survival times of

TABLE 1. Incidences of Tumors and Mean Survival Times of F344 Rats Treated with Low Doses of N-Ethyl-N-nitrosourea

Sex	Dose (ppm)	No. of rats		No. of rats with tumors (%)	Mean survival time (weeks)
		Initial	Effective		
Male	0	52	51	51 (100)	104 (70–111)
	0.3	52	52	51 (98)	103 (63–111)
	1.0	52	52	52 (100)	98 (61–111)
	3.0	52	51	51 (100)	94 (61–111)
	10.0	52	52	52 (100)	69 (41–87)
Female	0	52	52	38 (73)	103 (57–111)
	0.3	52	51	39 (76)	102 (63–111)
	1.0	52	52	44 (85)	104 (65–111)
	3.0	52	50	46 (92)*	95 (49–111)
	10.0	52	50	50 (100)**	77 (53–103)

* $p<0.05$; ** $p<0.01$ (χ^2 test).

rats in each group. An inverse dose-effect relation with the mean survival time was observed in both male and female groups. The incidence of tumors was about 100% in male groups including the control group and 70–100% in female groups. Most of these tumors were shown to occur in the testis, mammary gland, uterus, hematopoietic organs, lung, pituitary, thyroid, or adrenal, and their incidence were not significantly different between the treated groups and the control group.

In contrast, tumors of the nervous system and digestive tract were observed in ENU-treated groups of both sexes, and their incidences appeared to be dose-dependent (Table 2). The tumors of the nervous system which were found in the brain were mainly gliomas, whereas tumors in peripheral nerves were benign or malignant neurinomas. Of the gliomas, oligodendrogliomas were the most frequent, although a few astrocytomas, mixed gliomas, and glioblastomas were also observed. In the digestive tract, tumors were not restricted to the small intestine but were distributed along

TABLE 2. Incidence of Tumors in Nervous System and Digestive Tract in F344 Rats Treated with Low Concentrations of N-Ethyl-N-nitrosourea in Drinking Water

Location of tumors	Concentration of ENU (ppm)	No. of male rats		No. of female rats	
		Examined	With tumors	Examined	With tumors
Nervous system	0	51	0	52	0
	0.3	52	2	51	0
	1.0	52	2	52	0
	3.0	51	4	50	2
	10.0	52	18	50	20
Digestive tract	0	51	1	52	2
	0.3	52	2	51	1
	1.0	52	1	52	1
	3.0	51	4	50	2
	10.0	52	14	50	8

TABLE 3. Estimated VSD for Nervous and Digestive Tract Tumors in F344 Rats at a Risk Level of 10^{-6}

Location of tumors	Sex		Ind. Weibull[a]	Ind. Logit[b]
Nervous system	Male	Chi-square value		2.08
		p-value		0.35
		VSD[c]		0.935×10^{-3}
	Female	Chi-square value	0.22	0.18
		p-value	0.89	0.91
		VSD[c]	0.289×10^{-1}	0.404×10^{-1}
Digestive tract	Male	Chi-square value	0.76	0.74
		p-value	0.68	0.69
		VSD[c]	0.214×10^{-2}	0.332×10^{-2}
	Female	Chi-square value	0.54	0.53
		p-value	0.77	0.77
		VSD[c]	0.343×10^{-1}	0.376×10^{-1}

[a] Weibull model with independent background. [b] Logit model with independent background. [c] Dose (ppm) in the drinking water.

the whole digestive tract from the oral cavity to the colon. Histologically, tumors of the upper digestive tract were mainly papillomas or squamous cell carcinomas, and others were adenomas, adenocarcinomas, or mucinous carcinomas. In contrast to findings from a previous experiment using a high concentration of ENU (400 ppm) (*9*), myelogenic leukemia did not appear in this experiment.

On the basis of the dose-response data mentioned above, the VSD of ENU was estimated by application of 2 mathematical models, the Weibull model with independent background and the logit model with independent background. The estimated values at a risk level of 10^{-6} are shown in Table 3. The chi-square test indicated that both mathematical models fitted the experimental data well except for the tumors of the nervous system in the male groups. In this case, the Weibull model suited the dose-response data poorly and therefore the estimated value was excluded from Table 3.

Low dose effects of nitrosamines have been studied in several laboratories, mostly using liver tumors as the end point. Diethylnitrosamine (DENA) in the drinking water at a dosage of 0.075 mg/kg/day induced liver tumors in rats (*10*). Lijinsky *et al.* showed that DENA at a dose of 0.45 ppm in the drinking water resulted in significant occurrence of tumors in the liver and upper digestive tract in rats after 60 weeks or more of treatment (*11*). Terracini *et al.* reported that DMNA incorporated in the diet caused a small number of liver tumors after a 2-year exposure of rats to 2 ppm (*12*). Based on the dose-response data by Terracini, the VSD of DMNA at a risk level of 10^{-6} was calculated to be as follows: 1.9×10^{-2} ppm by the Weibull model and 7.7×10^{-2} ppm by the Multi-hit model (*13*). Compared with these data, the carcinogenic potency of ENU appears to be slightly stronger than that of DMNA and almost similar to that of DENA although the target organs are different among the compounds.

Dose-response Study of Potassium Bromate in F344 Rats

Potassium bromate ($KBrO_3$) has been widely used in the world as a food additive mainly in the bread-making process, utilizing its oxidizing properties. Previous results from a battery of short-term tests, including the Ames test, chromosome aberration test and micronucleous test, revealed this compound to be genotoxic. Subsequently Kurokawa *et al.* have conducted the carcinogenicity test for this compound in male and female F344 rats using oral administration at doses of 500 and 250 ppm in the drinking water for 110 weeks (*14*). As a result, significantly higher incidence of renal cell tumors in both sexes given 500 ppm and 250 ppm were observed. Also, weak carcinogenicity of potassium bromate in male Syrian golden hamsters, as evidenced by induction of renal cell tumors was noted after oral administration for 89 weeks at concentrations in the range of 2,000 and 125 ppm (*15*). At present, the mode of action of potassium bromate in renal carcinogenesis is unknown, but the oxidizing properties of this compound are thought to be responsible, at least in part, for its carcinogenic effects. Considering its widespread use and unique biological activity, it seemed important and informative to conduct the dose-response carcinogenicity studies at low concentrations for further risk assessment of this compound.

Seven groups of male F344 rats were given potassium bromate in drinking water at concentrations of 0, 15, 30, 60, 125, 250, and 500 ppm for 104 weeks (*16*). From the data of water consumption recorded twice weekly throughout the experimental period, the daily intakes of potassium bromate (mg/kg/day) were calculated to be as follows: group 1 (500 ppm)-43.4; group 2 (250 ppm)-16.0; group 3 (125 ppm)-7.3; group 4 (60 ppm)-3.3; group 5 (30 ppm)-1.7; and group 6 (15 ppm)-0.9. The mean survival time of the animals given 500 ppm (82.8 ± 11.7 weeks) was significantly shorter than that of controls (103.1 ± 3.3 weeks) while the survival rates of the lower dose groups were comparable to that of the control group.

Table 4 shows the incidences of renal cell tumors in each group. Renal adenocarcinomas developed only in 3 of 20 rats of the 500 ppm group. The incidence of renal adenomas was significantly increased in the 125, 250, and 500 ppm groups with a definite dose-response relationship. On the basis of combined incidences of renal adenocarcinomas and adenomas, the VSD at a risk level of 10^{-6} was estimated by application of 4 mathematical models: probit model, logit model, Weibull model, and gamma-multihit model (Table 5). The p-value indicates that the probit model best fits the experimental data, and therefore, it seems reasonable to adopt the value estimated by this model (0.95 ppm) as the VSD of potassium bromate.

Potassium bromate is known to be degraded both *in vivo* and *in vitro* to potassium bromide (KBr), which is much less toxic than $KBrO_3$. A previous study by Kurokawa *et al.* indicated that KBr has no effect on 2-stage renal carcinogenesis in rats treated in the initiation phase with N-ethyl-N-hydroxyethylnitrosamine (*17*). Pharmacokinetic studies by Fujii *et al.* showed that BrO_3^- could be detected in the urine of rats after a single oral administration of $KBrO_3$ at a dose of 5 mg/kg or more

TABLE 4. Incidence of Renal Cell Tumors in Rats Treated with Various Concentrations of $KBrO_3$ in Drinking Water for 104 Weeks

Group	Effective No. of rats	No. of rats (%) bearing		
		Adenocarcinoma	Adenoma	Renal cell tumors
500 ppm	20	3 (15)	6 (30)*	9 (45)**
250 ppm	20	0	5 (25)*	5 (25)**
125 ppm	24	0	5 (21)*	5 (21)**
60 ppm	24	0	1 (4)	1 (4)
30 ppm	20	0	0	0
15 ppm	19	0	0	0
0 ppm	19	0	0	0

* $p<0.05$ (by Fisher's exact probability test), ** $p<0.001$ (by Fisher's exact probability test).

TABLE 5. VSD of $KBrO_3$ for Renal Cell Tumors at a Risk Level of 10^{-6}

	Probit[a]	Logit[a]	Weibull[a]	Gamma-multihit[a]
Chi-square value	1.627	2.155	2.472	2.693
p-value	0.898	0.827	0.781	0.747
VSD[b]	0.950	0.160×10^{-1}	0.481×10^{-2}	0.182×10^{-2}

[a] All models are with independent background. [b] Concentration in ppm in the drinking water.

whereas in rats given 2.5 mg/kg or less, no BrO_3^- was detectable (*18*). In our study, the daily intake of $KBrO_3$ was larger than 5 mg/kg when the $KBrO_3$ concentration in the water was higher than 125 ppm. In these cases, the incidences of renal cell tumors were significantly higher than the control values. This fact suggests that the contact of BrO_3^- with the renal tubular epithelium is essential for induction of renal cell tumors by $KBrO_3$.

Meanwhile, it is known that almost all $KBrO_3$ added to flour is converted to KBr during the normal British baking process. When $KBrO_3$ was added to flour at 150 ppm the residual level was reported to be lower than 5 ppm in the final bread. In fact, no evidence of carcinogenicity and toxicity was reported in mice and rats given a bread-based diet made from flour treated with 50 and 75 ppm $KBrO_3$ (*19*). As a result of this scientific evidence, the joint FAO/WHO Expert Committee on Food Additives decided at the meeting in 1983 " to change the previous acceptance of potassium bromate for the treatment of flour used for baking products to a temporary acceptance with a maximum treatment level of 75 mg potassium bromate per kg of flour, provided that bakery products prepared from such treated flour contain negligible residues of potassium bromate " (*20*). It should be added that the detectable limit of potassium bromate by current analytical methods is below 0.95 ppm, the VSD of this compound at a risk level of 10^{-6}.

Comparison of TD_{50} *and VSD*

Quantitative measures of carcinogenic potencies have been proposed by various authors. Bryan *et al.* suggested the use of TD_{50} or $D_{1/2}$ which is the daily dose of test chemicals required to give a 50% incidence of tumors in animals after a long-term exposure. On the basis of TD_{50}, Meselson has defined potency as $K = \text{In } 2/D_{1/2}$ (*21*). Maugh has prepared a diagram compiled from reports by Ames *et al.*, which indicates the relative carcinogenic potencies of various chemicals by TD_{50} (*22*). The diagram clearly illustrates that there is a million fold difference in potency between aflatoxin B_1, one of the most potent carcinogens, and sodium saccharin, one of the weakest. Recently, Sawyer *et al.* proposed a potency index based on the TD_{50} (*23*), and according to the principle, Gold *et al.* have published a carcinogenic potency database of the results of standardized animals bioassays (*24*).

Table 6 presents the TD_{50} and VSD of 5 carcinogens or tumor-inducing chemicals. The values of these 2 indexes appear to be quite paralled among the compounds.

TABLE 6. TD_{50} and VSD at 10^{-6} Risk Level of Various Cancer-inducing Substances

Compounds	TD_{50} (mg/kg/day)	VSD (mg/kg/day)	VSD (ppm)	TD_{50}/VSD
Aflatoxin B_1	0.8×10^{-3} [a]	1.6×10^{-7}	4.0×10^{-5} [b]	5×10^{5}
ENU	0.2×10^{0}	3.5×10^{-6}	0.9×10^{-3}	6×10^{5}
DMNA	0.6×10^{0} [a]	3.5×10^{-5}	0.9×10^{-2} [b]	1.5×10^{5}
Potassium bromate	0.6×10^{1}	3.8×10^{-2}	0.95×10^{0}	1.6×10^{3}
Saccharin-Na	0.9×10^{4} [a]	2.1×10^{1}	5.3×10^{3} [b]	4×10^{2}

[a] Maugh, T. H., 1978 (*22*). [b] Food Safety Council Report, 1980 (*1*).

The TD_{50} reveals potencies varying over seven orders of magnitude while the VSD does so over five order of magnitude. However, it should be noted that the ratio of TD_{50} and VSD is variable for each chemical ranging from 10^2 for sodium saccharin to 10^5 for aflatoxin B_1. These variation may be mainly attributable to the differences in the steepness of the dose-response curve for each chemical. Therefore, it is said that the TD_{50} is a useful index of the relative potency of chemicals, but used alone, can not assess the risk of chemicals at low dose exposure.

Biological Implication of VSD

It is generally agreed that carcinogenic risk assessment consists of 4 steps or components, namely, 1) hazard identification, 2) exposure assessment, 3) hazard assessment or dose-response assessment, and 4) risk characterization (*25*). The hazard identification is a qualitative evaluation of data on the potential of chemicals to produce carcinogenic effects to humans. The exposure assessment is the process of measuring or estimating real or hypothetical human exposures to a chemical of interest. The hazard assessment or dose-response assessment is the evaluation of both hazard and exposure information to estimate the mathematical probability that the carcinogenic potential associated with an agent will be realized in the human population under defined conditions of exposure. At the final step, all relevant information from the first 3 steps is integrated to characterize the carcinogenic risk associated with the expected human exposure to the compound of interest. The estimation of VSD is regarded as a component of the third step of carcinogenic risk assessment.

The third step of risk assessment relies on data from exposed humans and also from long-term animal testings. The latter have traditionally focussed on the detection of carcinogenic potential of chemicals, and therefore, the methods are usually based on exposure at or near the level of maximum tolerated dose in the test animals which often reaches orders of magnitude higher than the human exposure levels. These situations require the application of mathematical models for low-dose extrapolation to estimate the expected response at the exposure levels by humans.

Mathematical models proposed for low-dose extrapolation can be classified into 3 categories: tolerance distribution models (probit, logit, and Weibull models), mechanistic models (one-hit, multi-hit, and multistage models), and pharmacokinetic models. Unfortunately, no single model is recognized as the most appropriate at present. The pharmacokinetic models appear to be the most biologically realistic of all, but, they are rarely used because of insufficiency of available information concerning the metabolic fate of the chemicals.

Carcinogenic risk assessment is still at an embryonal stage of development, and there remains a number of problems to be clarified. With regard to the estimation of VSD, it is an urgent task to establish the principle or scientific basis for selection of appropriate mathematical models. At the same time we must also take the following point into consideration: not one of these mathematical models includes procedures for extrapolation of animal data to human. This indicates that the estimation of VSD is merely one of the components for the process in risk assessment, not risk assessment *per se*. In conclusion, the estimated VSD should be utilized for carcinogenic risk assessment in combination with other data relevant to species difference between animals and humans.

REFERENCES

1. Food Safety Council. Quantitative risk assessment. *In;* Proposed System for Food Safety Assessment, Final Report of the Scientific Committee of the Food Safety Council, pp. 137–160, Food Safety Council, Washington D.C., 1980.
2. Magee, P. N. and Barnes, J. M. The production of malignant primary hepatic tumors in the rat by feeding dimethylnitrosamine. Br. J. Cancer, *10*: 114–122, 1956.
3. Hicks, R. M. Nitrosamine as possible etiological agents in bilharzial bladder cancer. *In;* P. N. Magee (ed.), Nitrosamines and Human Cancer, pp. 455–469, Cold Spring Harbor Laboratory, New York, 1982.
4. Fong, L.Y.Y. Possible relationship of nitrosamines in the diets to causation of cancer in Hong Kong. *In;* P. N. Magee (ed.), Nitrosamines and Human Cancer, pp. 473–485, Cold Spring Harbor Laboratory, New York, 1982.
5. Yang, C. S. Nitrosamines and other etiological factors in the esophageal cancer in the northern China. *In;* P. N. Magee (ed.), Nitrosamines and Human Cancer, pp. 487–501, Cold Spring Harbor Laboratory, New York, 1982.
6. Fine, D. H., Ross, R., Rounbehler, D. P., Silvergleid, A., and Song, L. Formation *in vivo* of volatile N-nitrosamines in man after ingestion of cooked bacon and spinach. Nature, *265*: 753–755, 1977.
7. Lakritz, L., Simenhoff, M. L., Dunn, S. R., and Fiddler, W. N-Nitrosodimethylamine in human blood. Food Cosmet. Toxicol., *18*: 77–79, 1980.
8. Maekawa, A., Ogiu, T., Matsuoka, C., Onodera, H., and Furuta, K. Carcinogenicity of low doses of N-ethyl-N-nitrosamine in F344 rats; a dose-response study. Gann, *75*: 117–125, 1984.
9. Ogiu, T., Nakadate, M., and Odashima, S. Rapid and selective induction of erythroleukemia in female Donryu rats by continuous administration of 1-ethyl-1-nitrosourea. Cancer Res., *36*: 3043–3046, 1976.
10. Druckrey, H., Schildback, A., Schmähl, D., Preussmann, R., and Ivankovic, S. Quantitative Analyse der carcinogenen Wirkung von Diäthylnitrosamin. Arzneimittel-Forsch., *13*: 841–851, 1963.
11. Lijinsky, W., Reuber, M. D., and Riggs, C. W. Dose response studies of carcinogenesis in rats by nitrosodiethylamine. Cancer Res., *41*: 4997–5003, 1981.
12. Terracini, B., Magee, P. N., and Barnes, J. M. Hepatic pathology in rats on low dietary levels of dimethylnitrosamine. Br. J. Cancer, *21*, 559–565, 1967.
13. Oser, B. L. The rat as a model for human toxicological evaluation. J. Toxicol. Environ. Health, *8*: 521–542, 1981.
14. Kurokawa, Y., Hayashi, Y., Maekawa, A., Takahashi, M., Kokubo, T., and Odashima, S. Carcinogenicity of potassium bromate administered orally to F344 rats. J. Natl. Cancer Inst., *71*: 965–972, 1983.
15. Takamura, N., Kurokawa, Y., Matsushima, Y., Imazawa, T., Onodera, Y., and Hayashi, Y. Long-term oral administration of potassium bromate in male Syrian golden hamsters. Sci. Rep. Res. Inst. Tohoku Univ. C, *32*: 43–46, 1985.
16. Kurokawa, Y., Aoki, S., Matsushima, Y., Takamura, N., Imazawa, T., and Hayashi, Y. Dose-response studies on the carcinogenicity of potassium bromate in F344 rats after long-term oral administration. J. Natl. Cancer Inst. (in press).

17. Kurokawa, Y., Aoki, S., Imazawa, T., Hayashi, Y., Matsushima, Y., and Takamura, N. Dose-related enhancing effect of potassium bromate on renal tumorigenesis in rats initiated with N-ethyl-N-hydroxyethylnitrosamine. Jpn. J. Cancer Res. (Gann), *76*: 583–589, 1985.
18. Fujii, M., Oikawa, K., Saito, H., Fukuhara, C., Onodaka, S., and Tanaka, K. Metabolism of potassium bromate in rats I. *In vivo* studies. Chemosphere, *13*: 1207–1212, 1984.
19. Fisher, N., Hutchinson, J. B., and Berry, R. Long-term toxicity and carcinogenicity studies of the bread improver potassium bromate. I. Studies in rats. Food Cosmet. Toxicol., *17*: 33–39, 1979.
20. Twenty-seventh Report of the Joint FAO/WHO Expert Committee on Food Additives, pp. 27–28, WHO, Geneva, 1983.
21. Meselson, M. and Russel, K. Comparisons of carcinogenic and mutagenic potency. *In;* H. H. Hiatt, J. D. Watson, and J. A. Winsten (eds.), Origins of Human Cancer, Book C, pp. 1473–1481, Cold Spring Harbor Laboratory, New York, 1977.
22. Maugh, T. H. Chemical carcinogenesis: how dangerous are low doses? Science, *202*: 37–41, 1978.
23. Sawyer, C., Peto, R., Bernstein, L., and Pike, M. C. Calculation of carcinogenic potency from long-term animal carcinogenesis experiments. Biometrics, *40*: 27–40, 1984.
24. Gold, L. S., Sawyer, C. B., Magaw, R., Backman, G. M. A carcinogenic potency database of the standardized results of animal bioassays. Environ. Health Perspect., *58*: 9–319, 1984.
25. Federal Register. Part II. Office of Science and Technology Policy. Chemical Carcinogenesis: A Review of the Science and its Associated Principles, pp. 78–84, 1985.

DIET, NUTRITION AND CANCER, Y. HAYASHI ET AL. (EDS.),
JAPAN SCI. SOC. PRESS, TOKYO/VNU SCI. PRESS, UTRECHT, PP. 305-321, 1986

Cancer, Diet, and Public Policy

Sanford A. MILLER and F. Edward SCARBROUGH

Center for Food Safety and Applied Nutrition, Food and Drug Administration, Washington, D.C. 20204, U.S.A.

Abstract: The response of Government to the relationship between diet and chronic disease is a legitimate public health policy issue, yet the proper role of Government is a subject of continuing debate. Regulatory mechanisms designed to address questions of food additives or contaminants in food are ill-suited to address questions of diet/disease relationships because of the different standards and types of evidence involved. In this paper we examine, through the use of several examples, the question of whether current evidence on the relationship of diet to cancer is sufficient to make changes in public helath policy. Finally, we discuss potential Governmental approaches, both short- and long-term, to implementing policy in the area of diet and cancer.

Profound changes are occurring in the lives of people throughout the industrialized world. Lifespan is increasing as a result of not only decreased infant mortality but also increased life expectancy at the other end of the scale. Morbidity and mortality rates associated with virtually all principal chronic diseases are on the downside of the curve. While much of this observed trend is thought to be the result of improved medical care and technology, it is generally accepted that a substantial portion is the result of a growing awareness of the importance of diet and nutrition in the maintenance of good health. Attention to the relationship of diet to chronic disease on the part of Federal Governments could make substantial contributions to the further improvement of national health. The benefits that could be derived are significant in terms of improved lifestyle and quality of life. Equally important is the relationship of such improvements to the cost of health care. The relationship between diet and chronic disease is a legitimate public health question of high priority which must be addressed as a policy issue.

The Government of the United States has been in the forefront of the consideration of these changes and has devoted considerable effort to seeking the proper role of Government in implementing public policies for the application of dietary modification to the maintenance of good health. One of the first public statements to address these issues by any Federal Government was the 1979 report from the Surgeon General of the United States (*1*) which recognized the emerging evidence suggesting that the major problems of heart disease, hypertension, cancer, diabetes, and other chronic disorders are significantly related to diet and called for the focusing of more attention on dietary habits as contributing causes.

In 1980 with the publication of "Dietary Guidelines for Americans," (*2*) by the U.S. Departments of Agriculture (USDA) and Health and Human Services, the concern of government for the role of nutrition in the prevention of disease entered a new phase, in which, for the first time, a Federal Government provided specific dietary advice to its citizens—advice designed to not only maintain but hopefully improve their helath.

Yet, there is a continuing and often heated debate over the role of contemporary science in assuring and maintaining public health and the special role that Government needs to play in such activities. And members of the international regulatory community are asking: "where do we go from here?"

Role of Government

It has always been a fundamental fact of American political and social philosophy that Government does have a special role in assuring the health and safety of its people. This is now even more true as contemporary science has moved into often arcane regions in which it is impossible for the non-specialist to stay abreast of the most recent information and to make appropriate choices based on this information. In fact, the information load has become so great that many in society now depend on Government to provide guidance.

When the decision matrix is relatively straightforward and the consequences catastrophic, generally Governments choose direct intervention. In the United States, the introduction of fortification and enrichment of foods in the 1940's to combat acute nutrient deficiency diseases, such as beriberi, pellagra, and iron-deficiency anemia, is an example of the direct approach (*3*). For more complex issues, an indirect process involving provision of information has been used. In the area of nutrition, this indirect approach has been much more widely employed. Nutrition information has been provided to consumers, in coordination with an educational process designed to teach them how to use this information to make appropriate dietary choices. In the United States, activities such as the programs to reduce dietary sodium intake have involved a judicious combination of governmental advice and food labeling initiatives (*4*).

There is a substantial body of data linking diet and nutrition to a wide range of chronic diseases. Yet, it is the growing evidence that dietary patterns significantly affect cancer incidence in humans that may have the most profound implications for public health officials. Why is this true with cancer and not, for example, with cardiovascular disease—the number one cause of death in the United States (*5*)? The answer lies in the very nature of the disease itself. Death by cancer is a particularly frightening prospect to people in that it is generally not sudden but a progressive deterioration accompanied by prolonged pain and suffering. Because of the great fear felt by everyone who has had to watch a loved one suffer through terminal cancer, Government will be required to assume a role in the implementation of dietary recommendations related to cancer. It is the possible shape and directions of this role that we want to briefly explore.

Regulation of Chemicals vs. Diets

Since at least 1958, when the Food Additives Amendment with its famous Delaney Clause was added to the Federal Food, Drug, and Cosmetic Act (FD and C Act) (*6*), the Food and Drug Administration (FDA) has focused its attention on controlling human exposure, through food, to suspected carcinogens. The agency has sought to identify food constituents that pose cancer risks and to eliminate them from food. For very practical reasons, the regulatory efforts have been directed toward food additives and carcinogenic residues from environmental contaminants or pesticides. Current regulations generally assess food constituents individually and in isolation. Data considered by regulators come from tests of that substance alone, and safety assessment focuses on the relationship between the capacity of the substance to cause harm to test animals and the levels at which humans are likely to be exposed. Again for practical reasons, almost no attention has been given to how the substance may interact with other constituents found in the food, because predictions of synergistic effects are virtually impossible and a systematic consideration of all possible interactions between the several thousand substances found in food is far beyond the capacities of the regulatory and scientific communities.

The convergence of data suggesting that diet is a major factor in the etiology of specific types of cancer is beginning to challenge this established regulatory framework. Theoretically, FDA could restrict or ban specific foods if they were proven to be unsafe (*7*). However, this is precisely part of the challenge we face, in that an implicit assumption underlying our regulatory philosophy has always been that traditional foods, especially those eaten with a minimum amount of processing, require no further regulatory consideration because of their inherent safety. This assumption is being challenged by evidence indicating that foods constituting major portions of traditional American diets, and thus important products of American agriculture, are being implicated as significant risk-increasing components of our diets. The greatest challenge to Government is posed by the fact that the risk of cancer is closely being linked to a lifetime dietary pattern and not to occasional exposure to individual foods or food constituents.

FDA's traditional regulatory responses are geared toward either prohibiting or permitting a substance in the food and not to altering dietary patterns of consumers. It is clear that the developing evidence of the capacity of the traditional American diet to contribute to the risk of cancer is a significant public health problem, incapable of being fully addressed by FDA's traditional regulatory strategies. Regulatory agencies must devise more effective means to deal with such concerns. At the same time, Government cannot abandon its traditional roles; rather, we must assume new burdens, while continuing to assure the safety of the food supply in the traditional sense.

Standards of Evidence

One result of the developing evidence of the relationship of diet to cancer and the increasing demands for action to be taken in this area has been a conflict in the scientific community which has led to difficulty in obtaining consensus on what should be done with the knowledge we are gaining from our research activities. To a large measure this conflict has revolved around the fundamental problem of the degree of scientific surety that should form the basis for deciding to apply contemporary nutrition knowledge to public health issues. The question being asked is what is the standard of scientific evidence that should be used to determine when it is appropriate for public health policy changes to be made? This directly raises questions of scientific absolutism *vs.* public health needs. The scientific method is designed to provide a system in which data and hypotheses are constantly being rechecked and reevaluated, an essential process for knowledge to increase and for science to advance. Scientists are trained to continually question conclusions. Scientific papers are larded with words such as "possible", "probable", or "likely"—words which recognize that scientific truth is an evolving idea, at best; yet words that irritate journalists, lawyers, and politicians and confuse the general public. Such qualifications are inherent in retrospective epidemiology and animal feeding studies, the very type of studies that constitute the bulk of the evidence we have on the relationship of diet to cancer. Thus, scientists are rarely able to say, without reservation, that any proposed relationship is unequivocally demonstrated. And certainly, no scientist would state that "beyond a reasonable doubt" (or even, perhaps, a "preponderance" of) the evidence supports the strongest of the causal relationships; yet these are the standards of evidence most familiar to, and sought by, the legal and political communities. Conversely, there is a tendency for most experimental scientists not responsible for public health issues to look for a degree of absolutism in relationships that, while appropriate for evaluating experimental hypotheses, may only serve to delay the implementation of important public health actions. There is a dichotomy between academic science and public health science and a consequent lack of agreement on the nature of the words "sufficient scientific evidence".

The tentative attitude of the academic scientist is not an acceptable position for the public health scientist. Persistent uncertainty should not impede government action. At some point in the process of collection of information concerning a public health problem, public health scientists must make what one author (SAM) has termed a "Leap of Faith" (*8*) and believe that the data are *sufficiently* convincing to take public health action, even though doubts may remain. If action is taken too soon, there is the danger of not only damaging the public trust but also of initiating unnecessary controversy in the scientific community. However, and more importantly, delay in taking action runs the danger of imperiling the health of substantial numbers of people. For the academic scientist, there is no such compelling responsibility or rationale, nor should there be, and any doubt is sufficient to withhold acceptance of a relationship. This resulting dichotomy between academic and public health science, on occasion, creates a conflict between those charged with protecting the public health and the professional community, which in turn leads to even greater confusion for the public.

What then should the standard of evidence be? For public health scientists, the standard is, to a significant extent, variable, depending on the particular situation. This is a fact of public health policymaking that is not always recognized. The appropriate standard depends not only on the strength of the evidence but also on the nature of the action contemplated. Perhaps to a larger measure, an appropriate standard depends upon how potentially important the action is in maintaining the public health as well as on the potential for counterproductive effects to occur. If the danger of doing something is considerably less than the possible benefits in terms of improved health, then it becomes an appropriate action for public health agencies to take. This often occurs under conditions where the academic scientist would correctly say that the data are insufficient to prove the hypothesis. And, in fact, an action undertaken may later be shown to be ineffective or less effective than predicted. Nevertheless, as long as the harm done by the action is minimal, and hopefully nonexistent, the fact that the selected action ultimately has little effect on public health does not necessarily mean that it was an inappropriate decision *at the time it was made.* Yet, it is this lack of guaranteed success that makes it essential that the potential benefits of public health actions not be overpromoted. Above all, rationality and moderation are required.

Types of Evidence

Evidence on the relationship between diet and cancer has been derived from a variety of types of studies, the major ones being animal experiments and epidemiology. A brief consideration of a few of these types of studies will help explain why doubts and uncertainty persist.

Studies in which a specific chemical is fed to experimental animals have formed the bedrock on which the regulation of food additives or environmental contaminants has been grounded. Questions have always persisted on the extrapolation from animal derived results to humans. Evidence of the relationships between diet and cancer is also being developed from animal experiments. But because the relationships appear to be multifactorial and to involve lifetimes of dietary patterns, questions about the applicability of the data are even more persistent. Clearly, we need better methods of extrapolating animal derived data to human experience. Also the establishment of the traditional dose / response relationship which then allows a risk calculation is not possible when the substances being investigated are major portions of the diet. With a food additive, for example, the dose can be exaggerated to several fold above expected exposures, whereas similar treatment of a dietary constituent such as fat is not feasible.

Intervention studies in a clinical setting have been used extensively in the investigation of drugs for the treatment of cancer. However, for a number of reasons, such studies are expected to yield relatively little information on the relationship between diet and cancer. Drugs are exogenous to the body and following their metabolic fate, for example, is much easier than investigating effects and associated mechanisms of nutrients, which in many cases are endogenous constituents of the body. For ethical reasons such studies must be limited to investigations of protective factors. Clinical studies would be generally ineffective for exploratory investigations and can thus only be considered when the supporting evidence from other studies is strong. All things considered, clinical studies will be infrequently used to develop data on the relationships between diet and cancer.

Much of what we know about the relationship of diet to cancer has been provided by epidemiological studies. It may be fairly said that epidemiology is a method rather than a basic science (*9*). However, as in other disciplines, solutions of significant problems demand imagination, curiosity, and ingenuity in the application of the scientific discipline. Epidemiological studies have the advantage of observing humans directly at normal levels of exposure and, thus, avoiding the necessity of extrapolating data from animals fed at exaggerated doses. Epidemiological investigation is based on the premise that disease in populations is nonrandom and that disease agents and factors can be identified. For example, in coronary heart disease, saturated fats and cholesterol may be considered agents; a high socioeconomic status may be an environmental factor that increases the likelihood that a person will be exposed to large quantities of these agents; and a variety of complex genetic factors determine which individuals on a high fat diet are most likely to develop the disease. There are several different types of epidemiological studies, which, for our purposes, can be characterized as falling into two basic categories: retrospective and prospective studies.

In retrospective studies, in general, population groups or groups of reliably diagnosed subjects are selected and their histories are reviewed in search of correlations between commonly shared exposures or characteristics and the occurrence of cancer. Establishing adequately precise food intake data is a major drawback to these types of studies. Not only is there a problem of inadequate recall of dietary history, but also average or representative food intakes calculated from other data sources may be inaccurate for discrete subgroups. Further, chance variation or multiple nondietary differences between the compared populations may account for observed small differences in risk. Another difficulty is that the period between exposure and development of cancer can often be decades, during which time other modifying factors are introduced. Despite these drawbacks, the advantages of retrospective epidemiological studies are likely to outweigh the disadvantages of laboratory studies.

The long term prospective epidemiological study seeks to answer questions by providing direct estimates of the risk of developing disease in an exposed group and in an unexposed group with otherwise similar characteristics. Both groups are followed through time and the development of disease is noted. It may be that only long term prospective studies are capable of exploring diet and cancer relationships. However, three basic difficulties tend to limit the use of the prospective approach. First, it may be difficult to identify the two populations to be sampled. Second, the time required for follow-up may be very long, especially when the study involves a chronic disease such as cancer. Investigators who initiate the study may fail to answer the questions within their lifetimes. Also, especially in a mobile society such as in the United States, study members may move away from easy observation. Third, both the original sampling effort and the long follow-up period mean that such studies are extremely expensive. For these reasons such studies are not often performed.

When all the above factors are considered it is evident that emphasis will have to be placed on animal and retrospective epidemiologic data in the development of public policy on diet and cancer. Also, because of the usual lack of complete data, the public health scientists must act on the weight, or confluence, of the evidence, rather than on the assurance of certainty demanded by academic sciences.

Weight of Evidence

It is surprising how difficult it has been to develop unequivocal data relating diet and cancer. In spite of retrospective epidemiologic and animal studies supporting many of the hypothesized relationships, focused studies have often been negative or, at best, equivocal. It has become a truism of nutrition to describe the relationship between diet and cancer as multifactorial, in an effort to indicate the complexity of these issues. In general, however, when an event is described as multifactorial, it means that many factors are involved in its etiology. Perhaps more importantly, it also means that each of the factors, in addition to its interaction with other components of the process, has a relatively independent role to play in the process. Thus, using the traditional experimental design of keeping as many factors constant as possible and varying only one, some response is expected if the test factor is related to the expected effect. In the case of the relationship between diet and cancer it may be that several factors must vary simultaneously in order for a response to be observed. The metabolic and physiologic relationships between nutrients suggest this is a possibility. Equally important may be the fact that the primary factors involved may not be nutritional in origin. If the primary factors are ubiquitous in the environment, then the role of diet may be only a modulating influence on the expression of these primary factors. These considerations may explain why, using traditional experimental approaches, the results obtained have been less powerful than expected.

Evidence on Diet and Cancer

With these remarks in mind, we must now ask the fundamental question, does current evidence on the relationship of diet to cancer meet the standards of evidence necessary to make changes in public health policy? Because, as we have stressed, cancer is probably associated with dietary patterns that extend over a number of years, causative agents are difficult to identify. The uncertainties that plague the efforts to quantify the role of other environmental factors affecting cancer incidence also encumber estimates of the contribution of diet. A 1981 study stated with some confidence that a substantial portion of cancers could be attributed to dietary factors, and found support in the literature for estimates ranging from ten to seventy percent (*10*).

To illustrate current levels of evidence and some of the conflicts involved, we have chosen three examples: a single nutrient—selenium; a dietary constituent—

dietary fiber; and interrelated components of the diet—lipids. Selenium is a nutrient for which establishing public dietary policy is particularly difficult because, although some data suggest that this mineral lowers the risk of cancer, selenium is known to be toxic at elevated intake levels. Dietary fiber illustrates the difficulty in forming a consensus in the biomedical community, especially for a dietary component which obviously requires considerable research to clarify its role and mechanisms of action. And the lipids, fat, and cholesterol, are dietary components for which potential dietary recommendations for lowering the risk of cancer may contradict accepted recommendations related to cardiovascular disease.

1. Selenium

Both epidemiological and laboratory evidence suggest that selenium may offer some protection against the risk of cancer at customary levels of intake, but as a dietary constituent it is known to be toxic at higher levels of exposure. The epidemiological evidence generally demonstrates an inverse relationship between the level of selenium intake and the risk of cancer, although it is not clear whether this applies to cancer at all sites or only to specific cancer sites. Numerous experiments in animals have demonstrated an antitumorigenic effect of selenium although the relevance to humans is not apparent since the levels of intake far exceed dietary requirements and border on levels that might be toxic (*11*, *12*).

Data suggest that selenium, through its function in glutathione peroxidase, an enzyme that protects the constituents of cells against free radical damage, could be involved in protecting cells against cancer induced by high intakes of fat (*13*). Glutathione peroxidase activity in human blood increases with increasing selenium intake, but reaches a plateau at intakes well below those customary in the United States (*14*). Thus attempts to increase the enzyme activity by selenium supplementation, superimposed on an adequate diet, would not be successful.

The toxicity of selenium to animals varies with the amounts and chemical forms of the selenium ingested, the duration and continuity of the selenium intake, the nature of the rest of the diet, and to some extent with the species (*15*). For example, in the rat selenium appears to be hepatocarcinogenic when fed in the diet at 5 to 10 ppm for many months (*16*). Until recently, there were no well-documented cases of human selenium toxicity. However, a 1983 study from China (*17*), as well as a report from the U.S. (*18*), have demonstrated definite toxic effects of excessive selenium intake. Consideration of these data led the National Academy of Sciences (NAS) to conclude that "although balanced diets in some areas furnish more than 200 μg selenium per day (*19*), the demonstrated chronic toxic effects of relatively low dietary levels in experimental animals suggests a maximal intake of 200 μg/day for adults, which should not be exceeded habitually if the risk of long-term chronic exposure is to be avoided (*20*)."

2. Dietary fiber

The effects of consumption of dietary fiber on the occurrence of colorectal cancer have been studied extensively, yet this is a dietary component for which a consensus in the scientific community continues to be extremely difficult to achieve. Both experimental studies in animals and epidemiologic surveys in humans have produced conflicting data (*11*). On balance, it appears that dietary fiber may play a protective role in the risk of colon cancer; but other factors are equally or more important. Not all types of fiber are effective, and those types that seem to exert some protective influence appear to work through different mechanisms. These significant conflicts in data presumably arise from methodologic problems and from the fact that different types of fiber and different basal diets are not equivalent. Also, there may be multivariant correlations with observed results related to two or more variables in combination, for example, "high fiber" diets tend to be low in fat and the two dietary constituents may have to be raised and lowered together.

Various mechanisms for the functioning of fiber have been proposed. Increased bulk and water retention may moderate development of colon cancer through dilution of carcinogens and promoters (*21*). Additionally, it has been suggested that dietary fiber modifies the metabolic activity of gut bacteria (*22*). Also, different types of fiber have been demonstrated to have binding capacities of varying degree for a number of carcinogens (*23*, *24*). Variability of this type could help explain the different findings with respects to diets and fiber types.

In the face of this sea of conflicting data, different authoritative bodies have reached different conclusions. For example, the Federation of American Societies for Experimental Biology (FASEB) in 1980 concluded: "Although epidemiologic studies suggest that low fiber intake is associated with increased risk of . . . colonic cancer, much of the evidence is faulted by genetic, environmental, cultural, dietary, and other uncontrolled variables. Conclusive evidence for a causal relationship between low-fiber intake and . . . colonic cancer is not available, and the fundamental question of whether dietary fiber protects against human colonic cancer . . . is not yet resolved (*25*).

NAS reached a similar conclusion in 1982 when they stated: " . . . that the epidemiological evidence suggesting an inverse relationship between total fiber intake and the occurrence of colon cancer is not compelling [and] . . . that it seem likely that further epidemiological study of fiber will be productive only if the relationship of cancer to specific components of fiber can be analyzed." Unfortunately, it is not yet possible to make firm scientific pronouncements about the association between diet and cancer (*26*).

However, in 1984, the National Cancer Institute (NCI) found the data sufficiently convincing to state: "Americans now eat about 10–20 g of fiber daily. Populations that consume diets containing twice this amount have a lower rate of cancers of the colon and rectum. So, the NCI recommends that you eat foods which provide 25–35 g of fiber a day (*27*)."

Finally, the American Council for Science and Health concluded in 1985 that " . . . review of the literature shows insufficient evidence to warrant establishment of a public policy of guidelines for diet modification for all Americans for the purpose of reducing the risk of cancer (*28*)."

Clearly, much additional research is needed to overcome these differences of opinion and to establish a consensus dietary recommendation.

3. Lipids (fat and cholesterol)

The NAS has concluded that, of all the dietary components studied, the combined experimental and epidemiological evidence is most suggestive of a causal relationship between fat intake and cancer occurrence (*25*). Yet, even in this case the Academy detailed the impossibility of establishing with certainty that fat *per se* is the causative factor. Epidemiologically, diets that are high in fat will most likely be relatively low in fiber, cereal grains, vegetables, and associated nutrients. Interpretation of the data is further complicated by the wide diversity of cancer sites for which epidemiological evidence of correlation with fat exists. For example, in addition to gastrointestinal, especially large bowel, cancers, there are even more consistent correlations with cancers of the breast and prostate, as well as relatively convincing evidence linking fat with cancers of reproductive organs—sites which have no direct exposure to dietary fat (*29, 30*). As a further complication, it appears that different types of fat show quite different relationships to cancer. In general, it is not possible to identify specific components of fat as being responsible for the effects observed in epidemiological studies, although total fat and saturated fat are most frequently associated. However, laboratory studies on animals suggest that, when total fat intake is low, polyunsaturated fats are more effective than saturated fats in enhancing tumor formation (*26*).

Dietary fat appears to act as a promoter of carcinogenesis, however it is likely that promotion occurs by different mechanisms in the various types of cancer. Organs of the gastrointestional tract are exposed to fat directly, and it may be that some component of fat is metabolized to an active carcinogen. On the other hand, hormone-related cancers require different types of mechanisms, which may vary with cancer site (*11*).

Elevated serum cholesterol levels are asociated with atherosclerosis (*5*). Therefore, there has been intense interest in studying dietary cholesterol intake and cardiovascular disease. However, these long-term studies have provided the opportunity to also study the relationship between cholesterol intake and cancer. The data are inconsistent, although several epidemiological studies suggest that a low serum cholesterol is associated with increased frequency of colon cancer in males (*31–33*). This may be coincidence rather than cause and effect. This is in contrast to certain animal studies which have indicated that the protective effects of dietary fiber can be largely nullified with increased cholesterol intake, suggesting a major association of colon cancer with dietary cholesterol, although extrapolation to humans is uncertain (*34*).

Possible Regulatory Approaches

Assuming that the relationship of a specific dietary constituent or pattern to cancer has been sufficiently established to justify the implementation of public health programs, many questions would still remain on the best methods to take advantage of these findings. FDA has several options within its arsenal of traditional regulatory approaches which, with modification in some cases, could be applied, in the short-term, to these issues. These options fall into three broad categories: education, labeling, and controlling food compostion (*35*).

1. Education

As mentioned earlier, the indirect approach of providing information has been widely used in the area of nutrition. Clearly, providing consumers with information, and instruction on how to use this information to deal with the complex issues being raised by scientific inquiry into the relationship between diet and cancer, is an essential first step. However education alone is not likely to be enough. For education to be effective, foods that have attributes consistent with the dietary principles being advocated must be available and must be identified in such a way that consumers can readily incorporate them into their dietary patterns. Also, foods are eaten for many reasons in addition to their contribution to nutrition and health. Foods are associated with pleasure, ethnic customs, and societal traditions, all of which are factors tightly woven into the cultural fabric of a country or of a people. Given the fundamental and often competing roles of food, Government faces a difficult task in using education alone to alter dietary patterns. If our experience with the educational campaign on smoking and health is any indication of the difficulty in changing detrimental habits, implementing changes in basic dietary patterns will require substantial innovation and understanding. Finally, achieving motivation through education, given the latent onset of cancer, will be difficult.

2. Labeling

Placing information on the labels of food would not seem to require the same level of evidence that is necessary to justify banning a substance. There are two types of labeling to be considered: voluntary and mandatory, that is, Government can either encourage manufacturers to place certain information on the food label or it can require that the information appear on the label. In either case, simple disclosure of the quantity of a particular substance, for example the grams of saturated fat or of fiber contained in a serving of the food, may not be sufficient. Discussion of why these constituents are important in the diet may be necessary for labeling to be effective in altering dietary patterns.

Government encouragement of voluntary discussion of diet / health relationships raises several problems. A major concern would be that there is no incentive for a manufacturer to make a candid, balanced presentation of the information. For example, a manufacturer would not be inclined to discuss risks as well as benefits of consuming the constituent being promoted. A case in point might be polyunsaturated fats which appear to be associated with a decrease in cardiovascular disease but may be associated with an increase in breast cancer (*26*). Also, manufacturers would have no inclination to indicate the lack of scientific consensus about the relationship being discussed. As a hypothetical example, should a label declaration discussing the relationship of dietary fiber to colorectal cancer reveal that, although NCI has determined that there is sufficient evidence to state without reservation that fiber has a protective influence, NAS has found that the same evidence is not compelling? This raises the additional issue of what constitutes the proper degree of consensus before a relationship can be discussed at all on a food label.

It is evident that Government must have guidelines or standards against which to judge the propriety of health-related discussions on food labels. Control of what appears on the label could be maintained in one of three ways: by dictating the language which may appear on the label, by premarket review of labels, or by active monitoring of the marketplace coupled with vigorous enforcement actions against offending labels. Regardless of the method of control adopted, an additional burden of deciding what is appropriate labeling is placed on Government resources. Guidelines for evaluating health-related information on labels are also essential to prohibit the overpromotion of products on the basis of overstated or speculative claims based on diet / cancer relationships that have not been accepted by the scientific community.

With respect to mandatory labeling requirements, the FD and C Act could be construed to empower FDA to require any food label to bear just about any information that the Agency believes necessary to inform consumers about the food's important characteristics, presumably including its capacity to reduce or increase the risk of cancer (*7*). However, if we again assume that simple disclosure is not sufficient, government will be in a position of having to dictate the language which would be allowed to appear on food labels. This is by no means a simple task. The suggested relationships between diet and cancer and the resulting dietary recommendations are complicated, detailed, and often encumbered with numerous conditions and caveats. They do not lend themselves to reduction to a few words that will conveniently fit on a food label while at the same time remaining accurate reflections of the scientific findings or understandable to the general consumer. This is especially true when the limited size of the food label is considered along with the large amount of information that is already required to appear on the label. It may be that we are simply asking the food label to do too much. FDA has long viewed the label as a regulatory tool, while manufacturers increasingly look at it as an advertising and marketing tool. We are now asking it to function as an educational tool, which may be the final straw.

The existing food label is already overcrowded to such an extent that information overload is occurring. A large body or research has shown that providing people with too much information can actually detract from their ability to make a decision because they lose the ability to focus on the key information. Once a person is overhelmed by the amount of information, he stops trying to process it and makes random decisions. FDA's own studies with food labels support this finding (*36*).

These considerations raise the issue of whether mandating food labeling is an effective approach to providing dietary information or is it, in fact, counterproductive. Government should consider methods of providing information on diet and cancer as it relates to specific products other than by using the traditional food label. For example, a computerized data base of diet/cancer information, keyed to the universal product code (or a similar code) could provide an information seeking consumer all pertinent information related to a product at a point-of-sale terminal activated by scanning product codes.

A problem with labeling, whether voluntary or mandatory, as a device to transmit information about dietary patterns and health is that many people simply do not read food labels or do not understand what they read (*36*). Unfortunately, in many cases, these are just the people who should be reached with the messages. This illustrates that educational efforts and labeling can not be separated—there must be education for labeling to be effective; and education is not effective in the absence of labeling.

Another problem with labeling is that a large portion of the typical diet is not labeled. For example, fresh foods, such as fresh fruits and vegetables, which may play pivotal roles in desired dietary patterns, are not usually labeled. Also, an increasingly large proportion of food is eaten in restaurants and food service facilities. Such unlabeled food contributes significantly to the dietary patterns of many individuals. To address this problem, FDA is encouraging several experiments in providing point-of-sale information to aid in the selection of foods for desired dietary patterns. These experiments include shelf labeling information in supermarkets and posters and booklets with nutrition information in fast food restaurants.

3. Controlling food composition

Government has at least two broad avenues of approach for influencing the composition of food. On one hand, Government can encourage the food industry to develop new products containing specific constituents or patterns of nutrients that are consistent with accepted diet/cancer relationships. These new products could be based on traditional foods to which nutrients have been added or the levels of certain naturally occurring constituents have been limited. Food manufacturers would most likely have to be given some incentives to develop these products. Perhaps, providing for special label statements could be an incentive. FDA has adopted a modification of this approach in its imitation/substitute policy (*37*) which perimits a fabricated food intended to replace a more traditional food to be labeled as a substitute rather than as an imitation—a more negative term from a marketing view point—as long as it is not nutritionally inferior to the product it is replacing. In determining nutritional inferiority, calories and fat are not considered, thus recognizing certain principles of good nutrition. We are now, however, talking about foods that, when viewed from the perspective of diet and cancer, the manufacturer may want to promote as better than their traditional counterparts, thus raising a new set of regulatory problems. Also, how are we to guard against, or even distinguish, so-called "quack" or fraudulent products which have no dietary rationale other than a manufacturer's desire to capitalize on consumers' concerns and fears about cancer? A product with added selenium might fall in this category because there is no demonstrated need for added selenium in the American diet even though, at customary dietary levels, this mineral may have a protective effect relative to cancer.

A second way of influencing the composition of food is through a variety of standards administered by the Government. These would include FDA's food standards—historically, recipes for food staples to prevent economic fraud—and USDA's grading standards—measures of quality to determine price received by the producer. For example, many of these standards are based on the level of fat as an indicator of quality, that is, the more fat, the better the product according to the standards. Now, with many professional groups calling for a reduction in overall fat intake, perhaps it is time to critically reexamine these standards and reassess measurements of quality in terms of current dietary recommendations. Again, this is an undertaking beset with problems. First, we can expect the industry to object because of vested economic interest in the status quo. Second, consumers may turn to better tasting, less healthy alternatives if the composition of basic foods is altered.

4. Long-term approaches

To be truly successful in modifying the dietary patterns, more long-term, far reaching approaches must be adopted. We believe that, despite the differences between public health and academic science and the difficulties in developing unequivocal data, scientific research on the relationship of diet to the risk of cancer has produced a convergence of data which is sufficient for a modification of national food and agricultural policies. These policies must now be formulated with their nutritional consequences in mind, and health must be factored in as a major component in our planning for agriculture and the food industry for the rest of this century. Changes in national policies will also involve such programs as the USDA food stamp and school lunch programs and feeding in institutional settings such as the Armed Forces. The planning for scientific research related to food production must be guided by this same convergence of data. We must begin to encourage manufacturers and growers to produce food products that are, to the best of our knowledge, useful in reducing the risk of cancer. However, this must be a long-term process because abrupt changes would have damaging consequences on agriculture.

As part of a long-term approach to modifying dietary patterns, Government must establish extensive monitoring systems which will allow a tracking of the effects of these profound shifts in food and agricultural policies. Systems which determine what is grown or manufactured have been in place for some time, but systems which monitor dietary patterns and health status of populations are in their relative infancies. What is needed are more sophisticated systems which, in effect, will provide post-market surveillance of food products incorporating dietary recommendations to determine how such foods are being adopted into dietary patterns and what, if any, changes in health status are occurring as a result of the national policies. Evaluations and analyses of data derived from such monitoring systems must be accurate and timely enough to permit adjustments in public health policies and strategies. Further, any monitoring system must be flexible to incorporate the results of new research or dietary recommendations as they become available. However, the full value of monitoring systems may not be realized until such systems are in place for many years.

In a related area, as the relationships between diet and health become more well understood, we are moving into an era in which diets or even individual foods may become the major treatment modalities for certain diseases. Already, many products exist in the market, and many more are under development, which are used in the dietary management of medical conditions and thus, from a regulatory point-of-view, fall in the grey area between food and drugs. We can only anticipate that this trend will continue to accelerate. Yet, the concept of a food as a medicine is difficult to incorporate into our existing regulatory framework. We must recognize that the categorization of products as either food or drugs may become inoperable and begin to devise appropriate regulatory strategies.

CONCLUSION

Any change in public policies in the area of diet and cancer will be difficult to implement and, as we have discussed, will present numerous problems to be overcome in order to be successful. However, there are certain actions that Government must now undertake to have any hope of significant accomplishments in this area. We suggest that the following five actions are appropriate for immediate consideration.

Government must take whatever actions are necessary to continue research, both basic and applied, on the role of diet as both causative and supportive factors in cancer and other chronic diseases found in the human population. Even greater support must be given to exploring the relationships between nutrients and other factors and modification of the disease process.

Government should develop and maintain sophisticated monitoring systems to ensure that public health officials have an adequate grasp of trends in nutrition behavior of the public.

Government should develop and advocate the use of standardized methodologies and protocols for the conduct of epidemiological, clinical, and animal research studies to ensure that the results of various studies are valid and relatively comparable.

Government should more fully develop educational campaigns to acquaint the public with the need for good nutrition, as a reflection of the convergence of data and the derived concepts about the association of dietary practice and cancer.

Finally Government must develop new and more effective regulatory strategies to deal with the challenge presented by our emerging understanding of the relationships between diet and health.

REFERENCES

1. Department of Health, Education and Welfare. Healthy People: the Surgeon General's report on health promotion and disease prevention. DHEW publication no. (PHS) 79–55071, U.S. Government Printing Office (USGPO), Washington, D.C., 1979.
2. U.S. Department of Agriculture, U.S. Department of Health, Education and Welfare. Nutrition and Your Health: Dietary Guidelines for Americans, Home and Garden Bulletin No. 232, USGPO, Washington, D.C., 1980. (Note: After review by an expert committee appointed by the Secretary of Agriculture a second edition of the Guidelines was released in September 1985 which is essentially identical to the first edition, suggesting that the Guidelines represent a reasonable compromise between what we know and what we suspect).
3. Federal Register. 1941 Title 21—Food and Drugs. Chapter 1—Food and Drug Administration, Part 15—Wheat flour and related products; definitions and standards of identity (6 FR 2574–2582).
4. Shank, F. R., Larsen, L, Scarbrough, F. E., Vanderveen, J. E., and Forbes, A. L. FDA perspective on sodium. Food Technol., *37*: 73–77, 1983.
5. National Institutes of Health. Consensus Development Conference Statement: Lowering Blood Cholesterol to Prevent Heart Disease, Dec. 10–12, 1984.
6. Federal Food, Drug, and Cosmetic Act of 1938, chapter 675, 52 Stat. 1040 amended by Food Additives Amendment Act of 1958, Pub. L. No. 85–929, 72 Stat. 1764 (codi-

fied as further amended at 21 U.S.C. §§ 301–392 (1982).
7. Merrill, R. A. Reducing diet-induced cancer through Federal regulation: opportunities and obstacles. Vanderbilt Law Rev., *38*: 513–538, 1985.
8. Miller, S. A. and Stephenson, M. G. Scientific and public health rationale for the Dietary Guidelines for Americans. Am. J. Clin. Nutr., *42*: 739–745, 1985.
9. Fox, J. P., Hall, C. E., and Elvebach, L. R. Epidemiology: Man and Disease, The Macmillan Company, Toronto, 1970.
10. Doll, R. and Peto, R. The causes of cancer: quantitative estimates of avoidable risks of cancer in the United States today. J. Natl. Cancer Inst., *66*: 1192–1308, 1981.
11. Creasey, W. A. Diet and Cancer. Lea & Febiger, Philadelphia. 1985.
12. Schrauzer, G. N. Selenium and cancer: a review. Bioinorg. Chem., *5*: 275–281, 1976.
13. Ip, C. and Sinha, D. K. Enhancement of mammary tumorigenesis by dietary selenium deficiency in rats with a high polyunsaturated fat intake. Cancer Res., *41*: 31–34, 1981.
14. Thomson, C. D. and Robinson, M. F. Selenium in human health and disease with emphasis on those aspects peculiar to New Zealand. Am. J. Clin. Nutr., *33*: 303–323, 1980.
15. Griffin, A. C. Role of selenium in the chemoprevention of cancer. Adv. Cancer Res., *29*: 419–442, 1979.
16. Nelson, A. A., Fitzhugh, O. G., and Calvery, H. O. Liver tumors following cirrhosis caused by selenium in rats. Cancer Res., 3: 230–236, 1943.
17. Yang, G., Wang, S., Zhou, R., and Sun, S. Endemic selenium intoxication of humans in China. Am. J. Clin. Nutr., *37*: 872–881, 1983.
18. Jensen, R., Closson, W., and Rothenberg, R., Selenium intoxication—New York. Morb. Mortal. Wkly. Rep., *33*: 157–158, 1984.
19. Sakurai, H. and Tsuchiya, K. A tentative recommendation for the maximum daily intake of selenium. Environ. Physiol. Biochem., *5*: 107–118, 1975.
20. National Academy of Sciences. Recommended Dietary Allowances, Ninth Edition, p. 163, National Academy Press, Washington, D.C., 1980.
21. Robertson, J. A. and Eastwood, M. A. An examination of factors which may affect the water holding capacity of dietary fibre. Br. J. Nutr., *45*: 83–88, 1981.
22. Stephan, A. M. and Cummings, J. H. Mechanism of action of dietary fibre in the human colon. Nature, *284*: 283–284, 1980.
23. Furda, I. Interaction of pectinaceous dietary fiber with some metals and lipids. *In;* G. Inglett and I. Falkehag (eds.), Dietary Fibers: Chemistry and Nutrition, pp. 31–48, New York, Academic Press, 1979.
24. Story, J. A. and Kritchevsky, D. Comparison of the binding of various bile acids and bile salts *in vitro* by several types of fiber. J. Nutr., *106*: 1292–1294, 1976.
25. Life Sciences Research Office, Federation of American Societies of Experimental Biology (FASEB). The role of dietary fiber in diverticular disease and colon cancer, October 1980.
26. National Academy of Sciences. Diet, Nutrition, and Cancer, Report of the Committee on Diet, Nutrition, and Cancer of the National Research Council, National Academy Press, Washington, D.C., 1982.
27. National Cancer Institute. Diet, Nutrition & Cancer Prevention: A Guide to Food Choices. NIH Publication No 85–2711. USGPO, Washington, D.C., 1984.
28. Parza, M. Diet and cancer. American Council on Science and Health, Feb. 1985.
29. Carroll, K. K. and Khor, H. T. Dietary fat in relation to tumorigenesis. Prog.

Biochem. Pharmacol., *10*: 308–353, 1975.

30. Erickson, K. L. and Thomas, I. K. The role of dietary fat in mammary tumorigenesis. Food Technol., *39*: 69–73, 1985.
31. Kagan, A. McGee, D. L., Yano, K., Rhoads, G. G., and Nomura, A. Serum cholesterol and mortality in a Japanese-American population. Am. J. Epidemiol., *114*: 11–20, 1981.
32. Garcia-Palmieri, M. R., Sorlie, P. D., Costas, R., and Havlik, R. J. An apparent inverse relationship between serum cholesterol and cancer mortality in Puerto Rico. Am. J. Epidemiol., *114*: 29–40, 1981.
33. Beaglehole, R., Foulkes, M. A., Prior, I.A.M., and Eyles, E. F. Cholesterol and mortality in New Zealand Maoris. Br. Med. J., *280*: 285–287, 1980.
34. Cruse, J. P., Lewin, M. R., Ferulano, G. P., and Clark, C. G. Cocarcinogenic effects of dietary cholesterol in experimental colon cancer. Nature, *276*: 822–825, 1978.
35. Hutt, P. B. Regulatory implementation of dietary recommendations. Food Drug Cosmet. Law J., *36*: 66–89, 1981.
36. For examples see: (1) Heimbach, J. T. and Stokes, R. C. FDA 1978 Consumer Food Labeling Survey, Department of Health, Education and Welfare, Washington, D.C., 1979; (2) Heimbach, J. T. Cardiovascular disease and diet: the public view. Public Health Rep., *100*: 5–12, 1985; (3) Heimbach, J. T. and Stokes, R. C. Nutrition labeling for today's needs. National Technical Information Sevice, Pub. no. PB8214402, Springfield, VA, 1981.
37. Federal Register. 1973 Title 21—Food and Drugs, Chapter 1—Food and Drug Administration, Part 1—Imitation Food; Application of the term "imitation" (38 FR 20702–20704).

CONCLUDING REMARKS AND FUTURE PERSPECTIVES

DIET, NUTRITION AND CANCER, Y. HAYASHI ET AL. (EDS.), JAPAN SCI. SOC. PRESS, TOKYO/VNU SCI. PRESS, UTRECHT, PP. 325-335, 1986

Diet, Nutrition and Cancer: Concluding Remarks and Future Perspectives

Lorenzo TOMATIS

International Agency for Research on Cancer, Lyon Cedex 8, France

Abstract: Several dietary exposures may increase the risk of cancer, while others may have a protective effect. The degrees of evidence for a causal relationship between these exposures and human cancer vary considerably, as do their suitability for prevention. Investigations on the mechanisms by which food components contribute to either increase or decrease cancer risks therefore deserve priority, as their results would allow a quantitative evaluation of risks and of their preventability. Dietary factors are likely to contribute directly or indirectly to the induction of cancer in a variety of organs—namely, oesophagus, stomach, colon and rectum, liver, breast and endometrium, as well as the oral cavity and larynx. In addition, certain dietary factors may contribute to the prevention of cancer at other sites, as, for instance, the lung and prostate. Dietary interventions may therefore have a very considerable impact on prevention. This is certainly not the least reason for the attraction that intervention studies exert on scientists. Attractive as they may be, however, they should not encourage short-cuts, in the belief that understanding of the means for the prevention of cancer might be easier than understanding of the mechanisms of its induction.

This very timely symposium has exposed the controversies that still exist about certain basic assumptions on the role of dietary factors in human cancer, and has underlined the importance of a multidisciplinary approach to understanding of the underlying mechanisms.

There is general consensus, although perhaps some exaggeration, about the importance of the relationship that may exist between what we eat and the risk of developing cancer (*1*). There has also been widespread consensus that the proportion of all cancer cases attributable to diet is rather high, even if the proportions proposed by different authors vary considerably, (*2–4*), a variance that has not aroused the debate that has instead occurred over the proportion of cancer cases attributable to occupation (*5*). In spite of the importance attributed to the role of diet, however, none of the risk factors for which there is currently sufficient evidence for a causal association with cancer in humans appear to be, besides alcohol (*6*), directly related to dietary exposure; most are related to occupational or therapeutic exposures (*7, 8*). This may indicate either an overestimation of the proportion of all cancer cases attributable to dietary factors, or a bias in the ways in which the causes of cancer have been investigated, which favours the identification of certain causes over others.

The reason may lie in both explanations. There is certainly a bias in favour of investigating causes that are easier to measure. It is easier, for instance, to carry out epidemiological studies or to obtain case reports on occupational and therapeutic exposures than on others; similarly, it is easier to test individual chemicals in isolation than to mimic complex human exposures experimentally. Groups exposed occupationally and iatrogenically somewhat resemble experimental groups in which high levels of exposure may generate effects sufficiently conspicuous to be detectable among a limited number of exposed individuals. Even in occupational settings, however, it is difficult, if not impossible, with the present methodology, to find evidence of an increased cancer risk when the exposure level is low. Similar difficulties are encountered in the investigation of dietary factors—not the least difficulty is that everybody is exposed to some food.

At the International Agency for Research on Cancer (IARC), great attention has been given to the risks produced by exogenous environmental factors. The IARC programme of the Evaluation of the Carcinogenic Risk of Chemicals to Humans has concentrated for years mainly, although not exclusively, on industrial chemicals, exposures to which was mainly occupational or iatrogenic. When the IARC programme was started, that was the obvious thing to do, as well as also the only direction in which, at that time, progress was envisageable. The programme has been expanded recently though, to include the evaluation of complex mixtures like tobacco, both tobacco as it is chewed and as it is smoked (*9, 10*). It will be further expanded and the criteria used to evaluate carcinogenicity will be reconsidered to enable to deal adequately within the programme with such complex situations as those related to nutritional factors and/or hormonal status.

The apparent contradiction—between the high proportion of cancer cases that have been attributed to diet and the low number of recognized causes of cancer that are related to the diet—may partially disappear, however, if it becomes possible to mount epidemiological studies on all those exposures through the diet for which there is some suspicion of carcinogenicity or for which there is sufficient evidence of carcinogenicity from experimental studies. Given the complexity of the interactions involving several variables and confounding factors, it is a little doubtful though, that many of such epidemiological studies will necessarily provide clear-cut results.

Sato (*11*) has suggested that exposure to the many carcinogens / mutagens present in our food might account, on the basis of an approximate estimate of their carcinogenic potency, for only a tiny fraction of human cancer cases. Low levels of exposure to carcinogens present in food may be only occasionally *per se* the cause of a cancer; in most cases, they probably contribute to increase the number of cells in the organism that are initiated, or are at least a step or two ahead of other cells towards neoplastic transformation. If there is a very large number of initiated cells, the probability is increased that some of them will be pushed one or more further steps in the direction of malignancy by promoters or by stimuli that have promoting activity, like bladder stones, inflammatory states, endocrine disorders and, in general, conditions that induce accelerated cell turnover.

The experiments that Berenblum and Shubik (*12–14*) performed in the late 1940s and 1950s on initiation and promotion showed, among other things, that low levels of exposure to a carcinogen do not necessarily produce cancer but may imprint cells to make them susceptible to subsequent action of a substance that does not *per se* produce cancer, but causes a large proportion of initiated cells to proliferate and finally give rise to cancer. It was decades before the importance of the two-stage carcinogenesis model was recognized in situations other than skin painting in mice. A further difference, however, is that we now speak more often of multi-stage, and not merely of two-stage, carcinogenesis, following a path indicated already in the 1960s by Leslie Foulds (*15*).

It also took a long time before the importance of Albert Tannenbaum's work (*16–18*) became recognized, probably because it provided disturbing evidence that the sequence of cellular and subcellular events leading to cancer can be substantially modified by "simply" varying the caloric intake. We know now that this is by no means a simple phenomenon, and that in order to follow the series of events and connections between what we eat and the risk of cancer one would need a sort of Ariane's thread, which according to the Greek mythology led Theseus out of the labyrinth.

Fortunately (but complicating the picture a bit more), we are also exposed to factors that may prevent one or more of the events necessary to reach complete malignant transformation (*19*, *20*). A further complication can derive from the fact that most of these factors may be bifunctional, that is, they may have opposite effects depending on the set they are put in (*21–23*). This does not favour the accuracy of either qualitative or quantitative predictions of risks. An understanding of the role as well as of the mechanisms of action of factors other than directly acting carcinogens/mutagens is therefore of the greatest importance. At the same time, we should not be under the illusion that accurate, quantified predictions will be easier to make just because we have discovered that there is not one but several variables to be measured. Assessment of human risks on the basis of an evaluation of the carcinogenicity of individual carcinogens, as if they acted alone and in isolation, is justified by the existence of certain extreme situations, in which exposure levels are so high as to prevail over any confounding factor. In situations that are not so extreme, evaluations of the carcinogenic risk to humans of individual chemicals, serve the purpose of indicating that a risk is likely to exist, without necessarily providing an idea of its degree.

The search for carcinogens/mutagens in food has undergone various phases—from the search for preformed carcinogens, added unintentionally to foods, like pesticide residues and other environmental pollutants, or added intentionally, like food additives; to the search for carcinogens formed endogenously, such as *N*-nitroso compounds, from the interaction of chemicals that may be added intentionally or unintentionally to the diet; to the search for polynuclear hydrocarbons formed during the process of cooking (*24*, *25*); to the findings, through the pioneering and extensive work carried out by Sugimura and his colleagues, of an entire series of carcinogens and/or mutagens that are natural components of our food or are originating from its pyrolysis (*26*, *29*). We can only speculate that before the era when humans learned the use of fire and started to cook their food, food-related cancer may have been related mainly to the ingestion of mycotoxins or of natural plants such as bracken fern, while, following the introduction of cooking, an entirely new series of carcinogens derived from the pyrolysis of food has appeared. It would obviously be interesting to document if, and the extent to which the frequency of cancer and the distribution of target sites in humans changed at that time.

There is little wonder that studies on protective factors exert a strong attraction for scientists, since they suggest the possibility of seeing a concrete, favourable outcome in a reasonably short time. This explains the rapid rise in enthusiasm for and expectations generated by studies on vitamin A and related compounds (*30*). Even if those expectations have not been fulfilled by the results of some recent studies (*31*), there is good evidence that certain components of the diet do indeed exert protective effects. The pioneering work of Hirayama (*32*) has not only demonstrated the protective effect of what he has called synthetically "green/yellow vegetables," but has also contributed to focusing attention on particular components of the diet, namely certain vitamins and minerals and the fibre content of our daily food.

Dietary habits are an important component of socio-cultural characteristics and, as such, they are closely interconnected with a variety of other risk factors. An example of the resulting complexity is well illustrated by the attempts made to assess the protective role of fibres in human cancer, in particular colon cancer. While most studies indicate a negative correlation between the intake of fibres and the risk of large-bowel cancer (*33*), there are populations at low risk for colon cancer even with low average intake of fibre, as shown by the situation in Japan (*34*, *35*). Fibre may have a protective role only when acting upon an already increased risk of cancer due to some other environmental (dietary or other) risk factor.

We do not yet know with sufficient accuracy whether it is the lack of certain food components, or their presence below minimal levels, that permits certain steps of the carcinogenesis process to occur more frequently; or if certain food components play a more active role, that is, if their presence actually prevents certain reactions or steps from taking place. In both instances, preventive intervention would be justified. In the first case, cancer risk would be decreased only in those individuals or population groups in which the risk is linked to improper nourishment; in the second case, it might be possible to intervene actively and produce an overall decrease in cancer frequency everywhere. In the first case, preventive intervention would consist of the addition of certain elements which are absent or present in minimal quantity, to the diet; in the second case, preventive intervention would consist of the provision of certain dietary elements over and above the normal level of intake. We have been warned about the possible inadequacy of assessing cancer risks for humans on the basis of the carcinogenicity of individual chemicals; it would probably also be inadequate to assess the preventive capacity of one or a few selected micronutrients or dietary components in isolation. In most cases, it is perhaps the relative imbalance between the various components of our daily diet that is as, or more, important than the amount of any one component. There is a large body of evidence, for instance, pointing to the role of fat, in increasing the risk of breast cancer, showing a better correlation with total dietary fat than with any particular type of fat. When dealing with dietary components, it is also important, in particular when planning long-term preventive interventions, not to disregard the significance of the diet components for other diseases (*36*).

It is well known that the incidence of stomach cancer is lower in Japanese who have emigrated to the U.S.A., and the decrease has been linked to changes in dietary habits. To a certain extent, the populations of most industrialized countries have undergone an experience similar to that of Japanese migrants to the U.S.A. The considerable decrease in the frequency of stomach cancer over the last 50 years coincides approximately with the increase in the incidence of lung cancer, their frequencies crossing over in about 1951 (*37*). In spite of this colossal natural experiment, we still do not know what it is that causes, or, equally importantly, what protects us from, stomach cancer, although many hypotheses have been advanced.

Gastric cancer has in fact decreased wherever socio-economic conditions have improved, and it has remained frequent—often the most frequent cancer—in countries where socio-economic conditions are and have remained poor (*38*). Within industrialized countries, the relationship between socio-economic conditions and gastric cancer persists, even with the general decline in the incidence of gastric cancer (*39*, *40*). However, we do not know how to translate a social condition into mechanistic terms, nor how to identify within the social condition the specific factor(s) responsible. It would seem evident though, that a good proportion of gastric cancer cases could be prevented by raising the living conditions of less favoured socio-economic groups. We also know, and this is confirmed by results presented at this Symposium, that frequent consumption of fresh vegetables and fruits and low intake of salty foods help to decrease the risk of gastric cancer.

It is tempting to assume that there was and perhaps is, a common risk factor across the great diversity in the composition of the diet which exist between, for instance, Japan and some mediterranean countries where incidence of gastric cancer is also high. Perhaps the unifying factor can not be found in what is eaten, but in what in particular in the past, was not eaten. The case of the Filipino ethnic group in Hawaii (*41*) would seem particularly interesting to follow, as this group which is among the socio-economically less favoured, experiences low incidence of both lung and stomach cancer.

The era in which the possible viral etiology of liver cancer was almost completely ignored was followed by a period in which only chronic infection with hepatitis B virus appeared to be important in the genesis of primary liver cancer. In accordance with the hypothesis of the multifactorial origin of most human cancers, results presented at this Sympositum (*42*) confirm the possible joint role of viral infection and other environmental factors, possibly aflatoxins, in its causation. In this context, it is important to discover whether any other mycotoxins to which we are exposed also play a role in the causation of human cancer. Good harvesting and storage conditions of most food commodities may guarantee protection against high levels of infestation, but they do not always protect against low levels of contamination. As for all carcinogens, we do not know the significance for humans of low levels of exposure to mycotoxins; however, in this case, the ignorance is accompanied by disturbing experimental evidence that even very low levels of exposure to aflatoxins are carcinogenic (*43*).

CONCLUSION

Several categories of dietary exposure may increase the risk of cancer (Table 1). They differ considerably both in terms of their suitability for prevention and in the degree of evidence for a causal relationship with the occurrence of cancer in humans. While it would be therefore injudicious to try to rank them according to importance in the causation of cancer in humans, we can assume that alcohol, dietary fats, and total caloric intake probably are the most important.

There are also a number of factors that may protect against human cancer (Table 2); however, a quantitative evaluation of the preventive capacity of each factor would be as unsound as a quantitative evaluation of the causative role of each factor listed in Table 1.

We can instead try to draw some conclusions on the basis of what we know

TABLE 1. Food Components and Intentional and Unintentional Food Additives That May Increase the Risk of Cancer

Alcohol	Coffee (?)
Total caloric intake	Natural plants (such as bracken fern, cycasin)
Dietary fat	Certain pesticide residues
Mycotoxins	Certain food additives
N-Nitroso compounds	Pyrolysis products (such as polycyclic aromatic hydrocarbons and heterocyclic amines)
Salty foods	

TABLE 2. Factors That May Protect against Cancer

Vitamin A and beta-carotene
Ascorbic acid
Vitamin E
Riboflavin
Selenium
Other anti-oxidants
Dietary fibre

TABLE 3 Dietary Factors for Which There Are Varying Degrees of Evidence of a Role in Increasing or Decreasing the Risk of Cancer

Increase	Decrease
Alcohol	Green-yellow vegetables (containing vitamin C, A, E and . . .?)
Aflatoxins	Citrus fruits (containing vitamin C and . . .?)
Total dietary fat	Fibre (grain fibre)
Total caloric intake	
N-Nitroso compounds	
Salty foods	

best. Table 3 lists those dietary factors for which there are varying degrees of evidence that they have a role in increasing or decreasing the risk of cancer. While it would seem justified, in the present state of our knowledge, to urge reduction of consumption of alcoholic beverages (particularly hard liquors), to encourage a decrease (certainly not the abolishment!) in the consumption of fat, an increase in the consumption of fresh vegetables and fruits, a reduction in the intake of nitrites and nitrates, as well as of salty foods, avoidance of the consumption of foods contaminated with mycotoxins and prevention for such contamination to occur, there clearly remains a need (and I don't know whether to call it "constant" or "urgent", or both) for investigations on the mechanisms by which food components contribute to cancer risk. Understanding of the mechanisms by which dietary fats modulate cancer risks, and the development of methods to assess the actual risks represented by the pyrolysis products in our daily food, are among the priorities for research.

The factors listed in Table 3 are those for which results of epidemiological surveys, or case reports, providing varying degrees of evidence for a causal association with human tumors, are available. The availability of results of human studies guarantees that a critical element for the evaluation of human risks is in our hands, but should not become the condition that determines by itself which risk or protective factor is really important, and which is only potentially important. Such an approach would in fact imply that epidemiology—even though the criticism is often made that it is intrinsically incapable of detecting low levels of risk which may nevertheless be of great importance to public health—is the only pivot around which prevention can be organized. One should not be encouraged to disregard the important warning that experimental results may provide, as if the existence of human data, independently from their quality, would indicate *per se* a higher probability that an exposure is carcinogenic to humans, or, by extension, that exposures for which only experimental results are available are not a human hazard.

Dietary factors are thought to contribute directly or indirectly to the induction of cancer in a variety of organs, in particular, oesophagus, stomach, colon and rectum, liver, breast, and endometrium; with oral cavity and larynx, if alcohol is also considered. We do not know, however, to what extent neoplasms occurring at these sites are due to dietary factors alone, to diet plus other factors, or to factors other than dietary ones. Any attempt to quantify the cancers that could be prevented is thus bound to be inaccurate. We know, however, that the cancers listed above represent 65% and 58% respectively in men and women of the total cancer cases in Japan, 28% and 60% of those in the U.S.A. and 34% and 54% of those in Western Europe (Fig. 1). If we also consider that certain dietary factors may contribute to the prevention of cancer at other sites, as for instance, the lung, and possibly the prostate without attempting a numerical assessment, we can confidently state that a considerable proportion of human cancers could indeed be prevented by intervening on the diet.

Besides what can be already today, rather cautiously recommended and put into practice (see Table 3), there are of course a variety of initiatives which can and should be taken; many are already under way and the results may become available soon. On the "passive" side these concern more detailed, more sophisticated surveys,

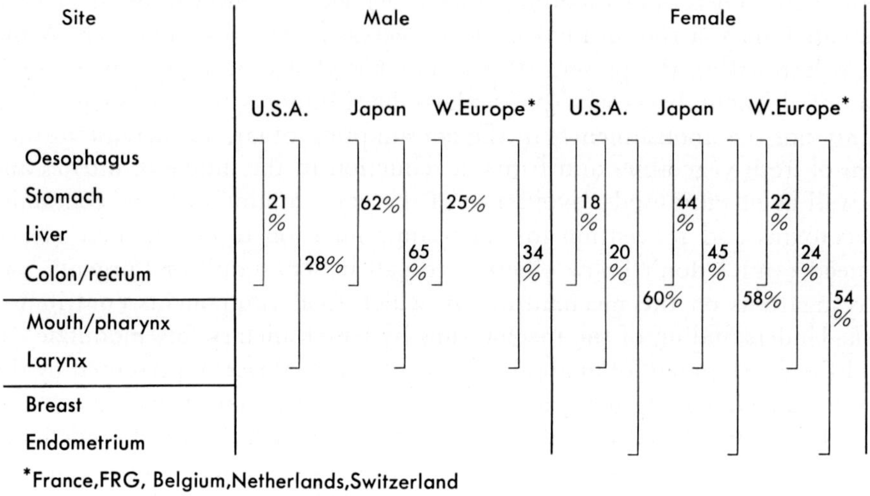

FIG. 1. Proportion of total cancer cases occurring at selected sites.

possibly prospective in nature of the correlation between nutrition and cancer. On the "active" side these are primarily intervention studies, that should, however, not be taken as possible short cuts toward primary prevention. The results of a recently terminated (*44*) intervention study, in an area of China at high risk of oesophageal cancer and with a very high prevalence of chronic oesophagitis, are a very useful warning that short cuts do not exist. After one year of administering riboflavin, retinol, and zinc, while the low serum levels of these micronutrients were found to be actually raised with subjects under treatment, there was no difference between treatment and placebo groups in the prevalence of chronic oesophagitis which, in certain instances, is assumed to be a precancerous lesion. Possibly the period of administration was too short, or given too late in life. We should, however, not be discouraged if the first attempts are not immediately successful. The intervention studies have not only a great potential for testing hypotheses but also provide the possibility of being of actual help to populations at high risk of cancer.

This Symposium has put in evidence the degree of controversy which still exists on certain concepts generally accepted as almost uncontroversial and that, during this Symposium, appeared rather like interesting assumptions needing further studies. While to have controversies as they exist, for instance over the role of dietary fats (*36*) is, at this stage, probably very healthy, if not yet for our colon, for science, it is hoped that results of studies of which we have heard during this Symposium (*11*, *21*, *35*, *36*, *45*) will become available soon to make some of the controversies obsolete.

ACKNOWLEDGMENTS

I would like to thank Mrs. E. Rivière and Mrs. W. Fèvre-Hlaholuk for their assistance in preparing the manuscript, Mrs. E. Heseltine for editing same and Dr. D. M. Parkin for providing Fig. 1.

REFERENCES

1. Abelson, P. H. Dietary carcinogens. Science, *221*: 1240, 1983.
2. Wynder, E. L. and Gori, G. B. Contribution of the environment to cancer incidence: an epidemiologic exercise. J. Natl. Cancer Inst., *58*: 825–832, 1977.
3. Higginson, J. and Muir, C. S. Environmental carcinogenesis: misconceptions and limitations to cancer control. J. Natl. Cancer Inst., *63*: 1291–1298, 1979.
4. Doll, R. and Peto, R. The causes of cancer. J. Natl. Cancer Inst., *66*: 1191–1308, 1981.
5. Peto, R. and Schneiderman, M. (eds.) Quantification of Occupational Cancer. Banbury Report No. 9, pp. 1–756, Cold Spring Harbor Laboratory, New York, 1981.
6. Tuyns, A. J. Alcohol. *In;* D. Schottenfeld and J. F. Fraumeni (eds.), Cancer Epidemiology and Prevention, pp. 293–303, W. B. Saunders Company, Philadelphia, 1982.
7. International Agency for Research on Cancer. IARC Monographs on the Evaluation of the Carcinogenic Risk of Chemicals to Humans, vols. 1–36, IARC, Lyon, 1971–1985.
8. Vainio, H., Kemminki, K., and Wilbourn, J. Data on the carcinogenicity of chemicals in the IARC Monographs Programme. Carcinogenesis, *6*(11): 1653–1665, 1985.
9. International Agency for Research on Cancer. IARC Monographs on the Evaluation of the Carcinogenic Risk of Chemicals to Humans: Tobacco habits other than smoking, betel-quid and areca-nut chewing and some related nitrosamines, vol. 37, IARC, Lyon, 1985.
10. International Agency for Research on Cancer. IARC Monographs on the Evaluation of the Carcinogenic Risk of Chemicals to Humans: Tobacco Smoking, vol. 38, IARC, Lyon, 1986.
11. Sato, S. (1986) The role of dietary mutagens/carcinogens in human cancer development. Presented at the 16th International Symposium of The Princess Takamatsn Cancer Research Fund, Tokyo, 1986.
12. Berenblum, I. and Shubik, P. The role of croton oil applications, associated with a single painting of a carcinogen, in tumor induction of the mouse's skin. Br. J. Cancer, 1: 379–382, 1947.
13. Berenblum, I. and Shubik, P. A new quantitative approach to the study of the stages of chemical carcinogenesis in the mouse's skin. Br. J. Cancer, *1*: 383–391, 1947.
14. Berenblum, I. Sequential aspects of chemical carcinogenesis: skin. Cancer, *1*: 323–344, 1975.
15. Foulds, L. Tumour progression: a review. Cancer Res., *14*: 327–339, 1984.
16. Tannenbaum, A. The genesis and growth of tumours. II. Effects of caloric initiation *per se*. Cancer Res., *2*: 460–467, 1942.
17. Tannenbaum, A. The genesis and growth of tumours. III. Effects of a high fat diet. Cancer Res., *2*: 468–474, 1942.
18. Tannenbaum, A. The dependence of tumor formation on the composition of the calorie-restricted diet as well as on the degree of restriction. Cancer Res., *5*: 616–625, 1945.
19. Wattenberg, L. Inhibition of carcinogens and toxic effects of polycyclic hydrocarbons by phenolic antioxidants and ethoxyquin. J. Natl. Cancer Inst., *48*: 1425–1430, 1972.
20. Wattenberg, L. Inhibition of chemical carcinogenesis by antioxidants. *In;* T. J. Slaga (ed.), Carcinogenesis—A Comprehensive Survey. vol. 5: Modifiers of Chemical Carcinogenesis: An Approach to the Biochemical Mechanism and Cancer Prevention, pp. 85–98, Raven Press, New York, 1980.

21. Sporn, M. B. and Roberts, A. B. Suppression of carcinogenesis by retinoids: interactions with peptide growth factors and their receptors as a key mechanism. This volume, pp. 149–158, 1986.
22. Wattenberg, L., Hanley, A. B., Barany, G., Sparnins, V. L., Lam, L.K.T., and Fenwick, G. R. Inhibition of carcinogenesis by some minor dietary constituents. This volume, pp. 193–203, 1986.
23. Willett W. Vitamin A and selenium intake in relation to human cancer risk. This volume, pp. 237–245, 1986.
24. Kuratsune, M. Benzo(a)pyrene content of certain pyrogenic materials. J. Natl. Cancer Inst., *16*: 1484–1486, 1956.
25. Lijinsky, W. and Shubik, P. Benzo(a)pyrene and other polynuclear hydrocarbons in charcoal-broiled meat. Science, *145*: 53–54, 1964.
26. Sugimura, T., Nagao, M., Kawachi, T., Honda, M., Yahagi, T., Seino, Y., Sato, S., Matsukura, N., Matsushima, T., Shirai, A., Sawamura, M., and Matsumoto, H. Mutagens-carcinogens in food, with special reference to highly mutagenic pyrolytic products in broiled foods. *In;* H. H. Hiatt, J. D. Watson, and J. A. Winsten (eds.), Origins of Human Cancer, pp. 1561–1577, Cold Spring Harbor Laboratory, New York, 1977.
27. Sugimura, T. Mutagens, carcinogens and tumor promoters in our daily food. Cancer, *49*: 1970–1984, 1982.
28. Ohgaki, H., Kusama, K., Matsukura, N., Morino, K., Hasegawa, H., Sato, S., Sugimura, T., and Takayama, S. Carcinogenicity in mice of a mutagenic compound 2-amino-3 methylimidazo [4-5-f] quinoline from broiled sardine, cooked beef and beef extract. Carcinogenesis, *5*: 921–924, 1984.
29. Takayama, S., Masuda, M., Mogami, M., Ohgaki, H., Sato, S., and Sugimura, T. Induction of cancers in the intestine, liver and various other organs of rats by feeding mutagens from glutamic acid pyrolysate. Gann, *75*: 200–213, 1984.
30. Peto, R., Doll, R., Buckley, J. D., and Sporn, M. B. Can dietary beta-carotene reduce human cancer rates? Nature, *290*: 210–218, 1981.
31. Kalache, A., Peto, R., and Doll, R. Case-control study on beta-carotene consumption and cancer risk (in preparation).
32. Hirayama, T. Does daily intake of green-yellow vegetables reduce the risk of cancer in man? An example of the application of epidemiological methods to the identification of individuals at low risk. *In;* H. Bartsch and B. Armstrong (eds.), Host Factors in Human Carcinogenesis, IARC Scientific Publication No. 39, pp. 531–540, IARC, Lyon, 1982.
33. Zaridze, D., Muir, C. S., and McMichael, A. J. Diet and cancer: value of different types of epidemiological studies. *In*; J. V. Joossens, M. J. Hill, and J. Geboers (eds.), Diet and Human Carcinogenesis, pp. 221–233, Elsevier Science Publishers, Amsterdam, 1985.
34. Kuratsune, M., Honda, T., Englyst, H. N. and Cummings, J. H. Dietary fibre in the Japanese diet. This volume, pp. 247–253, 1986.
35. Bingham, S. Non-starch polysaccharides as a protective factor in human large bowel cancer. This volume, pp. 183–192, 1986.
36. Stemmermann, G. N., Nomura, A.M.Y., and Heilbrun, L. K. Cancer risk in relation to fat and energy intake among Hawaii Japanese: a prospective study. This volume, pp. 265–274, 1986.
37. Cairns, J. The origin of human cancers. Nature, *289*: 383–387, 1981.

38. Parkin, D. M., Stjernsward, J., Muir, C. S. Estimates of the worldwide frequency of twelve major cancers. Bull. WHO, *62*: 163–182, 1984.
39. Logan, W.P.D. Cancer mortality by occupation and social class, 1851–1971. IARC Scientific Publications No. 36, IARC, Lyon, 1982.
40. Tomatis, L. The contribution of epidemiological and experimental data to the control of environmental carcinogens. Cancer Lett., *26*: 5–16, 1985.
41. Kolonel, L. N., Hankin, J. H. and Nomura, A.M.Y. Multiethnic studies of diet, nutrition, and cancer in Hawaii. This volume, pp. 29–40, 1986.
42. Sun, T., Wu, S., Wu, Y., and Chu, Y. Measurement of individual aflatoxin exposure among people having different risk to primary hepatocellular carcinoma. This volume, pp. 225–235, 1986.
43. Wogan, G. N., Paglialunga, S., and Newberne, P. M. Carcinogenic effects of low dietary levels of aflatoxin Bl in rats. Food Cosmetic. Toxicol., *12*: 681–685, 1984.
44. Munoz, N., Wahrendorf, J., Lu Jian Bang, Crespi, M., Thurnham, D. I., Day, N. E., Zheng Hong Ji, Grassi, A., Li Wen Yan, Liu Gui Lin, Lang Yu Quan, Zhang Cai Yun, Zheng Su Fang, Li Jun Yao, Correa, P., O'Conor, G. T., and Bosch, X. No effect of riboflavine, retinol, and zinc on prevalence of precancerous lesions of the oesophagus. Lancet, July 20, 111–114, 1985.
45. Bruce, W. R., Bird, R. P., and Rafter, J. J. The effect of calcium on the pathogenicity of high fat diets to the colon. This volume, pp. 291–294, 1986.

Author Index

Subject Index